SECOND EDITION

EXPLORING BIOINFORMATICS

A PROJECT-BASED APPROACH

Caroline St. Clair
North Central College

Jonathan E. Visick
North Central College

World Headquarters
Jones & Bartlett Learning
5 Wall Street
Burlington, MA 01803
978-443-5000
info@jblearning.com
www.jblearning.com

Jones & Bartlett Learning books and products are available through most bookstores and online booksellers. To contact Jones & Bartlett Learning directly, call 800-832-0034, fax 978-443-8000, or visit our website, www.jblearning.com.

Substantial discounts on bulk quantities of Jones & Bartlett Learning publications are available to corporations, professional associations, and other qualified organizations. For details and specific discount information, contact the special sales department at Jones & Bartlett Learning via the above contact information or send an email to specialsales@jblearning.com.

Production Credits
Chief Executive Officer: Ty Field
President: James Homer
Chief Product Officer: Eduardo Moura
Executive Publisher: William Brottmiller
Publisher: Cathy L. Esperti
Senior Acquisitions Editor: Erin O'Connor
Editorial Assistant: Rachel Isaacs
Production Editor: Leah Corrigan
Marketing Manager: Lindsay White
Manufacturing and Inventory Control Supervisor: Amy Bacus
Composition: Circle Graphics, Inc.
Cover Design: Scott Moden
Photo Research and Permissions Coordinator: Lauren Miller
Cover Image: © enot-poloskun/iStockphoto.com
Printing and Binding: Edwards Brothers Malloy
Cover Printing: Edwards Brothers Malloy

To order this product, use ISBN: 978-1-284-03424-0

Library of Congress Cataloging-in-Publication Data
St. Clair, Caroline.
Exploring bioinformatics : a project-based approach / St. Clair, Caroline & Visick, Jonathan E. — Second edition.
pages cm
ISBN 978-1-284-02344-2
1. Bioinformatics. I. Visick, Jonathan. II. Title.
QH324.2.S72 2014
572'.330285—dc23
2013015263
6048

Printed in the United States of America
17 16 15 10 9 8 7 6 5 4 3

Contents

Preface

A Project-Based Approach to Bioinformatics

Why We Wrote this Book

Exploring Bioinformatics: A Project-Based Approach arose from the bioinformatics course that we team-teach at North Central College in the western suburbs of Chicago. Our course, in turn, came from our realization that (1) every working biologist uses bioinformatics in some way to make meaningful use of the mountains of genetic and genomic data available today, and (2) a lot of computer scientists are finding jobs in bioinformatics, doing research on bioinformatic algorithms or collaborating with biologists to solve bioinformatic problems. We began meeting our students' need for bioinformatics in both our fields by incorporating hands-on computational exercises in biology courses at all levels and using biological programming problems in computer-science courses. Eventually, we decided to bring the biologists and the computer scientists together to explore bioinformatics in more depth in a bioinformatics course and later an interdisciplinary bioinformatics minor.

From the beginning, we wanted our course to be inquiry-based, giving students an opportunity to work hands-on with real data. And, we believed that both computer science and biology students would benefit from understanding how bioinformatics tools work: the algorithms that make them tick. Anyone can run a BLAST search (see Chapter 4) or do a pairwise alignment (see Chapter 3), but to understand how changing parameters might produce a more biologically relevant and thus more informative BLAST search or alignment requires knowing something about how the algorithms work. Additionally, we wanted to teach our course on a level appropriate for students who've had one or two programming courses and one or two courses introducing molecular biology, genetics, and/or biochemistry. We had difficulty finding a text fully compatible with these goals, and the "labs" that we wrote for our students first grew to replace the books we were using and then evolved into this book. We think that students elsewhere can also benefit from our approach, and we tried to make this book adaptable to a variety of audiences and to the needs of a variety of instructors.

What's Different About *Exploring Bioinformatics*?

Exploring Bioinformatics doesn't try to be encyclopedic. Each chapter covers a major area of bioinformatics, one which we think is fundamental. We don't try to cover every aspect of that topic, however; we instead focus on a specific biological problem and how bioinformatics can be applied to that problem. We try to include "just enough" biological background to understand the context of the problem, with a brief biological introduction in each chapter supplemented by a *BioBackground* box to help students with limited biology backgrounds. We take an "under the hood" look at an important algorithm used in computational solutions to the problem so that programmers and non-programmers alike can understand what the bioinformatics tools are doing. We then use web-based bioinformatics tools involving the algorithm (or related ones) in a hands-on *Web Exploration* project that explores aspects of the biological problem. For programming courses, a *Guided Programming Project* then offers an opportunity to actually implement the algorithm in code, with significant guidance and a pseudocode solution to use as a starting point. The third project in each chapter, the *On Your Own Project*, encourages students to work independently to understand (and, in programming courses, implement) a more advanced solution or to apply the solution to a more complex problem. The exercises and questions found throughout the book are intended not merely to test students' knowledge but also to develop it, as we feel that the best learning takes place when students are actively engaged in finding their own solutions.

Exploring Bioinformatics is intended to be flexible. Students can learn how computational algorithms are applied to a particular problem and find research-rich exercises to apply existing web-based bioinformatics tools to real data without having to do any programming. Students in programming courses will find pseudocode and discussions of algorithms that are not language-specific, allowing solutions to be implemented in any desired language. Although exercises and supporting data are provided, instructors could apply the same concepts to any desired dataset if they would like to incorporate their own research or interests into their courses. The text is closely tied to resources for both students and instructors available on a companion website, and instructors can, as desired, download working programs for students to use in lieu of writing their own solutions as well as test data, answers for exercises, and other features. *More to Explore* boxes suggest ways that students or instructors can readily take their inquiry beyond what is in the chapter.

What's New in the Second Edition

After we published our first edition, we started to hear reports of how instructors were using our text and what they wished they could do with it, and we realized that we needed to make the content more flexible in this second edition. To that end, we have removed its dependence on the PERL programming language (though PERL code along with solutions in other languages are still available on the website) in favor of flexible pseudocode solutions. We have also made our discussions of the bioinformatics algorithms more central to the chapters and more accessible to students in courses where no programming is required. We have added new material reflecting changes in how bioinformatics is used, such as next-generation sequencing, metagenomic analysis, and statistical methods like hidden Markov models. In the process, we re-wrote nearly all of the text with better and more applicable biological problems, more authentic data sets, more real-world problem solving with the web-based bioinformatics tools, clearer explanations of algorithms, better figures, more references, and more useful questions and exercises. We would be happy to hear your reactions and your suggestions for further improvement down the road.

How to Use *Exploring Bioinformatics*

The following table lists the main elements of each chapter, their intended use, and the exercises or other assessments that can be completed to reinforce concepts and skills. Each chapter focuses on a major area of bioinformatics and presents a biological problem to which that bioinformatic concept can be applied. Although later chapters do draw on material from

Using *Exploring Bioinformatics: a Project-Based Approach, Second Edition*

Chapter Element	Goal	Assessment
Introduction and Conceptual Development		
Understanding the Problem	Overview the biological problem to be addressed.	BioConcept Questions
BioBackground	As necessary (depending on biology background), learn key biological concepts relevant to the problem.	
Bioinformatics Solutions	Gain a broad understanding of how computational techniques can be applied to the biological problem.	
Understanding the Algorithm	Explore in detail how a computational algorithm relevant to the biological problem works "under the hood."	Test Your Understanding
Chapter Project		
Web Exploration	Apply existing, freely available web-based tools to an aspect of the biological problem, using real-world data.	Web Exploration Questions
Guided Programming Project	Implement an algorithmic solution to an aspect of the biological program in a desired programming language (programming courses only).	Putting Your Skills Into Practice
On Your Own Project	Work independently, using skills developed in the chapter, to further explore the biological problem either through more in-depth programming or more in-depth analysis.	Solving the Problem and/or Programming the Solution
Optional Elements		
Learning Tools	Download an exercise, spreadsheet, or other aid to assist in understanding a biological or computational concept.	
More to Explore	Continue investigating a topic beyond the exercises given in the chapter with additional tools or ideas for exploration.	
Connections	Applications of the chapter's ideas to additional real-world problems or current questions.	
References and Supplemental Reading	Sources for algorithms discussed in the text, background reading on the biological problems, and additional reading to better understand biological ideas or bioinformatics solutions.	

earlier chapters, where time is limited, an instructor could select a group of chapters to cover in a course or omit chapters s/he does not wish to cover.

The initial sections of each chapter lay out the biological and bioinformatic concepts needed to complete that chapter's projects. For students with less background in biology, the *BioBackground* boxes provide "just enough" introduction to the topic. *Understanding the Algorithm* is the heart of the chapter's conceptual foundation, walking students through a key computational algorithm to provide the basis either for using existing bioinformatic tools or implementing the algorithm in their own code. *BioConcept Questions* and *Test Your Understanding* exercises provide a convenient means of assessing student learning; instructors can assign a selection of these questions or devise exercises of their own.

Students will spend most of their time for each chapter working on the *Chapter Project*, which has three parts. In the *Web Exploration*, web-based bioinformatic tools are used to approach the biological problem, with *Web Exploration Questions* providing exercises ranging from simple to more sophisticated to test students' mastery of the programs. For programming courses, the *Guided Programming Project* helps students implement the algorithm in whatever programming language is used in their course, and *Putting Your Skills Into Practice* requires them to use their programming ability to extend or improve the solution. The *On Your Own* project, which has options for both programming and non-programming courses, gives an opportunity for more independent investigation. Each chapter lists specific learning objectives and offers suggestions and options for effective use of the projects in programming and non-programming courses. Many chapters have additional downloadable *Learning Tools* to help students grasp the problem or algorithm under discussion.

We intend for this book to be usable in a variety of ways to suit the interests and goals of instructors and their students. Instructors can select chapters, project segments, and exercises that suit their courses best, add their own exercises, or use their own data. We hope this book will be an effective tool to help build undergraduates' understanding of this new and essential field at the interface of biology and computer science. We welcome your comments and ideas as you explore bioinformatics.

Resources

For the Student

Hosted on the Navigate Companion Website (go.jblearning.com/Bioinformatics2ecw), students will find the following resources available for download:

- **Web links**: Links to access the tools and websites used in the chapter projects or referenced elsewhere in the text.
- **Learning Tools**: Visual demonstrations, simulations, and hands-on exercises to help students understand important concepts.
- **Python and Perl syntax guides**: A breakdown of key syntax needed for each chapter's programming exercises for both PERL and Python.
- **Sequences and Test Data** used in the chapter projects.

Access to this companion site is included with each new, printed copy of the text.

For the Instructor

The following resources will be made available to all instructors who adopt this text:

- **Keys for chapter exercises**: Answer keys for the Web Exploration, Test Your Understanding, and BioConcept Questions should the instructor choose to assign these questions as homework.
- **Programming solutions**: Complete, executable code corresponding to the Putting Your Skills Into Practice exercises and the On Your Own Project for each chapter will be provided in both PERL and Python; these files can be used as keys or could be downloaded and given to students in non-programming courses for use in completing the exercises.
- **Additional Sequences and Data Files**: Where appropriate, sequences that students will download or generate as part of the chapter's exercises are provided for the instructor's convenience.
- **Updates**: If additional tools are added in the future (e.g., in response to instructor feedback) or if corrections need to be made, updated files will be added to the website.

The instructors' resources are available via secure download; please contact your sales representative for more details.

Acknowledgments

We are grateful for the support of our faculty colleagues and acknowledge the North Central College biology and computer-science students who "test-drove" the exercises and commented on the text. We particularly thank Michael Holler for his invaluable contributions. We have also benefited from feedback from the participants in a 2010 HHMI faculty bioinformatics workshop at Lewis and Clark College who worked through early versions of many exercises used in this edition, contributed to the development of the concepts, and provided valuable suggestions.

Caroline St. Clair
Jonathan E. Visick

Jones & Bartlett Learning Titles in Bioinformatics

Biomedical Informatics: A Data User's Guide
Jules J. Berman

Medical Informatics 20/20: Quality and Electronic Health Records through Collaboration, Open Solutions, and Innovation
Douglas Goldstein, Peter J. Groen, Suniti Ponkshe, and Marc Wine

Perl Programming for Medicine and Biology
Jules J. Berman

Python for Bioinformatics
Jason Kinser

R for Medicine and Biology
Paul D. Lewis

Ruby Programming for Medicine and Biology
Jules J. Berman

Chapter 1

Bioinformatics and Genomic Data: Investigating a Complex Genetic Disease

Chapter Overview

This is a skill-development chapter: it does not address a specific bioinformatic algorithm. The goal of this chapter is to build skills in retrieving information from genomic databases; it is appropriate for both programming and nonprogramming courses but could be skipped if students are already familiar with genomic databases and genome browsing. For nonbiologists, the BioBackground box at the end of the chapter provides a basic tutorial on genes and genomes.

Biological problem: Genes associated with Parkinson disease

Bioinformatics skills: Searching genome databases and metadatabases, genome browsing

Bioinformatics software: Entrez interface to NCBI databases, UCSC genome browser

Programming skills: None

Understanding the Problem: Parkinson Disease, A Complex Genetic Disorder

At least seven million people are currently living with Parkinson disease (PD), a severe, progressive, and incurable disorder of the central nervous system. Actor Michael J. Fox's candid discussion of his condition has helped to focus attention on this disease, which begins with shaking, stuttering, difficulty walking, or involuntary muscle movement and worsens over time. PD is a ***complex*** *or* ***multifactorial*** *disorder: It has a heritable component but cannot be attributed to any single gene. Like autism, type 2 diabetes, asthma, or obesity, it likely results from the interaction of multiple genes as well as environmental or developmental components, complicating research into causes and treatments. We know that PD symptoms result from the death of a population of brain cells (neurons) residing in an area called the substantia nigra that normally produce the neurotransmitter dopamine. However, the cause of these cells' demise is unclear, and whereas a small percentage of PD patients have clearly defined defects in one of several specific genes, most do not, leaving biomedical researchers with no specific target toward which therapy can be directed. A genetic component of the disease is suggested by the fact that close relatives of Parkinson patients are at higher risk, but there is no clear pattern of inheritance; understanding of the disease is further complicated by environmental risk factors, including tobacco smoke.*

Parkinson's disease is just one example of a biological problem whose solution will depend heavily on bioinformatics. The cause of many well-studied genetic diseases can be narrowed down relatively easily to the inheritance of a dysfunctional allele of a single gene from one (in the case of **dominant** genetic disorders such as Huntington disease or hypercholesterolemia) or both (for **recessive** disorders such as cystic fibrosis, sickle-cell anemia, phenylketonuria, and Tay-Sachs disease) parents. Identifying the genetic factors in PD, however, is a more difficult undertaking that requires computational tools to facilitate the analysis of large amounts of genomic data.

A working draft of the nucleotide sequence of the entire human genome was completed in 2001, adding nearly 3 billion nucleotides to publicly accessible databases such as GenBank. Today, thousands of different genomes from plants, animals, bacteria, archaea, fungi, viruses, and protists have been completely sequenced, and new data continue to be added rapidly as the cost of genome sequencing projects continues to fall (**Figure 1.1**). These genomic data are used not only to investigate diseases but are also used by research scientists in areas as diverse as molecular biology, physiology, evolutionary biology, immunology, and ecology and by doctors, pharmaceutical companies, public health officials, animal breeders, food scientists, sociologists, law enforcement agencies, and many others. Diagnosing genetic diseases, determining evolutionary relationships, understanding metabolic functions, designing new drugs, investigating forensic evidence, improving food supplies, tailoring medical treatments to individuals, and reversing environmental degradation are just a few of the numerous current and potential applications of sequence data.

Making sense of these hundreds of billions of base pairs of genetic information is a daunting task. Simply printing out the human genome sequence would require some 250,000 pages, double-sided and single-spaced. **Bioinformatics** is the new science at the interface of molecular biology and computer science that seeks to develop better ways to

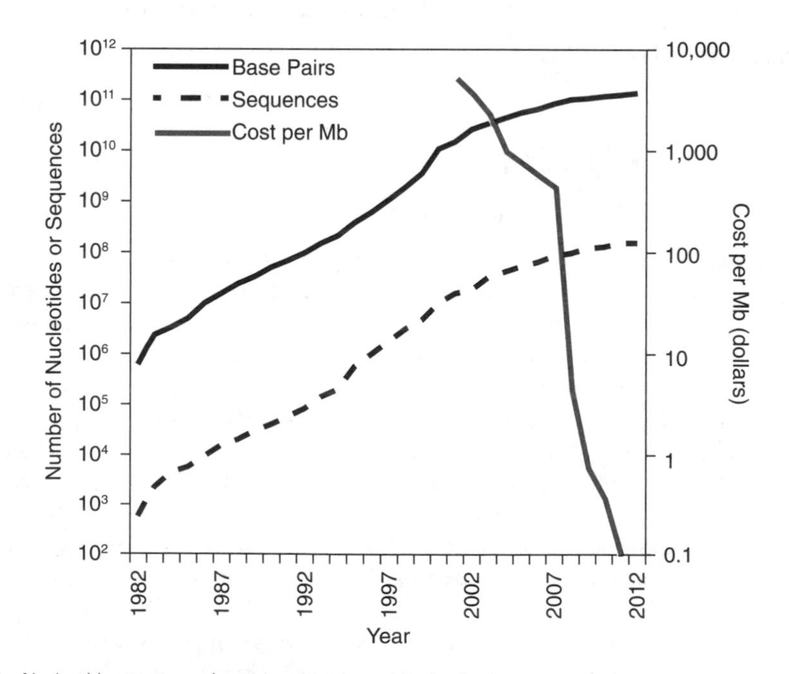

Figure 1.1 Nucleotide sequence data stored in the public GenBank database has grown exponentially for three decades, while the cost of large sequencing projects has declined dramatically. Data from: National Center for Biotechnology Information.

explore, analyze, and understand this vast wealth of genomic data. It is a branch of **computational biology** (the application of mathematical methods and computer algorithms to biological problems) that focuses specifically on the storage, retrieval, manipulation, and analysis of DNA and protein sequences as well as the information on structure, expression, and so on that can be derived from them. Bioinformatic tools provide one avenue for better understanding complex disorders such as PD, increasing our ability to work toward improved treatments or cures. Our goal is to help students become familiar with the algorithms and software used to address key questions in bioinformatics through the use of existing Web-based tools and/or individual programming solutions to investigate genuine biological questions. In this chapter, we explore how databases that store and link genomic information can allow us to investigate and better understand complex biological problems like Parkinson's disease.

Bioinformatics Solutions: Databases and Data Mining

Although most programs and algorithms discussed in this text involve ways that new data are generated, analysis—often referred to as "**data mining**"—of existing information in genomic databases can also yield valuable new insights. The nucleotide sequence of the entire human genome has been determined and is publicly accessible, but we have not yet unambiguously identified all of the genes within the genome (see the chapters on gene prediction), let alone determined their functions (see the chapter on protein alignment). However, in addition to sequence information, we have access to data about phenotypes, expression, intron/exon prediction, transcription factor binding, and more. "Mining" databases that bring together these different types of information can allow us to make new discoveries about known sequences, especially when the power of the bioinformatic data is combined with experimental results.

Except for the small proportion of PD patients whose disease clearly results from mutation in one of several specific genes, the genetic contribution to PD has generally been thought of as relatively small: Only about 15% of PD patients have a parent, sibling, or child with PD. However, a number of regions of the human genome have been correlated with PD through experiments such as **genome-wide association studies** (**GWAS**) in which a large number of genetic differences known to occur among individuals (often identified through genome sequencing) are examined to identify possible links to a specific genetic disease. A 2011 study by Do et al. (see References and Supplemental Reading at the end of this chapter) identified several new regions of the genome associated with PD and suggested that genetic changes contribute to a much larger proportion of both early- and late-onset cases of PD than was previously believed.

GWAS, however, can only identify genome *regions* that may be associated with a disease, not specific genes or specific mutations. Further analysis is necessary to determine what gene or genes are encoded in the identified regions, which of those genes are likely to be involved in the disease of interest, and what specific mutations may account for the disease phenotypes; this is where bioinformatics comes in. In this chapter's projects, we use genomic databases and metadatabases and the University of California Santa Cruz (UCSC) genome browser to examine the PD-associated genome regions identified by Do et al. and see how mining these databases using bioinformatic techniques can enable us to formulate hypotheses about which specific genes in the identified regions could be involved in PD. Such hypotheses can guide future research into the causes, prevention, and treatment of Parkinson's disease.

BioConcept Questions

BioConcept questions test your understanding of the biological ideas needed for each chapter. If you find that you need some help with these concepts, the BioBackground boxes are intended to provide a working knowledge sufficient to complete the chapter projects; these can be skipped if you are already clear on the concepts tested here.

1. In thinking about genetic diseases, it is common to refer to a "disease gene," such as the cystic fibrosis gene or the Tay-Sachs gene. A geneticist, however, would insist that we should refer to a "disease *allele*" instead. Why is this seemingly subtle difference an important one?
2. A recessive genetic trait (such as cystic fibrosis, sickle-cell disease, or simply short eyelashes) is only shown phenotypically if an individual inherits the recessive allele from *both* parents. If a mutation results in an allele that does not encode a functional protein, it is generally true that the allele is recessive. Can you explain why nonfunctional alleles are most often recessive and not dominant?
3. If every cell in the human body has the same DNA, how can they have such different structures and functions?
4. Briefly outline the pathway of gene expression, starting with DNA and ending with the protein product of a gene.
5. Which is longer, a gene's transcript or its coding sequence? Why?
6. When you see DNA represented as a string of letters (such as AACGATCC . . .), what do those letters represent? Why isn't it necessary to write out both strands?
7. Does the sequence `CAGCCUCCGA` represent a DNA sequence or an RNA sequence? How do you know?
8. When you see a protein represented as a string of letters (such as `MFRVAMP` . . .), what do those letters represent?

Understanding Genomic Databases

Learning Tools

From the text's website, you can download an HTML file, `DNASidebar.htm`, containing links to all Web databases, tools, and other sites mentioned in this text. Firefox users can load this list into the sidebar to provide a "bookmark" list that is always visible while working on other things. To do this, save the file to your computer and then open it in Firefox. Create a bookmark to the page and then edit the bookmark properties and check `Load this bookmark in the sidebar`. When you access this bookmark, the link list appears as a sidebar alongside whatever page you are viewing. Throughout the text, the link icon (the wrench shown to the left) denotes a website whose URL can be found in the list in `DNASidebar.htm` or at the *Exploring Bioinformatics* website. Websites that correspond to the link icon will be in bold blue type.

Primary Databases

Since molecular geneticists began to acquire significant DNA sequence information in the late 1970s, they have emphasized the importance of making these data widely available. The vast majority of all known gene and genome sequences have been deposited in databases accessible to scientists worldwide—and to the general public—via the Internet. The three major databases for DNA sequence information (each of which mirrors the others) are (1) GenBank, maintained by the National Center for Biotechnology Information (NCBI), a unit of the National Library of Medicine, which in turn is a branch of the U.S. National Institutes of Health; (2) the European Molecular Biology Laboratory (EMBL) database; and (3) DNA Data Bank Japan, maintained by the National Institute of Genetics of Japan. On the protein side, the UniProt (Universal Protein Resource) database is considered the most comprehensive; it combines data formerly maintained in three distinct repositories. These are called **primary databases**, because this is where "raw" nucleotide or amino acid sequence information is deposited. They are also **annotated databases**, because database records contain additional information about the sequences, such as the locations of protein-coding regions, introns and exons or other genetic features, as well as references to the scientific literature. Additional primary databases have also been created to store specific kinds of data such as the results of gene **expression** experiments, known **polymorphisms**, and so on. **Table 1.1** shows a list (by no means comprehensive) of some useful primary databases.

When we retrieve a DNA or protein sequence from a database, typically we see it laid out in some clear, readable format, usually within a Web browser. Consider, however, the problem of how the raw data should be stored. It is in fact not very practical to store formatted data, because this reduces the likelihood that a different application can effectively access it or that it can be readily repurposed as needs and software change. Like any database, then, genomic databases are divided into records (sequences) and then into fields. There are fields for the raw sequence itself; for the locations of features such as coding sequences, promoters, or introns; and for annotations such as references or additional information. Formatting is left up to the software that retrieves the sequence. Ideally, it should be easy for another database to retrieve a desired piece of information and connect it to related information stored elsewhere to generate a metadatabase.

Metadatabases

Along with the enormous growth of sequence information has come a need for additional resources that assist researchers in finding and retrieving data and, increasingly, finding the interconnections among the various kinds of stored data. **Secondary databases** or **metadatabases** (as the term is used by bioinformaticians) select and combine data from other databases (**Figure 1.2**). For example, **NCBI's Gene database** pulls together the DNA sequence, the protein sequence, references and information on expression, alleles and phenotypes, genomic location, and much more for genes that have been well studied. The **OMIM (Online Mendelian Inheritance in Man) database** brings together a wealth of information about all the known human genetic diseases and the genes that contribute to them. Or, one might choose the **KEGG Pathways database** to focus on metabolic pathways and the genes that encode metabolic proteins. Table 1.1 also lists a few useful metadatabases.

Database Searching

Searching a database requires some form of user interface. Today, this is most often a Web-based interface, with the data stored on a remote server. The search interface is not part of the database itself, and multiple Web-based interfaces can be found that all search the same underlying database. It is of course impossible to comprehensively describe these search interfaces here, but **Box 1.1** provides syntax information for the Entrez search interface. Entrez is the common interface for all databases maintained by NCBI—databases that are heavily used in this text and by actual researchers.

Table 1.1 Summary of some useful databases and resources.

Resource and Location	Description
	NCBI Databases
Nucleotide www.ncbi.nlm.nih.gov/nuccore	Main interface to annotated nucleotide sequences in GenBank and other major repositories
Protein www.ncbi.nlm.nih.gov/protein	Main interface to annotated protein sequences and translated DNA sequences from GenBank and other major repositories
Gene www.ncbi.nlm.nih.gov/gene	Metadatabase compiling information on genes from well-annotated genomes, including maps, sequences, functions, expression, etc.
OMIM www.ncbi.nlm.nih.gov/omim	Database of human diseases and genes associated with human disease; includes entries for both the diseases and the genes
Map Viewer www.ncbi.nlm.nih.gov/mapview	Genome browser: graphical view of genes and sequences in the context of chromosome, phenotype, marker, single-nucleotide polymorphism (SNP) and other maps
Gene Expression Omnibus (GEO) www.ncbi.nlm.nih.gov/geo	Composite of results from a large number of gene expression experiments
HomoloGene www.ncbi.nlm.nih.gov/homologene	Metadatabase focused on showing conservation of genes and identifying their orthologs
dbEST www.ncbi.nlm.nih.gov/dbEST	Database of expressed sequence tags: sequences of cDNAs representing fragments of genes expressed in some tissue or condition
dbSNP www.ncbi.nlm.nih.gov/dbSNP	Database of known SNPs
	Other Databases
KEGG www.genome.jp/kegg	Metadatabase focused on protein structure and function
Human Gene Mutation Database www.hgmd.cf.ac.uk/ac	Database of known human mutations (registration required)
PDB www.pdb.org/pdb	Database of protein structures and nucleic acid secondary structures
KEGG Pathways www.genome.jp/kegg/pathway.html	Database of metabolic pathways linked to known genes and proteins
UCSC Genome Browser genome.ucsc.edu	Genome browser: graphical view of genome sequences, with tools for mapping sequences to the genome, visualizing expression, etc.
STRING string.embl.de	Database of protein interactions
Pfam www.sanger.ac.uk/resources/databases/pfam.html	Database of protein families and domains
Wormbase www.wormbase.org	*Caenorhabditis elegans* genome and associated resources
Flybase www.flybase.org	*Drosophila melanogaster* genome and associated resources
SGD www.yeastgenome.org	*Saccharomyces cerevisiae* genome and associated resources
Colibri genolist.pasteur.fr/Colibri	*Escherichia coli* genome and associated resources

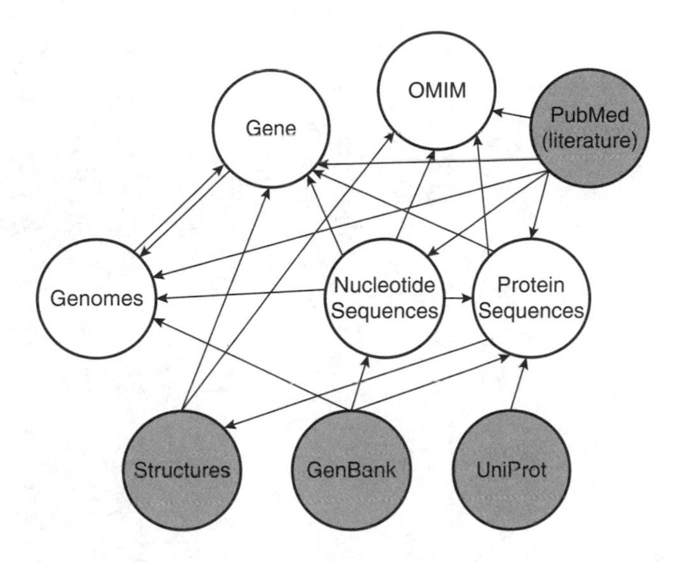

Figure 1.2 Example showing how primary genomic databases (filled circles) and metadatabases (open circles) might be interrelated.

Genome Browsers

A genome browser acts like a metadatabase in that it brings together information from many genomic databases, but it does so in a graphical form (**Figure 1.3**). A computer scientist might think of a genome browser as a graphical user interface (GUI) for genomic databases. Typically, a genome browser shows a graphical representation of the chromosomal position of a specified gene or genome segment. This view can be zoomed out to show more

Box 1.1 Syntax for the Entrez Search Interface to NCBI Databases

- **AND, OR,** and **NOT** must be capitalized.
- **Quotes** can be used to search for "exact phrases"; however, use quotes with caution, because the phrase will only be found if it occurs in the database's phrase list; the complete text of database entries is not searched.
- **Parentheses** can be used for grouping: `(CFTR AND complete) NOT human.`
- The asterisk is the wildcard character, representing any characters; therefore, `cys*` would find cystic fibrosis but also cysteine.
- The search can be limited to a particular database field using square brackets. For example, `smith [Author]` will limit the search to records with authors named Smith. Other useful fields are [Organism], [Title], [Text Word], and [Gene Name].
- Searching for a name followed by initials has the same effect as limiting to the author field: `smith j` or `smith je` would search for the author J. Smith or J. E. Smith, respectively.
- Click `Limits` on a search result page to include or exclude sequences by date or by other criteria appropriate to the particular database; the `Filter` option on the results page is another way to limit search results.
- Click `Advanced` to build a query using drop-down field lists and other tools or to run an additional query on a previous search result.

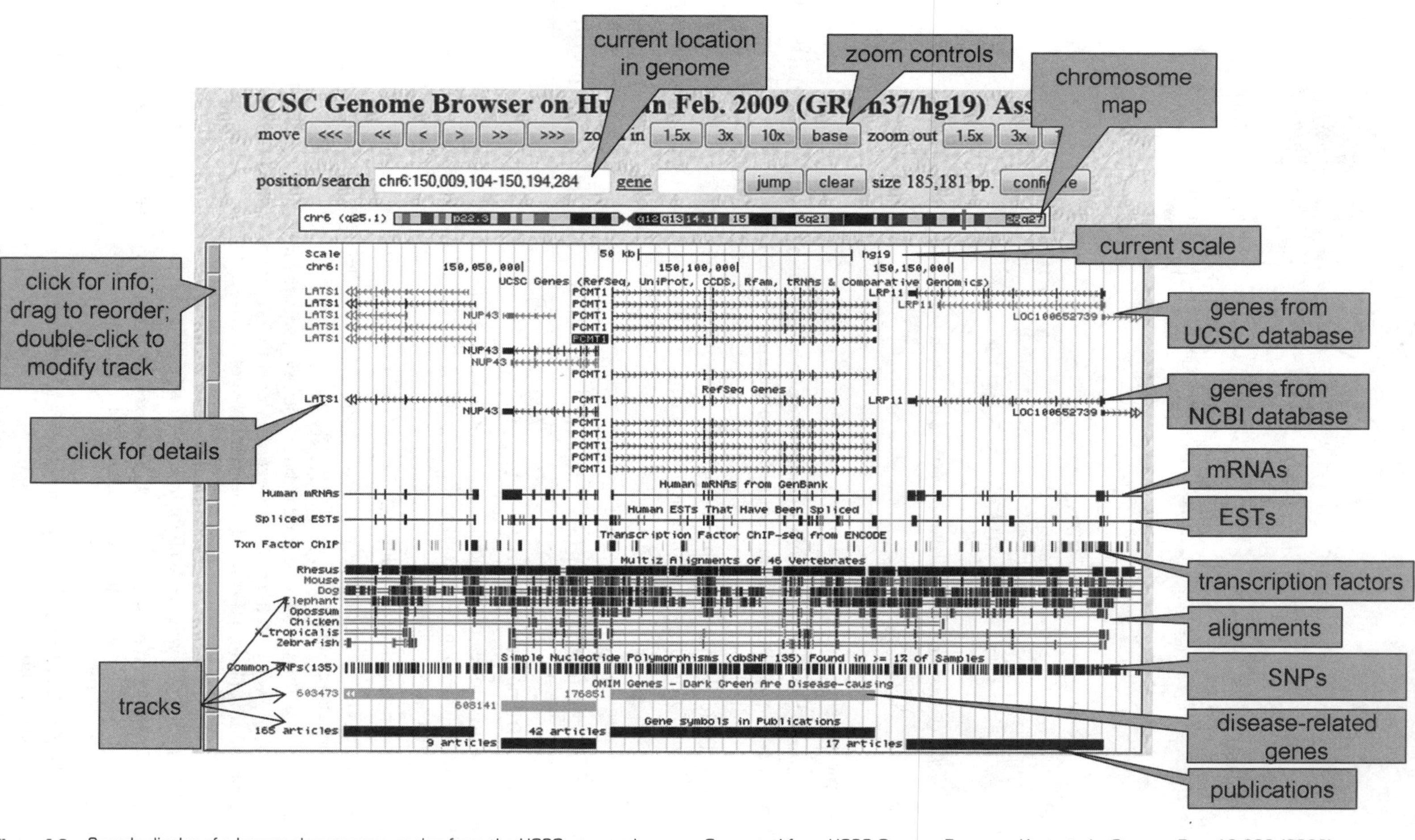

Figure 1.3 Sample display of a human chromosome region from the UCSC genome browser. Generated from UCSC Genome Browser: Kent et al., *Genome Res.* 12:996 (2002).

of the chromosome or zoomed in to show particular regions of a gene or even the actual DNA sequence. There are then various **tracks** that can be shown, hidden, or in some cases modified by the user. By default, one or more tracks representing genes are usually shown; "genes" in this case are typically defined as transcribed regions, so this track may show multiple known or hypothesized transcripts, and several different gene tracks can represent different database sources or types of evidence for transcription. Introns and exons are also represented in this view (if you are unfamiliar with concepts such as transcripts, introns, or coding sequences, see BioBackground at the end of this chapter), and each gene can be clicked to view more detailed descriptions, lengths, and references or to retrieve sequences or link to additional databases.

Additional tracks show binding sites for transcription factors, expression in specific tissues, the locations of known genetic variations (mutations or polymorphisms), methylation sites, repeated sequences, and comparisons with the genomes of other organisms. Users can even add custom tracks showing their own data. Genome browsers usually also have integrated tools for functions such as aligning a gene of interest with the genome or with the genomes of related organisms or predicting the sizes of **polymerase chain reaction (PCR)** products. Although all these data can also be accessed by other means, genome browsers have become popular due to the vast wealth of information consolidated in one location. Currently, the most-used genome browser is the **UCSC genome browser** maintained by the Genome Bioinformatics group at the University of California Santa Cruz. NCBI has its own genome browser, **Map Viewer**, as does EMBL, the **ENSEMBL genome browser**; many other genome browsers can be found online, some that are generally useful and some with specialized functions.

Test Your Understanding

1. Classify each of the following databases as primary databases or metadatabases.
   a. GenBank
   b. Gene
   c. Protein Data Bank, a database in which structural biochemists deposit newly determined, three-dimensional structures of proteins
   d. Colibri, a database providing information about each gene within the genome of the bacterium *Escherichia coli*
   e. Stanford Microarray Database, a repository of the results of microarray experiments
   f. TreeBASE, a database containing phylogenetic trees constructed based on nucleotide or protein sequence data
2. Using a Web search engine, identify two primary databases and two metadatabases not mentioned in this chapter and list the main goals or functions of each.
3. Although most sequence information has been deposited into one of the public databases, for-profit companies sometimes sequence genes or genomes and keep the information private, at least for a period of time. Discuss the value of public sequence data versus the need for industry to control access to information that may affect their ability to produce their products.
4. Navigate to the **UCSC genome browser** and choose any chromosome region to display. Identify two tracks not shown by default and list their functions.
5. In the sample genome browser display in Figure 1.3, several different bars are labeled as representing the same gene. These bars are sometimes, but not always, the same length. If these bars all represent one gene, why do you believe there are different bars, and why are they not identical?

CHAPTER PROJECT:

Genomic Regions Associated with Parkinson Disease

Do et al. (see References and Supplemental Reading) used a GWAS to look for genetic determinants of PD. GWASs, however, do not identify specific genes but rather use markers that permit the identification of genome regions correlated with the trait of interest. The researchers must then begin identifying candidate genes in the regions found, using bioinformatics to generate hypotheses and then testing them with laboratory research. In this project, we use genome databases and the UCSC genome browser to develop testable hypotheses about genes that might be involved in PD based on the regions identified by Do et al.

Learning Objectives

- Understand the kinds of information stored in genomic databases and metadatabases
- Develop skill in using various interfaces to find and retrieve genomic information
- Gain familiarity with the use of a genome browser to examine genomic regions and identify genes and other features
- Appreciate the value of bioinformatic data mining to develop hypotheses for further research

Suggestions for Using the Project

This chapter includes a Web Exploration Project but no Guided Programming Project. Its intent is to develop skills in the use of Web-based genomic databases and tools that will be needed in future chapters. This project is recommended both for courses that require programming skills and those that do not because of the familiarity the students will gain with how genomic information is stored and used and how computers represent and manipulate DNA and protein sequence information. Instructors should feel free to develop exercises of their own that use these same skills. An optional On Your Own Project is included for instructors who would like their students to practice their skills further.

Web Exploration: Data Mining in the Genome Databases

A GWAS requires a method to determine many individuals' **genotypes** for a large number of sites in the genome where genetic variation is known to occur. This is often done by means of a DNA microarray using allele-specific oligonucleotides: Short, single-stranded DNA segments with sequences matching known polymorphic sites are allowed to base pair with DNA from a particular individual under conditions where even a one-base mismatch would prevent binding. Do et al. genotyped 3,426 PD patients and 29,624 healthy control individuals for 522,782 known **simple nucleotide polymorphisms (SNPs)**, which are places in the genome where variation occurs among individuals in the form of a single nucleotide change or a one- or few-base insertion or deletion. They identified 11 SNP sites where one allele was correlated with PD with a statistically significant frequency.

Part I: Genome Browsing to Identify Possible Disease-Associated Genes

Exploring Bioinformatics on the Web

Find quick links to Web-based tools (updated as URLs change) at the *Exploring Bioinformatics* website.

Genome browsing for SNPs. Known SNPs in the human genome are deposited in the primary genomic database dbSNP; each has a unique **accession number** that identifies it. One SNP identified by the Do et al. GWAS is rs11868035; we start by investigating the genomic neighborhood of this SNP, using the **UCSC genome browser**. Note the following:

- If you use a Web search to find the genome browser, you may land on the UCSC Genome Bioinformatics main page; if so, find the genome browser in the list on the left side.
- The genome browser saves your preferences; if you have used this program before (or someone else has used it on the same computer), click the `Click here to reset` link before continuing to be sure your experience matches the discussion that follows.

You can use the UCSC genome browser to browse a variety of genomes; by default, the search parameters should be set to the latest version ("assembly") of the human genome, which is what we want in this case. You should see an input box where you can enter a specific position in the genome or a search term: gene names, key words, and accession codes all work here. Using rs11868035 as your search term, search the genome for the SNP identified by Do et al.

From the results page, you can see various data sets that include this SNP. Of specific interest to us is the listing for the NHGRI Catalog of Published Genome-Wide Association Studies, because this means the SNP has been identified as a point of interest in at least one GWAS. Because the SNP occurs at a specific point within the genome, any of these links will show you that genome region in the browser; however, different links will turn on different tracks that may provide different kinds of information. (You may wish to review Figure 1.3 to familiarize yourself with the track interface.) For our purposes, click the NHGRI link to find the SNP in the genome.

Examine the resulting browser view and get a feel for the kinds of information displayed. Notice from the scale bar (see Figure 1.3) that you are zoomed in to see a very small region of the genome: About how many nucleotides are displayed in this view? Just below the scale bar, you should see a track labeled NHGRI Catalog of Published Genome-Wide Association Studies; this track was turned on because of the link you selected and shows that the genome view is centered on the rs11868035 SNP. The point where the SNP occurs is shown by a bar, and its accession number is highlighted. Just above this display, you should also see an **ideogram**, or schematic drawing of the chromosome where this SNP occurs; the dark and light regions represent bands that are visible when real chromosomes are stained. Bands along the p (short) and q (long) arms of the chromosome are numbered, so that a chromosome position can be described by notation like 6q25.1. What is the chromosome position for the rs11868035 SNP?

Near the bottom of the track display, you should see a track labeled Simple Nucleotide Polymorphisms. Here, you should see bars for other SNPs in this region. To see their labels, right-click the track title and choose `full`. The rs11868035 SNP should again be highlighted. If you click its accession number in this track, you will jump to a page of information from the dbSNP database where you can see what nucleotides have been observed in human alleles at this position: In this case, the SNP is a base substitution, and either A or G can occur at this position, giving two known alleles. You can also click the SNP's accession number in the NHGRI Catalog track; you should see that it has been identified in the Do et al. study, as expected. Has it also been identified by other GWASs?

Returning to the track display, try zooming out by clicking on the `10× zoom out` button. As you zoom out, you will see more SNPs and begin to see the boundaries of genes (change the display of the SNP track back to `dense` to keep it from getting excessively long). At the top of the track display are three tracks that show different views of genes in this chromosome region: UCSC Genes, RefSeq Genes, and Human mRNAs. The labels on the left give the official gene symbol ("name") for each gene. Thick lines represent segments of coding sequence, called **exons**, whereas thin lines indicate **introns**, which are sequences that interrupt the coding sequence and are spliced out after transcription. In the UCSC Genes track, introns are shown with arrows to show the direction of transcription and transcribed, but untranslated regions (UTRs) flanking coding sequences are shown by medium-width lines. Does the rs11868035 SNP occur within a gene? If so, what is the name of the gene?

Now, zoom in far enough to read the actual nucleotide sequence by clicking on the `base` button. The display should still be focused on your highlighted SNP; if you have changed the focus (such as by clicking on a gene), you can always zoom out, find your SNP in the NHGRI or SNP track, right-click it, and choose `Zoom to ...` to recenter. With the display zoomed in to this level, you should be able to see the position of the SNP within the gene: Notice in this case (as shown by the Human mRNA track) that the SNP is within an intron, not within the coding sequence itself. Mutations within introns—especially if they occur near the boundary between an intron and exon—can affect the expression of a gene by impeding the splicing process. However, this could also indicate that the SNP identified in the GWAS occurs within a chromosome region important in PD but not within the actual *gene* correlated with the disease. We need additional evidence before we decide to focus on this particular gene.

Evidence for possible disease involvement. One line of evidence that a particular genomic region is important is **conservation**, which is the maintenance of the region in the genome over evolutionary time. A track labeled Multiz Alignments should be visible by default; this track shows an **alignment** between the human genome region you are browsing and the genomes of other sequenced vertebrates. Is the gene in which the rs11868035 SNP is found conserved? Is the specific *sequence* where this SNP occurs conserved? What animals might be appropriate model systems in which to conduct further research into the role of this gene in PD?

Although the Do et al. GWAS identified the rs11868035 SNP, this was just a marker in the study, and in fact any genes in its neighborhood could be responsible for the correlation between this region and PD. To analyze additional evidence, we need to turn on some tracks that are not displayed by default. First, zoom out so you can see a larger genome region surrounding the SNP of interest: Get a view showing approximately 1,000 kilobases (1,000 kb, or 1,000,000 base pairs) of the genome. This should bring the entire length of several genes into view, still centered around your original SNP. Now, scroll down below the track display, where you should see a number of drop-down boxes allowing you to turn tracks on and off. Under Phenotype and Disease Associations, find a drop-down list labeled `GAD View` and choose `pack` from the list. Click `refresh` to display the result of your change. This track will show previously described associations of genes with complex disorders. Similarly, turn on the track labeled `OMIM Genes`, showing genes listed in the OMIM database associating human genes with phenotypes, including disease phenotypes. Also, under Expression, turn on the `GNF Atlas 2` track, which summarizes experiments testing expression of the genes in various tissues. Finally, under Mapping and Sequencing Tracks, turn on the `Publications` track if it is not already on (you may want to choose the "dense" option for this one). Refresh the display.

Now, what kind of evidence would suggest the possible involvement of a particular gene in PD? Because PD is a disorder of the central nervous system, genes expressed in the brain or in nervous system tissue are good candidates, as well as any genes previously associated with some type of neurological disorder. Genes shown in red in the GAD View track have been associated with some kind of disease; click them to see if any seem relevant to PD. The OMIM

Genes track uses a color scale to show how clearly a gene has been associated with a disorder or phenotype; click one of the green or gray bars for a page that describes this scale. Then click the OMIM entry number link to see a summary of information from the OMIM database. OMIM catalogs both genes and disorders, so from the summary page, you can click either the gene link or the disorder link to retrieve entries from OMIM. By mining the data others have already found, can you strengthen the case for involvement in PD for any of these genes?

In the GNF Expression Atlas track, red bars represent genes that were strongly expressed in the indicated tissues (see labels at left): the brighter red, the more expression. Black represents a gene that is neither over- nor underexpressed, and green represents a gene that is underexpressed in the given tissue. A gene that is expressed in brain or nervous system tissue would certainly suggest a possible link to PD, but so might a gene normally underexpressed in these tissues turned *on* as the result of a mutation. Is there evidence that any of these genes are expressed in appropriate tissues?

Finally, click on each gene's bar in the Gene Symbols in Publications track to see a summary of published papers that refer to it. If you are in dense view, you will need to click twice, once to expand the gene list and then again to choose a particular gene. Is there any experimental evidence to suggest possible involvement of this gene in a neurological disorder? Taking together various pieces of evidence such as this, Do et al. concluded that the genes in this region most likely to be involved in PD were *SREBF1* and *RAI1*. Do you agree with their analysis? What evidence did you find that would support this conclusion?

Web Exploration Questions

You have seen how the UCSC genome browser can be used to expand on a PD-associated SNP found by a GWAS, leading to the identification of one or more candidate genes for further study. Now, use the skills you have learned to investigate a second SNP reported by Do et al.: rs34637584. Answer the following questions about that SNP:

1. On which human chromosome is this SNP located and at what position? (Use the conventional chromosome position notation described earlier.)
2. Does rs34637584 occur within a gene or between genes? If it occurs within a gene, is it within an exon or an intron?
3. Describe the alleles that occur through variation at this site.
4. List the genes found within approximately 500 kb on either side of the SNP.
5. Has this site or any of the nearby genes been associated with PD previously?
6. What evidence did you find to support the identification of one or more of the genes in this region as a candidate for a PD-associated gene?

Part II: Retrieving Sequences and Examining Genes in Detail

As you have seen, the UCSC genome browser is a powerful tool for genome analysis, bringing together a vast wealth of data that can be mined to answer many different kinds of questions. Similar analysis could be done by using **NCBI's Map Viewer**, the **ENSEMBL genome browser**, or any of a variety of related tools. In fact, without leaving the genome browser, you could retrieve the DNA or protein sequence of a gene you are interested in and go on to the kinds of analysis described next. For the purposes of this exercise, however, we will retrieve sequences by searching GenBank (a primary repository of nucleotide sequences) directly, so that we can learn to use the NCBI Entrez interface. We will then learn more about them through the use of metadatabases.

You found that the rs11868035 SNP was located within the gene *SREBF1*. Suppose you now want to download the sequence of this gene and/or the protein it encodes, perhaps to clone the gene for further experimentation or to perform a more detailed bioinformatic analysis such as alignment, structure prediction, or phylogenetic analysis. To accomplish this, go to the **NCBI Nucleotide database**. All NCBI databases can be searched using an interface called Entrez; once you find the NCBI site, start with a simple search by typing `SREBF1` into the search box at the top of the page and choose `Nucleotide` as the database to search using the drop-down box.

You will quickly see that this search may have been a little too general. Although you may see useful sequences in the result list, you will also recognize that you have more than 100 total results, only some of which represent human genes. A narrower search might be easier to interpret; one way to limit the search to human genes would be to use `SREBF1 AND homo sapiens` as your search term; note the AND must be capitalized (see search tips presented in Box 1.1). Another way to accomplish this is to look at the list of matched organisms on the right side of the results page (a variety of tools and links are listed here) and click on the link next to *Homo sapiens.* You should also see a list of "filters" that allow you to pare down your results further: Clicking on the `RefSeq` link, for example, will leave only those that are also found within NCBI's Reference Sequence database, which is a subset of GenBank containing nonredundant, well-characterized sequences. Note also that some of the results listed are not actually for *SREBF1* but for some nearby gene (whose record probably mentions *SREBF1* somewhere in its text). To eliminate these, you could use `SREBF1 [Title]` or `SREBF1 [Gene Name]` as search terms to require a match to a particular field.

Even with these limits, you will still see more than one result. Some matched records include "mRNA" in their titles, indicating they are sequences of cDNAs—mRNAs copied to DNA by the enzyme reverse transcriptase and then sequenced—rather than sequences of genomic DNA. For our purposes, look for an entry that includes "RefSeqGene" in its title; it should have the accession number NG_029029.1. Click on this sequence.

By default, you will see the complete GenBank record for this gene, including description, references, and a listing of a variety of features. The actual nucleotide sequence is at the bottom of the record: Are you surprised by its length? To see only the nucleotide sequence, click on the link at the top labeled `FASTA`; this will show the sequence in a conventional format called FASTA (introduced in a popular alignment program of the same name) in which the sequence (with no spaces or numbers) is preceded by a one-line descriptive comment marked by the > symbol (**Figure 1.4**).

Return to the GenBank display and consider the features list. This list shows the location of important features within the sequence, like the coding sequence (designated CDS in a

```
>gi | 28302128 | ref | NM_000518.4 | Homo sapiens hemoglobin, beta (HBB), mRNA
ACATTTGCTTCTGACACAACTGTGTTCACTAGCAACCTCAAACAGACACCATGGTGCATCTGACTCCTGA
GGAGAAGTCTGCCGTTACTGCCCTGTGGGGCAAGGTGAACGTGGATGAAGTTGGTGGTGAGGCCCTGGGC
AGGCTGCTGGTGGTCTACCCTTGGACCCAGAGGTTCTTTGAGTCCTTTGGGGATCTGTCCACTCCTGATG
CTGTTATGGGCAACCCTAAGGTGAAGGCTCATGGCAAGAAAGGTCTCGGTGCCTTTAGTGATGGCCTGGC
TCACCTGGACAACCTCAAGGGCACCTTTGCCACACTGAGTGAGCTGCACTGTGACAAGCTGCACGTGGAT
CCTGAGAACTTCAGGCTCCTGGGCAACGTGCTGGTCTGTGTGCTGGCCCATCACTTTGGCAAAGAATTCA
CCCCACCAGTGCAGGCTGCCTATCAGAAAGTGGTGGCTGGTGTGGCTAATGCCCTGGCCCACAAGTATCA
CTAAGCTCGCTTTCTTGCTGTCCAATTTCTATTAAAGGTTCCTTTGTTCCCTAAGTCCAACTACTAAACT
GGGGGATATTATGAAGGGCCTTGAGCATCTGGATTCTGCCTAATAAAAAACATTTATTTTCATTGC
```

Figure 1.4 The gene encoding the beta chain of human hemoglobin shown in FASTA format.

GenBank entry), the mRNA, and the various exons. Clicking on the links associated with these features alters the sequence display to show only the desired feature. Or, the locations of the features within the sequence can be readily visualized by choosing `Highlight Sequence Features` from the list of links on the right side of the page and then choosing the desired feature from the drop-down box that appears at the bottom of the page. Try highlighting the gene, then the mRNA, then the CDS, and compare the results.

The list on the right also links to other resources with additional information about this gene. You will recognize, for example, the link to OMIM as a database where you could learn about disease or phenotype associations of this gene. PubMed is a database of scientific publications where you could look up what has been published. For each protein-coding gene indexed in Nucleotide, there is a corresponding entry in Protein for the amino acid sequence that can be linked from this list. Another valuable choice from this list is the Gene metadatabase, where NCBI has compiled details about well-studied genes from a variety of sources. Here, you will see a graphical display of the gene in its genomic context and a mini genome browser showing a close-up view of the gene with its introns and exons. Many kinds of information are available on this page or through links or roll-overs, some of which should be familiar to you from the previous section of this project.

Web Exploration Questions

Mining the resources in the GenBank entry and the Gene metadatabase provides valuable information to a researcher seeking to better understand this gene or its physiological roles and should enable you to answer the questions below.

7. How long is the entire *SREBF1* gene (in bp or kb)?
8. When you click on the CDS link in a GenBank entry, only short segments of the gene sequence are highlighted. What do these highlighted segments represent?
9. How long is the spliced mRNA for *SREBF1*? What fraction of the gene is thrown away in the form of spliced-out introns?
10. What accounts for the difference between the sequence segments that are highlighted when you click on the `mRNA` link versus when you click the `CDS` link?
11. How long is the SREBF1 protein in amino acids? What are the first 10 amino acids in the protein sequence?
12. What is known about the function of this gene?
13. Besides its hypothetical association with PD, what are two other known connections of *SREBF1* to disease?

On Your Own Project: Clues to a Genetic Disease

Now that you have seen some genomic databases and worked with sequence information in a variety of ways, you should be able to apply these skills to a new project. Choose a genetic disease of interest to you (your instructor may require that you choose a single-gene disease or a complex disease), identify a gene associated with this disease, and summarize in one or two typed pages basic information about the gene, the protein it encodes, its genomic location, its expression, its known function(s), and its association with the disease you chose. OMIM would be a great place to start. If it is a gene that has been clearly identified as the specific cause of the disease, tell how the disease allele differs from the normal allele, if that is known. Document the sources of your information and give at least two references to published papers where your reader could learn more.

BioBackground: Genomes, Genes, Alleles, and DNA

An organism's **genome** is the complete set of all its **genes**, each of which can in turn be thought of as the encoded instructions for synthesizing one protein. All members of a species have the same set of genes—genes encoding hemoglobin and digestive enzymes and hair proteins and eye lens proteins—every protein that any cell in the organism can possibly make. Every cell within the organism carries this same genome. Each gene is a segment of **DNA** (deoxyribonucleic acid), and the genes are joined together to make up a set of very long DNA molecules called **chromosomes** that reside within the **nucleus** of every cell.

Individuals vary because although they all have the same set of genes, they do not all have the same **alleles**. An allele is a specific form of a gene: If a gene is thought of as a sequence of the DNA nucleotides A, T, C, and G, then one individual might carry the allele of the hemoglobin gene that begins ATGGTGCATCTGACTCCTGAGGAG and another individual might carry an allele beginning ATGGTGCATCTGACTCCTGTGGAG. In both cases, this is the same gene, encoding the hemoglobin protein, but different alleles (ultimately arising from mutations) encode slightly different variants that may function differently. Genetic variations are also known as **polymorphisms**.

Actually, each individual—and each cell within that individual—carries *two* complete sets of genes: one inherited from the mother and one from the father. So, you might inherit an allele of the "dimple" gene that produces dimpled cheeks from your mother and an allele of the *same* gene from your father that does not. In this case, you would have dimpled cheeks: The dimpled allele is said to be **dominant** over the nondimpled allele—it only takes one dimpled allele to produce the dimpled **phenotype**, or observable characteristic. On the other hand, having round eyes is a **recessive** trait: It is necessary to inherit a round-eye allele of the eye-shape gene from *both* parents to show this phenotype; the almond-eye allele is dominant. Simple genetic diseases result from alleles that encode an abnormal version of a protein; these alleles can be recessive (as in the case of sickle-cell anemia, caused by a misfunctioning version of the hemoglobin protein: see **Figure 1.5**) or dominant. An individual randomly inherits one of the mother's two alleles and one of the father's two alleles; thus, if the parents' genetic makeup (**genotype**) can be determined, the laws of probability can be used to determine how likely it is that their children will show a particular trait.

The cells within one individual vary not because they carry different genes or alleles but because different types of cells *express* different genes. We can think of a gene as a segment of DNA

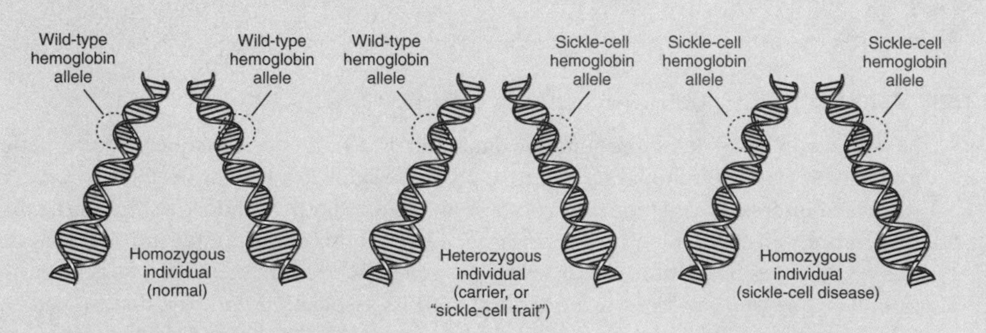

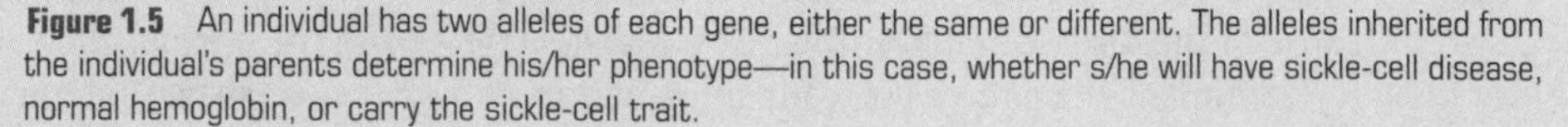

Figure 1.5 An individual has two alleles of each gene, either the same or different. The alleles inherited from the individual's parents determine his/her phenotype—in this case, whether s/he will have sickle-cell disease, normal hemoglobin, or carry the sickle-cell trait.

that encodes a protein (although some genes encode RNAs that are never translated to proteins), and **gene expression** is the process of actually making that protein. This is a two-step process: **Transcription** (carried out by **RNA polymerase**) copies nucleic acid information for one gene from DNA to **messenger RNA (mRNA)**, and **translation** (carried out by the **ribosome**) decodes that information to make the protein (**Figure 1.6**). This DNA → RNA → protein process is so fundamental to all of biology that Francis Crick dubbed it the "central dogma of molecular biology."

In bioinformatics, DNA, RNA, and proteins are all represented as simple strings of letters, making them easy to manipulate computationally. A closer look at their structures reveals why this works. DNA (**Figure 1.7**) is a nucleic acid molecule consisting of two very long chains of **nucleotides** or "**bases**" (an average human chromosome is more than 100 million nucleotides long). The four nucleotides, A, T, C, and G, can occur in any order, and it is the specific sequence of nucleotides that identifies a gene, protein binding site, or other functional feature of DNA. The two chains are held together by **base-pairing**: A always pairs with T and C always pairs with G. Base-pairing means that it is only necessary to write out the nucleotide sequence of one DNA chain, such as ATGGTGCATCTGACTCCTGAGGAG, to represent a double-stranded DNA molecule that really looks like ATGGTGCATCTGACTCCTGAGGAG
TACCACGTAGACTGAGGACTCCTC. **Figure 1.8** represents the same process of gene expression as Figure 1.6, but now a string of letters representing nucleotides takes the place of the double-stranded DNA molecule.

Where a gene occurs within the DNA sequence, a nucleotide sequence known as a **promoter** indicates the site where RNA polymerase should begin transcribing. RNA polymerase synthesizes a single-stranded RNA using the same complementary base-pairing rules as for DNA, except that the nucleotide T is replaced by U; this process continues until a **terminator** sequence is reached. Within the resulting mRNA **transcript** is the **coding sequence** of the gene (Figure 1.8), beginning with a **start codon** (AUG) and ending with a **stop codon** (UGA, UAA, or UAG). The ribosome uses these sequences as its start and stop signals for translation; each three-nucleotide **codon** between them represents an amino acid. The protein is then a chain of amino acids (which later folds into a three-dimensional structure), each of which can be represented by a three-letter or one-letter code; thus, like the nucleic acids, proteins can be simplified for bioinformatic purposes to a string of letters. DNA and protein sequences retrieved from genomic databases use these strings as shorthand representations of complex biological molecules.

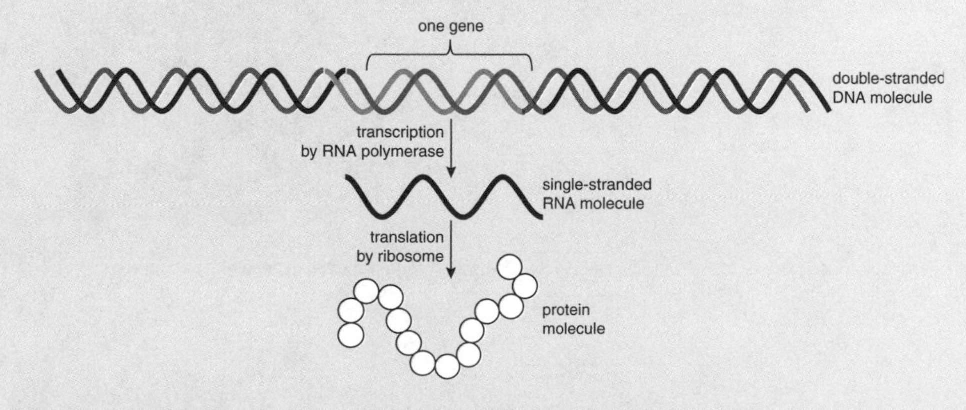

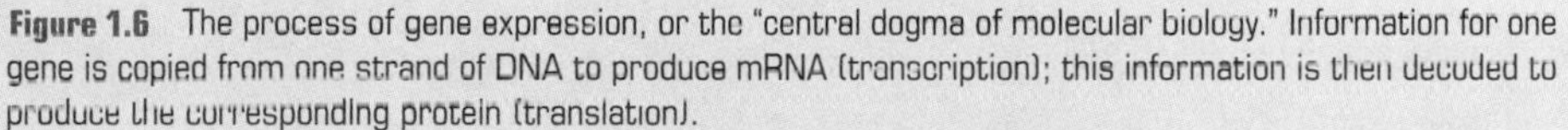
Figure 1.6 The process of gene expression, or the "central dogma of molecular biology." Information for one gene is copied from one strand of DNA to produce mRNA (transcription); this information is then decoded to produce the corresponding protein (translation).

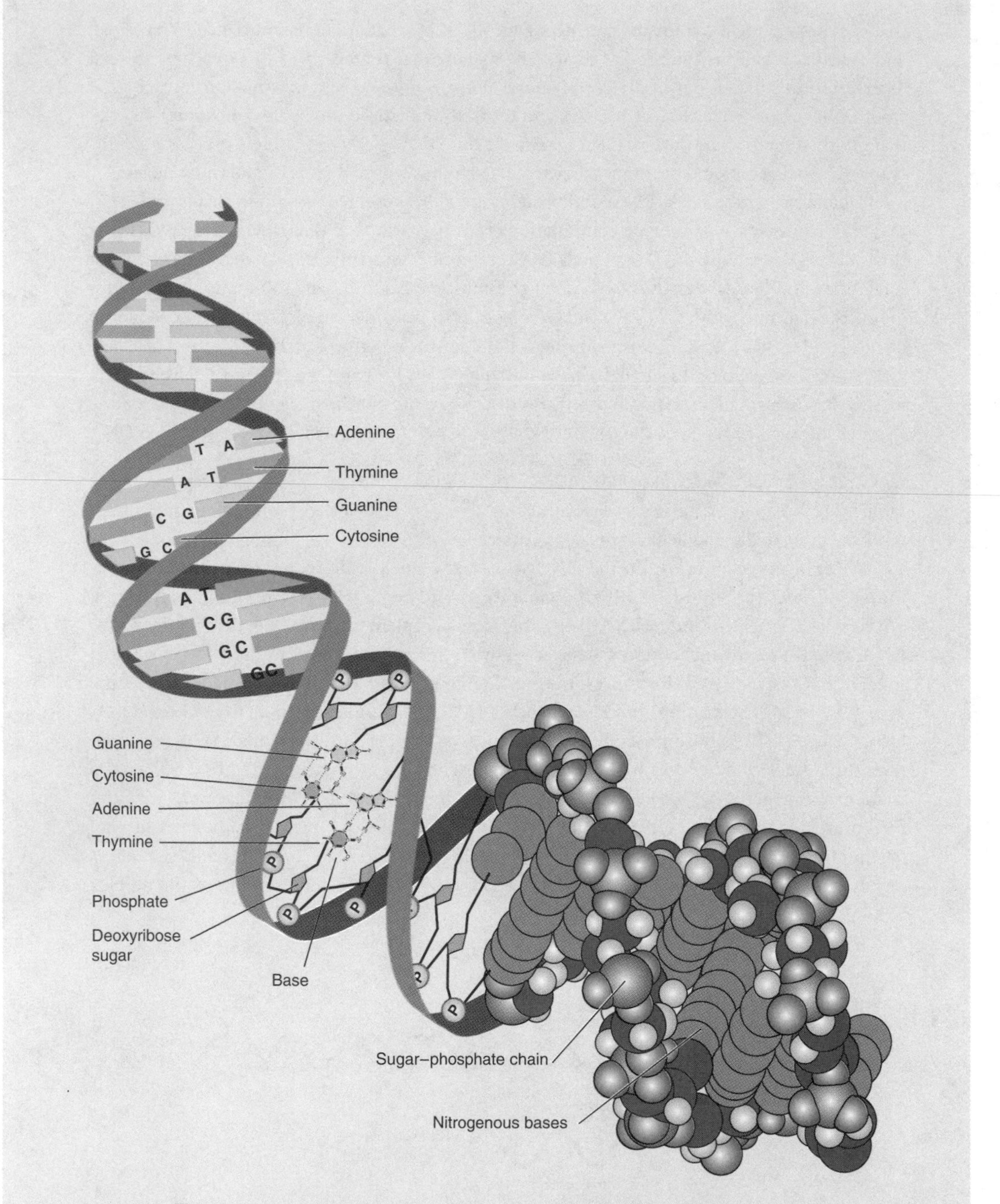

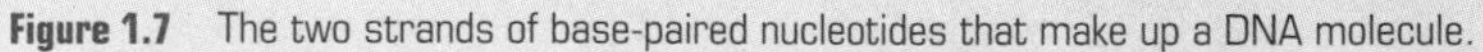
Figure 1.7 The two strands of base-paired nucleotides that make up a DNA molecule.

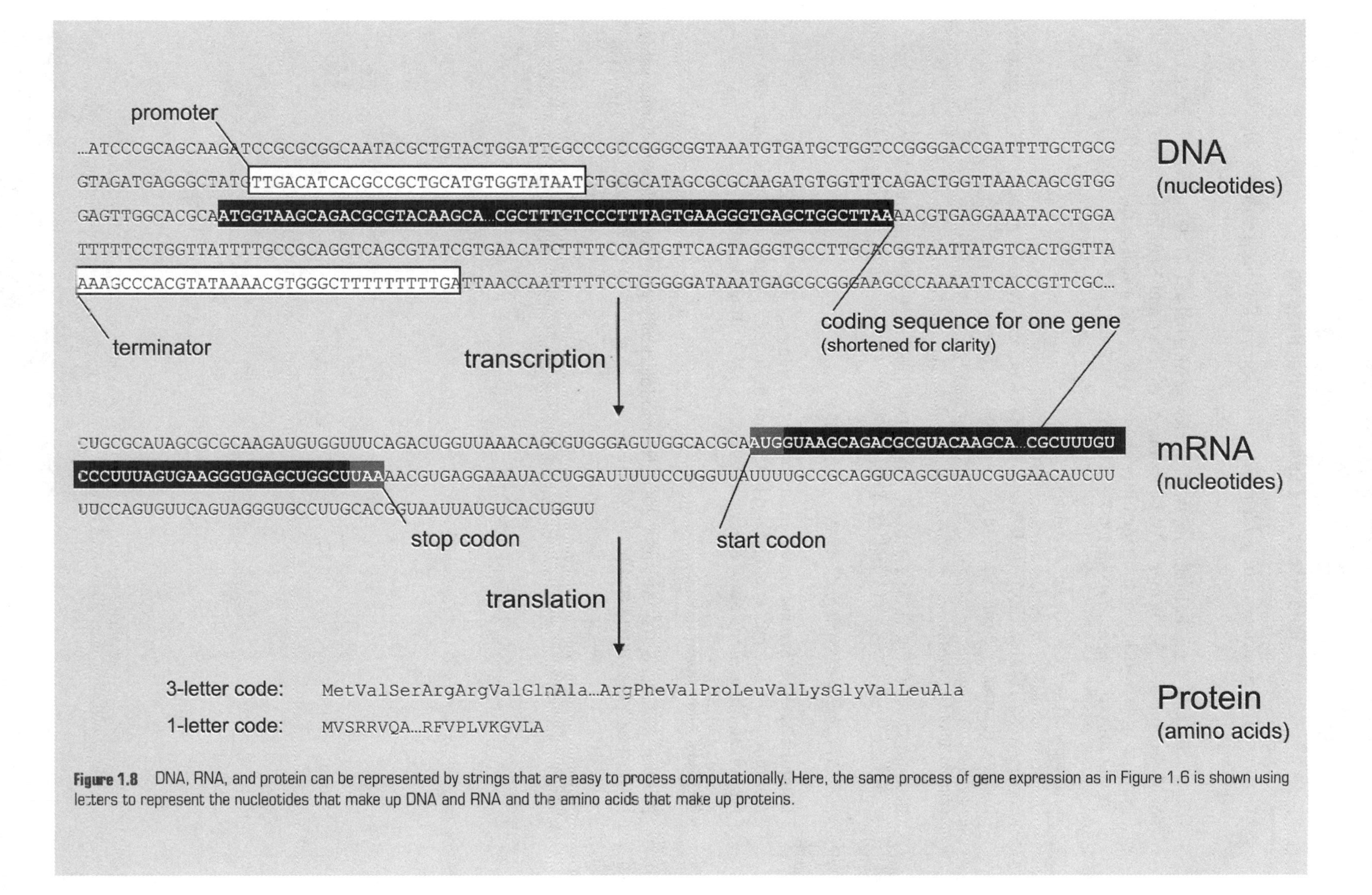

Figure 1.8 DNA, RNA, and protein can be represented by strings that are easy to process computationally. Here, the same process of gene expression as in Figure 1.6 is shown using letters to represent the nucleotides that make up DNA and RNA and the amino acids that make up proteins.

References and Supplemental Reading

Genome-Wide Association Study Identifying New Genome Regions Associated with Parkinson Disease

Do, C. B., J. Y. Tung, E. Dorfman, A. K. Kiefer, E. M. Drabant, U. Francke, J. L. Mountain, S. M. Goldman, C. M. Tanner, J. W. Langston, A. Wojcicki, and N. Eriksson. 2011. Web-based genome-wide association study identifies two novel loci and a substantial genetic component for Parkinson's disease. *PLoS Genet.* **7**:e1002141.

Overview of the UCSC Genome Browser

Zweig, A. S., D. Karolchik, R. M. Kuhn, D. Haussler, and W. J. Kent. 2008. UCSC genome browser tutorial. *Genomics* **92**:75–84.

Overview of GenBank

Benson, D. A., I. Karsch-Mizrachi, K. Clark, D. J. Lipman, J. Ostell, and E. W. Sayers. 2012. GenBank. *Nucleic Acids Res.* **40**:D48–D53.

NHGRI Catalog of Published GWAS

Hindorff, L. A., P. Sethupathy, H. A. Junkins, E. M. Ramos, J. P. Mehta, F. S. Collins, and T. A. Manolio. 2009. Potential etiologic and functional implications of genome-wide association loci for human diseases and traits. *Proc. Natl. Acad. Sci. U.S.A.* **106**:9362–9367.

To Learn More About GWAS

Pearson, T. A., and T. A. Manolio. 2008. How to interpret a genome-wide association study. *J. Am. Med. Assn.* **299**:1335–1344.

Chapter 2

Computational Manipulation of DNA:

Genetic Screening for Disease Alleles

Chapter Overview

The primary goal of this chapter is to give students in programming-oriented courses experience with manipulating strings in the language selected by the instructor by writing relatively short and straightforward programs to computationally "transcribe" and "translate" DNA and to compare the resulting strings (representing the amino acid sequences of proteins). The BioBackground box should build nonbiologists' foundational knowledge of the structure of DNA, RNA, and proteins and the use of the genetic code to decode a gene to understand how these processes can be modeled computationally. In a nonprogramming course, this chapter could be used in one of three ways. First, using the basic guide to programming in the Appendix and the **Perl or Python syntax guides** available at the *Exploring Bioinformatics* website, this could be an opportunity for nonprogrammers to try their hands at writing some relatively straightforward programs to better understand the challenges of bioinformatics software development. Second, the Web Exploration Project in this chapter can be used to reinforce students' understanding of the basics of molecular genetics. Finally, this chapter could be skipped if students already have a strong background and no programming experience is desired.

Biological problem: Genetic screening for cystic fibrosis

Bioinformatics skills: Manipulating DNA, RNA, and protein sequences

Bioinformatics software: Sequence Manipulation Suite

Programming skills: Computational algorithms, string manipulation, string comparison

Understanding the Problem:

Genetic Screening and the Inheritance of Cystic Fibrosis

Mary and her husband Tom would like to start a family. However, Mary's mother has cystic fibrosis (CF), and Mary carries the allele for this fatal genetic disorder. Although new therapies have extended the life expectancy and quality of life for CF patients, Mary knows that if her child inherits this incurable disease, he or she would need intensive therapy, would likely be

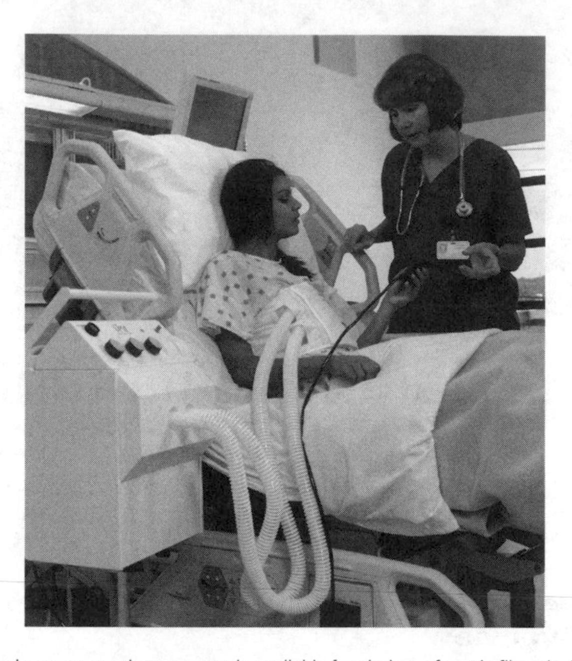

Figure 2.1 Currently, only symptomatic treatment is available for victims of cystic fibrosis. This CF patient is being treated by the use of a vest to mechanically enhance lung function.

*hospitalized frequently for respiratory infections, and would be unlikely to reach 40 years of age (***Figure 2.1***). Tom was adopted at a very early age, and little is known of his biological parents. However, CF is the most common fatal genetic disorder among Caucasians, and about 1 in 25 individuals carries the allele. Mary and Tom have sought the assistance of a genetic counselor, who has recommended genetic screening for Tom.*

Cystic fibrosis is a recessive, single-gene genetic disorder that affects the ability of epithelial cells to secrete fluids. Mothers are often the first to recognize something is amiss when they kiss their babies and notice their sweat tastes too salty. Indeed, not only is there not enough water in the sweat of a CF patient, but the secretion of digestive enzymes from the pancreas and, most importantly, the secretion of mucus in the respiratory system are affected. Failure of the vas deferens to develop properly commonly renders CF males sterile as well. All this results from the inheritance of two **mutant** alleles (one from the mother and one from the father) of a gene called *CFTR*. The product of this gene, the CF transmembrane conductance regulator, normally allows the regulated movement of chloride ions out of an epithelial cell, causing water to follow by osmosis and thus producing a watery secretion. In the absence of a functional copy of *CFTR*, thick mucus builds up in the lungs, making breathing difficult and providing a perfect environment for otherwise benign bacteria such as *Pseudomonas aeruginosa* to gain a foothold and cause chronic lung inflammation. Death may result from suffocation due to extensive scarring in the lungs or from an infection that cannot be managed by the combination of antibiotics and the patient's own immune response.

Pedigree analysis can determine the probability that an individual carries a CF allele but can decide with certainty whether he or she is a carrier only in some cases. However, now we can do more: We know the nucleotide sequence of *CFTR* and have sequenced many different CF alleles, and we have the technology to take a DNA sample from an individual and determine directly whether he or she is a carrier. Furthermore, we are beginning to use this

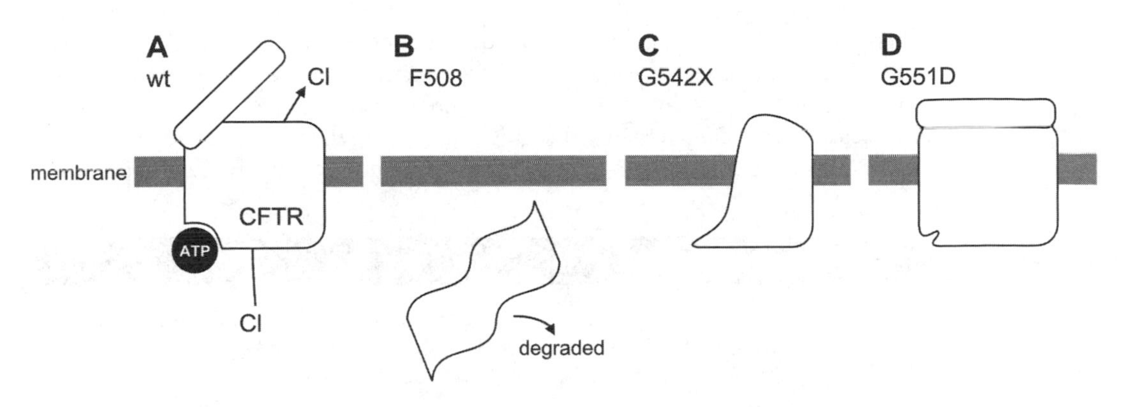

Figure 2.2 The wild-type allele (A) of the CFTR gene produces a chloride transport protein localized in the membrane; three different common CF alleles illustrated here result in variant proteins that are folded incorrectly (ΔF508; B), truncated (G542X; C), or unable to transport chloride (G551D; D).

information to personalize treatment for someone who suffers from CF. The various CF alleles all have in common the failure to encode a functional CFTR protein, but they result from different changes in DNA that have distinct effects on the protein. Some cause cells to be unable to make any CFTR protein at all, whereas others allow protein production but result in aberrant proteins that fail to fold correctly or fail to insert into the cell membrane (**Figure 2.2**; see References and Supplemental Reading at the end of this chapter for more information). Of course, someone who inherits *any* two nonfunctional alleles from his or her parents will have CF. However, it is becoming increasingly important to know *which* allele(s) a CF patient carries, because pharmaceutical companies are developing new drugs that target specific defects. For example, Kalydeco was approved by the U.S. Food and Drug Administration in January 2012 specifically for CF patients with a change at the 551st amino acid in CFTR from glycine to aspartate (G551D), a mutation that does not affect synthesis or localization of the protein but prevents opening of the channel. Thus, genetic testing is becoming increasingly important and taking on new roles.

Bioinformatics Solutions:
Computational Approaches to Genes

Bioinformatic tools have become essential in the analysis of genetic data: finding genes, identifying mutations, predicting the sequence, structure and function of proteins, and many other tasks. Suppose, for example, that you had the DNA sequence of Tom's *CFTR* gene, represented by a simple string of letters. You then might want to convert the DNA sequence to its corresponding mRNA sequence, decode the mRNA to obtain the amino-acid sequence of the protein, and compare that sequence to the sequence of the wild-type protein to identify any changes resulting from Tom's allele. By hand, this would be a tedious task: The coding sequence of *CFTR* is nearly 4,500 base pairs long, and the entire gene, including introns and exons, exceeds 180,000 base pairs. Using appropriate algorithms, however, a computer can automate the process of transcribing, translating, comparing, and otherwise manipulating sequences.

This chapter's projects focus on basic manipulation of DNA and protein sequences by computational means, looking "under the hood" to see how a computer program can convert among DNA, RNA, and protein sequences and make simple comparisons. For pedagogical purposes, we assume that we have chosen to actually sequence the coding region of Tom's *CFTR* gene, which was common practice at about 4% of genetic testing labs in a recent survey. Most labs use a faster, cheaper method to detect common *CFTR* mutations, such as PCR-based methods or ASO arrays. These methods, however, can only detect a particular set of alleles: typically, a CF carrier test looks for between 23 and 50 of the most common alleles. Sequencing remains the only truly comprehensive test, because it can detect any mutation, common or rare, including even previously unidentified mutations.

BioConcept Questions

Programming a computer to model manipulation of DNA requires that the programmer first have a solid understanding of how real DNA is manipulated within the cell. If you understand how DNA is transcribed and decoded, you should be able to answer these questions; use the BioBackground section to fill gaps in your understanding if needed.

1. How can Mary be sure she is a carrier of CF (one functional allele, one CF allele) without genetic testing? If Tom is also a carrier, what is the probability that their child would be born with CF?
2. One strand of a segment of a DNA molecule has the nucleotide sequence 5′ ACGTAGCAGATCAT. Show the double-stranded DNA molecule; don't forget to label the ends.
3. The *template* strand of a short DNA molecule has the nucleotide sequence 5′ GATGAGACTCCCATG. What would be the sequence of an mRNA molecule produced using this template? Write your mRNA sequence starting with the 5′ end, which is conventional whenever nucleotide sequences are written out.
4. The *nontemplate* strand of a short DNA molecule has the nucleotide sequence 5′ CGACCTATGCAGACT. What would be the sequence of an mRNA molecule produced from this DNA segment?
5. If your mRNA sequence for question 4 represented the first part of an mRNA encoding a protein, where could the coding sequence possibly start? Underline this start point and list the first two amino acids that would be found in the protein, using the three-letter code.

Understanding the Algorithm: Decoding DNA

Learning Tools

It is easier to understand how a computer would decode DNA if you have a clear understanding of how this process works. From the *Exploring Bioinformatics* website, you can download a short exercise, `DecodingDNA.pdf`, that asks you to analyze a DNA sequence by hand as an exercise to test your understanding.

Given a DNA sequence, such as might be generated by sequencing Tom's *CFTR* gene, how could a computer determine whether Tom carries a CF allele? Because some mutations are silent (see BioBackground), it would make the most sense to determine the amino-acid sequence of the CFTR *protein* and then compare it with a reference CFTR amino-acid sequence (neglecting for the moment that mutations outside the coding sequence can affect gene expression or processing). A DNA, RNA, or protein sequence can be represented by a string. In programming terminology, the term *string* is used to represent any sequence of characters, such as the nucleotide sequence AGCAT or the amino acid sequence KMVDR. In fact, we saw that only one string is needed to represent a DNA molecule, because we can deduce the sequence of the second strand from that of the first. However, to obtain a protein sequence, we need to know whether the DNA string we have represents the template or non-template strand. We then have to generate the matching mRNA sequence and then apply the

Box 2.1 What Is an Algorithm?

A computer **algorithm** is a set of specific steps that describes how a problem can be solved. The algorithm is the basis for any computer **program** or ("software"); the program is really just the steps of the algorithm converted into the syntax of a particular programming language. Although a computer may be able to solve a problem that a human cannot solve (for example, one that would take an impossible amount of time), a computer cannot solve a problem that a human does not know *how* to solve, because it can only execute the steps of a human-written algorithm.

To better understand the idea of an algorithm, you may want to try the following exercise. Lay out some playing cards on a table face up. Ignoring the suit, place the cards in numerical order. Easy, right? Now go back and do it again, but this time try to list the sequence of steps that you followed, as if you were going to teach someone how to put cards in order, step by step. This is a card-sorting *algorithm*, which you could convert into a program to instruct a computer how to sort something. Finally, can you think of a different algorithm that would accomplish the same thing? Which is more efficient?

genetic code to find the amino-acid sequence. Finally, we can compare this sequence with that of the wild-type protein. Programming a computer to perform these manipulations requires that we develop an algorithm for each one, as discussed next (to better understand the concept of an algorithm, see **Box 2.1**). These algorithms depend heavily on the ability to manipulate strings; one reason Perl and Python are popular languages for bioinformatics applications is that they have many convenient string manipulation functions built in.

A DNA Manipulation Algorithm

Although we only need one string (e.g., AGCAT) to represent a double-stranded DNA molecule, transcribing and translating DNA requires that we know whether that string represents the template strand or the nontemplate strand and its orientation (note that by convention, unlabeled single-stranded sequence strings are assumed to have their 5′ ends on the left). For our purposes, we will assume that we always want to output the template strand written from 3′→5′ (although these same manipulations could be used in many other contexts, as well), as diagrammed in **Figure 2.3**. The steps of one algorithm that would accomplish this computationally are shown below.

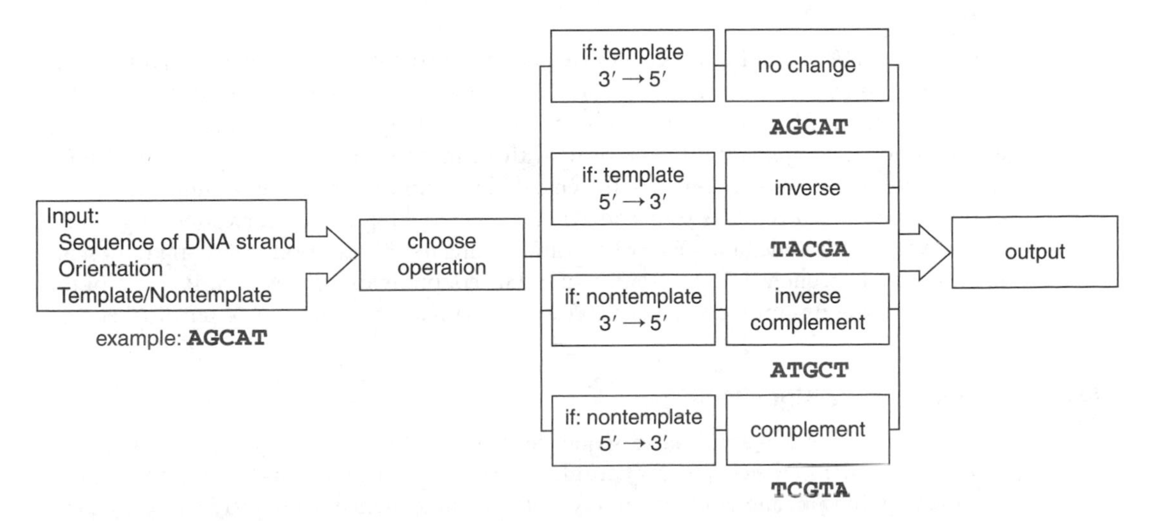

Figure 2.3 Illustration of an algorithm for manipulating a DNA sequence.

Algorithm

DNA Manipulation Algorithm

1. *Input a DNA sequence, whether it is the template or nontemplate strand, and which end is 5′.*
2. *Convert to all capital letters (so we don't later have to look for both C and c, for example).*
3. *Choose the appropriate operation:*
   a. *If it is the template strand and the 5′ end is on the right, simply output the same sequence directly.*
   b. *If it is the template strand and the 5′ end is on the left, generate the* ***inverse*** *sequence: traverse the string from right to left and add each character to the output string.*
   c. *If it is the nontemplate strand and the 5′ end is on the right, generate the* ***inverse complement*** *sequence: (1) traverse the string from right to left and (2) for each character (base), add the complementary character to the output string: for A, output T; for C, output G; for G, output C and for T, output A.*
   d. *If it is the nontemplate strand and the 5′ end is on the left, generate the* ***complement*** *sequence: (1) traverse the string from left to right and (2) for each character, add the complementary character to the output string as above.*
4. *Output the completed sequence, indicating the 5′ end.*

A Transcription Algorithm

By starting with the DNA manipulation algorithm above, we can ensure that the input for our transcription algorithm is a template strand in the 3′→5′ orientation. This greatly simplifies the next steps, because it is now only necessary to traverse the string from left to right, adding the complementary base to the output string for each character as in step 3c of the algorithm. Because the output string represents mRNA, however, the one key difference is that U (rather than T) is output for each A in the input sequence. The resulting mRNA sequence should be written 5′→3′, because that is the direction in which it would be read by the ribosome. Note that this is not the only possible algorithm: We could also have chosen to have our DNA manipulation program output the *non*template strand, in which case we would only have to replace all the T's with U's.

A Translation Algorithm

Given an mRNA strand in the proper orientation, translation is also quite straightforward. The most important complication is the identification of the start codon. In this chapter, we assume the mRNA starts with the start codon, although this is not the case in reality and additional complexity must be dealt with in more sophisticated algorithms. Determining the amino-acid sequence of the encoded protein then becomes a simple matter of examining three nucleotides (one codon) at a time, looking up the corresponding amino acid in the genetic code table (**Figure 2.4**) and adding its abbreviation to a string representing the protein sequence. In translation, the start codon defines the N-terminal end of the protein, and the amino acid sequence is conventionally written with the N-terminal end on the left.

A Mutation Detection Algorithm

Assuming we have two amino-acid sequence strings of the same length (see the On Your Own Project later in the chapter for consideration of comparing strings that are not equal in length), searching for alterations in Tom's CFTR protein sequence is now just a matter of comparing each amino acid in the amino-acid sequence we generated with the corresponding

	G	A	C	U
G	GGG, GGA, GGC, GGU: glycine (Gly, G)	GAG, GAA: glutamate (Glu, E) GAC, GAU: aspartate (Asp, D)	GCG, GCA, GCC, GCU: alanine (Ala, A)	GUG, GUA, GUC, GUU: valine (Val, V)
A	AGG, AGA: arginine (Arg, R) AGC, AGU: serine (Ser, S)	AAG, AAA: lysine (Lys, K) AAC, AAU: asparagine (Asn, N)	ACG, ACA, ACC, ACU: threonine (Thr, T)	AUG: methionine (Met, M) (start) AUA, AUC, AUU: isoleucine (Ile, I)
C	CGG, CGA, CGC, CGU: arginine (Arg, R)	CAG, CAA: glutamine (Gln, Q) CAC, CAU: histidine (His, H)	CCG, CCA, CCC, CCU: proline (Pro, P)	CUG, CUA, CUC, CUU: leucine (Leu, L)
U	UGG: tryptophan (Trp, W) UGA: stop UGC, UGU: cysteine (Cys, C)	UAG, UAA: stop UAC, UAU: tyrosine (Tyr, Y)	UCG, UCA, UCC, UCU: serine (Ser, S)	UUG, UUA: leucine (Leu, L) UUC, UUU: phenylalanine (Phe, F)

Figure 2.4 The genetic code. Codons are shown as they appear in mRNA (or in the noncoding strand of DNA, which would have T's in place of U's) with their 5′ ends to the left.

amino acid in the wild-type protein sequence and storing any changes detected. Steps for a mutation detection algorithm are shown next. Of course, the same algorithm could be used equally well to compare two DNA or RNA sequences, and in fact it is likely that we would want to go back to the DNA to see the actual mutation that led to the change in the protein.

Algorithm

Mutation Detection Algorithm

1. *Input two strings representing amino-acid sequences: the wild-type sequence and the patient's sequence.*
2. *Traverse each string from left to right, one character at a time.*
3. *If the character at a given position in string 1 fails to match the character at the same position in string 2, store the position and nature of the mutation for later output. For example, G551D is a shorthand notation meaning that the 551st amino acid in the protein, normally glycine (G), was changed to aspartate (D) as a result of a mutation.*
4. *Output a list of any differences between the sequences. If there are no differences, output a message stating the sequences are equivalent.*

Test Your Understanding

1. The DNA sequence shown here represents the first part of the wild-type human *CFTR* coding sequence. The sequence shown is the nontemplate strand and is written from 3′→5′. Write out the complement of this sequence, which would be the template strand. For convenience in doing the next step, write this sequence from 3′→5′ as well. Then use this to generate the mRNA sequence, written from 5′→3′. Label at least one end of all sequences.

   3′ CTTTTTTTCAAACCTCTGTTGCGACCGGAAAAGGTCTCCGCTGGAGACGTA

2. Using the genetic code table in Figure 2.4, translate your mRNA using the one-letter amino acid code.

3. The two sequences below represent matching DNA sequences from two patients (A and B) who are concerned they might be carriers of CF alleles. (Like the original sequence shown previously, they are nontemplate strands written 3′→5′; feel free to change them to a more convenient format if you like.) Compare their DNA sequences to the wild-type allele, note any mutations (changes in DNA), and then translate the sequences and determine the effects of any mutations on the CFTR protein. Should either patient be concerned about the possibility of having a child with CF?

```
A: CTTTTTTTCAAACCTCTGTTGCGACCGGAAAATGTCTCCGCTGGAGACGTA
B: CTTTTTTTCAAACCTTTGTTGCGACCGGAAAAGGTCTCCGCTGGAGACGTA
```

4. The previous case is slightly oversimplified: When these patients' *CFTR* genes were sequenced, there was actually not one DNA sequence recovered, but two. How could there be two different sequences? And, would a patient *always* have two different sequences?

CHAPTER PROJECT:

Genetic Screening for Carriers of CF Mutations

Mary and Tom's genetic counselor would likely send them to a genetic testing laboratory who would take a DNA sample from Tom. Because every cell has the same DNA (see BioBackground in Chapter 1), this can be done noninvasively by simply using a swab to remove some epithelial cells from the inside of the cheek. If the lab chose to sequence Tom's *CFTR* gene, their software would use algorithms similar to those we have seen to look for mutations in the DNA and their effects on the protein. In this project, we examine real *CFTR* sequences and arrive at a genetic diagnosis.

Learning Objectives

- Become familiar with tools that can be used to manipulate sequences in a variety of ways and make basic comparisons between sequences
- Understand the structure and orientation of string representations of DNA and protein sequences
- Gain experience with string manipulation in a chosen programming language and its application to DNA and protein sequence data
- Understand how genetic information is computationally decoded and appreciate important complications in working with sequences (introns/exons, start codons, template/nontemplate strand orientation, etc.)

Suggestions for Using the Project

The project in this chapter is primarily designed to build string-manipulation skills in courses that include programming, with a straightforward Web Exploration section to demonstrate existing tools and how they handle input and output of sequence information. For nonprogramming courses, these straightforward algorithms would provide an excellent opportunity for students to try their hand at writing code and appreciate the challenge of converting an algorithm into software, using the Guided Programming Project along with the programming information in the Appendix and the language-specific guides available on the *Exploring Bioinformatics* website. Alternatively, a nonprogramming course can use the Web Exploration as a stand-alone exercise: Although students with biology backgrounds are likely familiar with transcription and translation, seeing these processes represented computationally can be a valuable means of reinforcing these key concepts.

Web Exploration

The earlier Test Your Understanding exercise likely increased your appreciation for bioinformatics software: Although there is nothing at all difficult about manipulating and comparing DNA and protein sequences, the process can be very tedious even for a short sequence, let alone for an actual gene the size of *CFTR*. The ability to reverse, complement, translate, or otherwise manipulate DNA sequences is built into many bioinformatic programs for their users' convenience, but we can also find dedicated tools to carry out these tasks. For this project, we will use the **Sequence Manipulation Suite** (SMS), a set of tools written in JavaScript that run in any Web browser (see References and Supplemental Reading).

Obtaining the Coding Sequence of *CFTR*

For the purposes of this chapter, we limit ourselves to the analysis of the sequence corresponding to the spliced *CFTR* mRNA; analysis of the much longer *CFTR* genomic DNA sequence is complicated by the interruption of the coding sequence by introns, a topic that will be discussed in detail in later chapters. You should already have some familiarity with searching genomic databases. There are many *CFTR*-related sequences in GenBank, so to minimize confusion, try searching the Gene database at the NCBI site for *CFTR*. Open the Gene record for the human *CFTR* gene, then scroll down to find and retrieve the GenBank record (with the entire 189-kb sequence). Within the Gene record, you should also be able to find an accession number for the *CFTR* mRNA that links to this much shorter sequence (also available on the *Exploring Bioinformatics* website). Although you know how to obtain this sequence in FASTA format directly from GenBank, this time copy the entire mRNA sequence, including numbers and spaces, and save it in a text file for convenience.

Tools for Manipulating DNA

Navigate to **SMS**. Notice in the left-hand column the many useful tools that are brought together in a single Web interface. Some are very simple text manipulations but are very useful in working with real sequences—for example, often you have a sequence that is not in FASTA format but the programs you are working with require that format. Use the `Filter DNA` tool on the mRNA sequence you saved (also note that you could use the `GenBank to FASTA` tool to extract the DNA sequence in FASTA format from the complete GenBank record). Save the resulting FASTA-formatted mRNA sequence as a text file for future use (change the comment line to something more useful). Then use this sequence to try the `Reverse Complement` tool; notice the drop-down below the input box allowing you to choose reverse (that is, inverse), complement or reverse-complement output. Because the sequence you saved corresponds to the *CFTR* mRNA (but note that U's have been converted to T's), which option would give you the sequence of the *template* strand of the DNA?

Translating the *CFTR* mRNA

To obtain the amino-acid sequence of the CFTR protein, you need to translate the mRNA. Click the `Translate` tool in SMS. Notice there is some complexity here: The drop-down menus below the input box allow you to choose a reading frame and also a strand to translate. An mRNA can be broken into codons in three ways: The sequence ACUGCCAC . . . could be read as ACU | GCC | AC . . . (giving Thr-Ala . . .) or as A | CUG | CCA | C . . . (giving Leu-Pro . . .) or as AC | UGC | CAC . . . (giving Cys-His . . .). We say there are three **reading frames**. Worse, if we did not know our sequence represented mRNA, it could be either a template or a nontemplate strand of DNA, so we could either translate the strand as written (SMS calls this the "direct" strand) or generate its inverse complement and translate that ("reverse" in this program), which would give us *six* reading frames altogether. Here, at least we know that we want to translate the direct strand. Paste your FASTA sequence into the input box and examine the output for each of the reading frames.

You can quickly see that as soon as we start looking at real sequences, things get a lot more complicated than they might have seemed based on short test sequences. Indeed, you may initially have no clue what to consider as the amino-acid sequence for CFTR. To make the problem easier to visualize, try the `Translation Map` option in SMS instead; this gives the same results but shows all three reading frames at once and how they relate to the DNA sequence; a sample is shown in **Figure 2.5**. Note that here and in the Translate output, stop codons are indicated by asterisks in the amino-acid sequence. We would expect the *CFTR* coding region to take up most of the mRNA and not be broken up by stop codons (we call this a long **open reading frame**, or **ORF**); this consideration should make it easy to identify the correct amino-acid sequence in the correct reading frame. However, remember that real coding sequences start with AUG start codons, and AUG encodes methionine: This amino acid sequence does not start with a methionine. This probably means there are some potential codons *before* the start codon (there is no reason to expect a stop codon immediately before the start codon) that would simply be ignored by the ribosome. The real start of the amino-acid sequence is therefore most likely the first methionine (M) in the ORF, encoded by the first AUG (we further explore how start codons are found later).

To facilitate sequence comparison, you should save a file with just the coding region of the mRNA and one with just the CFTR amino-acid sequence. In the Translation Map output, note the nucleotide position of the A in the ATG start codon and the position of the TAG stop codon at the end of the open reading frame. Then use the `Range Extractor DNA` tool to get a new FASTA-formatted sequence representing just these nucleotides (read the information on how to specify a range carefully and check your output to see if you got what you expected). This is your coding region sequence. Now, use `Translate` (set for reading frame 1 and the direct strand) to get the corresponding amino-acid sequence. If you have done all this correctly, you should have 1,480 amino acids in your final sequence. To demonstrate one more SMS tool, try running `One to Three` on your amino-acid sequence. Although the one-letter amino-acid code is much easier to use in computer programming, many people find the less-cryptic three-letter code easier to work with.

Detecting Mutations

We can use the alignment tools in SMS to look for mutations in *CFTR*. Sequence alignment is a key concept in bioinformatics and becomes complex when sequences are not extremely similar; how alignments are made will be treated in detail in the next chapter. For our application, examining *CFTR* sequences for SNPs, however, we can think of alignment as simply lining up two nearly identical sequences and looking for their differences. To use SMS for this purpose, make a copy of your *CFTR* coding sequence and create a substitution "mutation" by changing one base to a different one. Now, go to the `Pairwise Align Codons` tool in SMS and paste your wild-type sequence into the first input box and your mutant sequence into the second. Don't worry about the parameters for now. Submit the sequences and examine the result. Not terribly revealing, is it? SMS simply returns two FASTA sequences, but what you cannot see using this example is that the sequences contain information about how they should be lined up. In this case, they line up base-for-base, but if that were not true, you would see dashes representing places where one sequence should be shifted over. To visualize the results, copy both sequences, including their comment lines, from the results page and paste them into the input box in the `Color Align Conservation` tool. On the results page, you should see a black background wherever

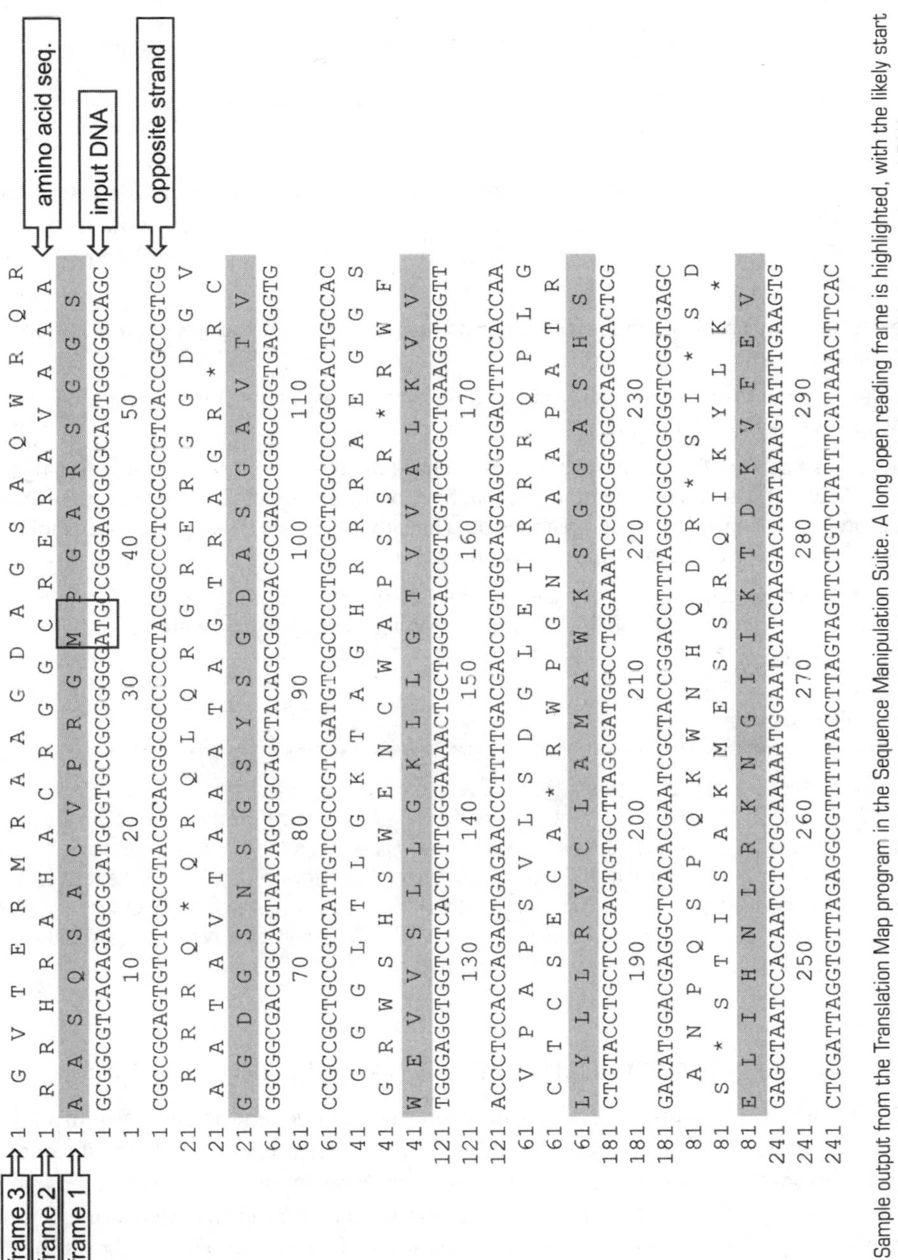

Figure 2.5 Sample output from the Translation Map program in the Sequence Manipulation Suite. A long open reading frame is highlighted, with the likely start codon boxed. Data from: Stothard P (2000) The Sequence Manipulation Suite: JavaScript programs for analyzing and formatting protein and DNA sequences. Biotechniques 28:1102–1104.

the two sequences have the same nucleotide and a white background where they differ; can you find your mutation? Now, translate your mutant coding sequence and align the result with the wild-type amino-acid sequence, this time using `Pairwise Align Protein` (and again `Color Align Conservation`). Did you make a missense mutation, a nonsense mutation, or possibly a silent mutation?

Web Exploration Questions

You should now be prepared to use SMS to get some answers for Mary and Tom. From the *Exploring Bioinformatics* website, you can download the file `CFScreening.txt`, which contains four FASTA sequences representing the coding regions of each of Mary's two alleles and the coding regions of each of Tom's two alleles.

1. Compare each of Mary's alleles with the wild-type coding sequence. Can you identify a mutation in either or both?
2. Now, translate each of Mary's alleles and compare them with the wild-type amino-acid sequence. What differences can you detect?
3. We know Mary is a carrier of the CF allele but does not have the disease. Summarize your findings for Mary's *CFTR* alleles: Describe the mutation(s) that have occurred, discuss how (if at all) they affect the CFTR protein, and explain how your genomic data fit with what Mary already knows, including which allele she must have inherited from each of her parents.
4. Repeat your analysis for each of Tom's two alleles. Is he a carrier of CF? Summarize your findings as in question 3 and determine the probability that Tom and Mary will have a child with CF.

More to Explore

CFTR is one of the best-studied single-gene diseases. The *CFTR* sequences have now been determined for a large number of individuals, and this information has been gathered into a single database, the **Cystic Fibrosis Mutation Database**, providing a valuable resource for CF researchers. Nearly 2,000 different mutations in *CFTR* (not all of which affect the CFTR protein or cause CF) have been catalogued to date. If you would like to explore further, note the position and description of each mutation you identified in Mary and Tom's *CFTR* sequences. Then choose Graphic Search on the Cystic Fibrosis Mutation Database page and examine the *CFTR* exons to find the ones in which your mutations occurred. Look at the specific mutations shown in that region and see if yours are known mutations or new mutations.

Guided Programming Project: Working with DNA and Protein Strings

If your job was to sequence DNA from people seeking genetic screening information and identify mutations, you probably would not want to use the SMS. You would want to have a single tool that manipulates, translates, and compares input sequences and reports on any mutations found. You learned earlier (review Understanding the Algorithm before continuing if you need to) about string manipulation and string comparison algorithms. In this project, you will develop a program that can identify mutations given a pair of equal-length input sequences in text files.

Initially, we simplify our algorithm by placing some constraints on it. We assume that the input sequences will be of equal length, that they are the sequences of coding regions (they begin at the start codon and end at the stop codon), and that they are DNA sequences of nontemplate strands in 5′→3′ orientation. We also assume all nucleotides will be in capital letters and the sequences will contain no white space.

Our algorithm allows the user to compare sequences stored in text files in FASTA format. After retrieving the sequences, the first step in our algorithm is "transcription," converting the DNA sequence to an RNA sequence (because the genetic code is written in RNA form). Next, the RNA string is translated. Recall that translation is the process of reading the genetic information encoded in mRNA and producing an amino acid sequence by "decoding" each codon and adding the appropriate amino acid to an output string. Then, we are ready to compare the amino-acid strings and report any characters that differ between them as the result of mutations, along with the position at which the change occurred. The results should be displayed to the user and also saved to an output file.

The pseudocode (that is, generic programming steps that do not show the syntax of any specific programming language) presented next shows how the algorithms discussed previously can be implemented in a mutation detection program as described. The Putting Your Skills Into Practice exercises will ask you to write a program based on this pseudocode in the language your instructor has chosen for your course; your instructor may then choose to assign additional exercises from this section, allowing you to refine your program further. This application, as well as all the other bioinformatics programs described in this text, could be written in almost any programming language; each has its own set of string manipulation functions, operators, and syntax. You will find a general discussion of key programming concepts in the Appendix to this text, and you can download guides to relevant Perl or Python syntax from the *Exploring Bioinformatics* website. Write a program to implement the mutation detection algorithm using the following pseudocode as a guide.

Algorithm

Mutation Detection Algorithm

Goal: Identify the location of all differences (mutations) between two strings.

Input: Two equal-length DNA sequences, representing the nontemplate strands for protein coding regions in 5′ to 3′ orientation (each in its own input file in FASTA format)

Output: Description and location of all mutations or a message indicating the sequences are identical.

```
// STEP 1: Read in sequences
// create I/O variables infile1, infile2 and outfile
open input file 1: infile1
open input file 2: infile2
open output file: outfile

// initialize variables
seq1 = seq2 = " "
aminoseq1 = aminoseq2 = " "

// read in data
read and discard first line of data in infile1
for each line of data in infile1
    concatenate line of data to seq1
read and discard first line of data in infile2
for each line of data in infile2
    concatenate line of data to seq2
```

```
// STEP 2: Transcription
// Transcribe sequence 1 – repeat steps for sequence 2
for each i from 0 to length of seq1 – 1, inclusive
    if (seq1[i] == 'T')
        seq1[i] = 'U'

// STEP 3: Translation
// create a map (dictionary or hash table) of key/value pairs
//representing each of the 64 possible codons and the amino acid
//(in one-letter code) corresponding to each
codes = map of all codon/amino acid pairs

//Translate sequence 1 – repeat for sequence 2
for each i from 0 to len(seq1)-3, inclusive, iterate by 3
    aminoseq1 = aminoseq1 + codes[seq1.substring(i,3)]

// STEP 4: String Comparison
mctr = 0 // mismatch counter
for each i from 0 to len(aminoseq1)-1, inclusive
    if (aminoseq1[i] != aminoseq2[i])
        //output mutation shorthand, e.g., K136R
        output aminoseq1[i] + (i+1) + aminoseq2[i]
        write aminoseq1[i] + (i+1) + aminoseq2[i] to outfile
        increment mctr
if (mctr==0)
    output "no mismatches found – strings are identical"
```

Putting Your Skills into Practice

1. Write a program in the language used in your course to implement the above pseudocode. Test your program initially with the short wild-type sequence from the Test Your Understanding section and a missense mutant version of this sequence. You can also test it with the mutant B sequence from Test Your Understanding. Be sure you know the expected results for these test sequences before you run the program. Then test it with your wild-type CFTR sequence and Mary's alleles (not Tom's—this is discussed in the On Your Own Project that follows) from the Web Exploration section.
2. Modify your program so it is more flexible, allowing the input sequences to contain either lowercase or uppercase letters by simply converting all input sequences to uppercase.
3. Modify your program so that stop codons are included in your codon table (if you haven't already included them). Then devise a way to represent a stop when you translate your DNA sequences and to show this nonsense mutation in your mutation list. Test your program with mutation A from the Test Your Understanding section.
4. This program could just as well be used to compare two DNA sequences. Modify it so the user can choose whether to compare the DNA sequences directly, translate them before comparing, or both.
5. Allow the user to input either the template or nontemplate strand of DNA *or* an mRNA sequence, in either 5′→3′ or 3′→5′ orientation. Request input from the user to identify the sequences, and then manipulate them as necessary to produce an mRNA sequence before translating. Use the short sequence in Test Your Understanding to generate test sequences for your program.

■ **On Your Own Project**: Identifying Insertions and Deletions

Understanding the Problem: Insertions and Deletions

A major constraint in our mutation detection program was that the strings had to be the same length so they could be compared amino acid for amino acid or base for base. However, this is unrealistic, because real mutations are often insertions or deletions (generically, **indels**) of one or more nucleotides, and these cannot be handled properly by our program. Consider, for example, the simple DNA strings `ACGTTA` and `ACTA`, where it seems a two-nucleotide deletion event has occurred. Comparing nucleotide 1 with 1, 2 with 2, and so on, our program would report the mutations G3T and T4A and would then encounter an error condition when it runs out of characters on the shorter string. What it should do is recognize that the second string is really `AC--TA` and report something like del3GT. How can we properly detect and report mutations when the strings are not the same length?

Solving the Problem

If the sequence strings are of different lengths, then they must be aligned so that the appropriate characters match up. You saw an example of this when you used SMS to align sequences in the Web Exploration section. The difficulty lies in deciding *how* to align the strings: Our two previous sequences could be aligned in five different ways, assuming a two-base deletion:

```
ACGTTA    ACGTTA    ACGTTA    ACGTTA    ACGTTA
ACTA--    --ACTA    A--CTA    AC--TA    ACT--A
```

Of course, this neglects the possibility that the two deletions did not occur together and that the second sequence might thus be represented, for example, as `A-C-TA`. In fact, how do we know this mutation is a deletion at all; couldn't it be a two-nucleotide *insertion* into the second sequence to produce the first? We keep our task manageable for now by adding some constraints: We assume that the first input sequence is the reference or wild-type sequence and that only a single mutational event (substitution, insertion, or deletion) has occurred at one site to produce the mutated sequence. We can then develop a more sophisticated mutation detection algorithm that aligns unequal-length sequences to minimize the number of mismatches. In the example, you can see that the fourth alignment is the most satisfactory, with only two mismatches, whereas the others have three or four.

Programming the Solution

One trick to solving new problems is to find similarities with previously solved problems. This problem is similar to the problem solved in the Guided Programming Project earlier in the chapter, except now you need a slightly more sophisticated algorithm to account for the possibility of different length sequences due to indels. Your program should do the following:

1. Input and manipulate DNA sequences as before. Designate the first strain entered as the wild-type or reference strain. Transcription and translation can be omitted for now; to avoid further complexity, we compare only the nucleotide sequences.
2. Determine the lengths of the two sequences. If they are equal, proceed to compare the two strings for substitutions as before.

3. For strings of unequal length, determine the number of nucleotides deleted from or inserted into the second string relative to the first. This is the number of gaps that need to be inserted into the shorter sequence to align it with the longer.
4. Construct all possible alignments by placing the desired number of gaps at each possible position in the shorter sequence and determining which alignment has the fewest mismatches. Based on this alignment, report the position and nature of the insertion/deletion.

Develop a step-by-step algorithm to solve this new problem by extending the algorithms used in the Guided Programming Project. Write a program in the language of your choice to implement your algorithm. Test your program using the short sequence given in the Test Your Understanding section as your reference sequence and sequences with a substitution or with one, two, or three nucleotides inserted or deleted as your mutant sequences. If your program behaves as expected, test it using the full-length *CFTR* coding sequence and see if it can correctly identify the mutations in Mary's and Tom's *CFTR* alleles.

When you run your program using full-length sequences with indels, pay attention to its execution time. Do you notice a distinct increase in the time it takes to complete the task when alignment of the two sequences is necessary? How many individual alignments does the program have to test for these sequences? Our alignment method is sufficient for this task, but this brute-force method is not very efficient and certainly not efficient enough to choose the best possible alignment when the constraints are relaxed. In the next chapter, we will consider more sophisticated methods to identify optimal alignments that are far less computationally intensive.

More to Explore

A useful modification of your program is to always align the DNA sequences and output the best alignment, showing the user directly where and how the mutation affects the sequence. An even more challenging modification is to deal with how indels affect the amino-acid sequence of the protein. Notice that indels whose length is a multiple of three look like insertions or deletions of one or more amino acids, whereas those of other lengths introduce frameshifts that need to be represented to the user in some clear format.

Connections: Future of Genetics

The explosive growth of genetic and genomic information also brings new questions and uncertainties. Could Mary's husband face higher medical insurance premiums if found to be a carrier of CF? If they have a CF child, could the child's insurability be questioned? What information does Tom have to disclose to his employer or insurance company? If he were a carrier, what would be the chances that their child would have the disease? Is it acceptable for the couple to decide to take this risk? Can screening be compelled to reduce the incidence of genetic disease? Many consider genetic testing to be both a blessing and a curse. The value of knowing that one is genetically destined or predisposed to develop a particular disease is questionable if the disease cannot be cured. Yet, understanding these genetic risks may lead to wiser family planning or better lifestyle choices, and advances in personalized medicine depend on genetic testing to select the best available therapy for a particular individual.

As these questions are debated, genetic research continues to advance, including development of possible genetic treatments. **Gene therapy** is the idea of curing a genetic disease by changing an individual's DNA. This might be done by changing DNA only in the cells affected by the disease (**somatic cell** gene therapy) or by making a change that would affect the sperm or egg cells so the cure would be passed on to future generations (**germ-line** gene therapy, which is far more controversial). So far, only somatic cell gene therapy has been tried, and although some successes have been achieved in clinical trials, many technical hurdles remain before gene therapy becomes safe, effective, and inexpensive enough for wide use. Also daunting are questions about the appropriateness of such therapy. Some issues are ethical in nature (are we "playing God" by making these genetic changes? What kinds of deliberate genetic change are allowable—changes to appearance, behavior, intelligence, or ability?), but others are scientific. CF is a case in point: Although there is no doubt we would like to end the suffering resulting from CF, there is also evidence that carrying one CF allele can be beneficial, possibly reducing the likelihood of cholera. Although cholera may be under control today, these and other "bad" alleles may well have positive effects of which we are presently unaware, and the potential implications of gene therapy must be considered very carefully.

BioBackground: The Genetic Code and Decoding DNA

DNA is a double-stranded molecule, with the two strands in **antiparallel** orientation: If one strand is thought of as running left to right, the other runs right to left. The two ends of a DNA strand are biochemically distinct, with one end terminating in a phosphate group and the other in a hydroxyl (-OH) group. Based on chemical naming conventions, we call these the **5′** ("five prime") and **3′** ("three prime") ends, respectively. Antiparallel orientation then means the 5′ end of one strand is adjacent to the 3′ end of the other (**Figure 2.6**).

RNA is a single-stranded molecule, and in the cell, it is made by unzipping the two DNA strands for a short distance and base-pairing RNA nucleotides with one of the DNA strands, the **template strand**. Identification of the template strand for a particular gene depends on the location of the promoter and other signals. The RNA is therefore **complementary** to the template strand, but its sequence reads the same as the opposite DNA strand, the **nontemplate strand** (Figure 2.6)—except that in RNA, the nucleotide U pairs with A and is used everywhere that T would be used in DNA.

To decode the information in the mRNA and obtain the amino-acid sequence of the corresponding protein, it is necessary to know the **genetic code**. The mRNA is read in three-base groups called codons; thanks to the work of Francis Crick, Marshall Nirenberg, and others, we know the amino acid represented by each of the 64 possible codons (Figure 2.4). In the cell, the ribosome identifies the start codon, AUG, within the transcript. Starting at that point, short

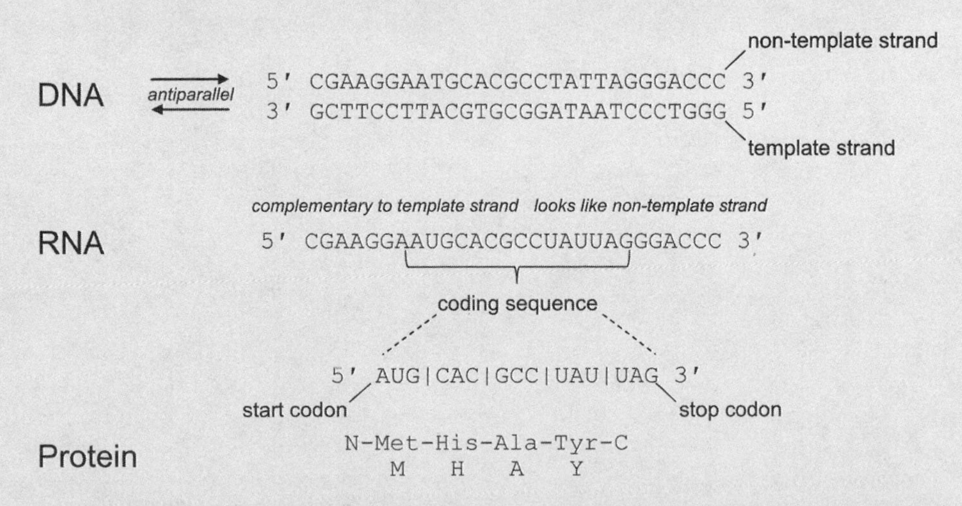

Figure 2.6 Important features of DNA, RNA, and protein sequences that must be dealt with in computational manipulation algorithms.

transfer RNA molecules with three exposed nucleotides at one end and the corresponding amino acid at the other base-pair one by one with the mRNA. As each tRNA brings an amino acid into the ribosome, the ribosome joins the amino acids together.

Codons are read in the 5′ toward 3′ direction without overlapping until a stop codon (UAG, UGA, or UAA) is reached. The start codon encodes an amino acid, methionine, but the stop codon does not. Like nucleic acid chains, proteins have two distinct ends: the **amino terminal** or **N-terminal** end (represented by N in Figure 2.6), which is made first and corresponds to the 5′ end of the mRNA coding sequence, and the **carboxyl terminal** or **C-terminal** end (C in Figure 2.6), corresponding to the 3′ end of the coding sequence.

When a mutation changes a DNA sequence, it may replace one nucleotide with another: a **substitution**. You can see from the genetic code (Figure 2.4), however, that not all substitutions change the protein: A mutation that changes GGG to GGA is a **silent** mutation, because both codons represent glycine. If a substitution mutation results in changing an amino acid, then it is a **missense** mutation, and if it changes an amino-acid codon to a stop codon (terminating the protein early), then it is a **nonsense** mutation. Nucleotides can also be added to or removed from the DNA: **insertion** or **deletion** mutations. The insertion or deletion of one or two nucleotides will cause a **frameshift**: *all* codons after the mutation point is misread.

References and Supplemental Reading

Cystic Fibrosis Alleles and Their Consequences

Rowntree, R. K., and A. Harris. 2003. The phenotypic consequences of *CFTR* mutations. *Ann. Hum. Genet.* **67**:471–485.

Sequence Manipulation Suite

Stothard, P. 2000. The Sequence Manipulation Suite: JavaScript programs for analyzing and formatting protein and DNA sequences. *BioTechniques* **28**:1102–1104.

© Sashkin/ShutterStock, Inc.

Chapter 3

Sequence Alignment:

Investigating an Influenza Outbreak

Chapter Overview

This chapter focuses on algorithms for optimal alignment of DNA sequences. Students in both programming and nonprogramming courses will understand how dynamic programming techniques can be used to make the complex problem of gene alignment tractable and, through the use of Web-based tools, how the choice of alignment parameters can influence the biological relevance of the results. Students will also consider how a basic algorithm can be modified to answer different biological questions. Students in programming courses will develop their own solutions that implement these algorithms.

Biological problem: Origin of new influenza virus strains

Bioinformatics skills: Optimal global, semiglobal, and local alignments of DNA sequences; gap penalties and alignment parameters

Bioinformatics software: EMBOSS implementations of pairwise alignment algorithms

Programming skills: Two-dimensional arrays, dynamic programming, backtracking

Understanding the Problem:

The 2009 H_1N_1 Influenza Pandemic

In March 2009, epidemiologists responsible for influenza surveillance at the Centers for Disease Control and Prevention (CDC) and the World Health Organization (WHO) were surprised by an outbreak of influenza in Mexico City. Because influenza virus mutates rapidly, the strains that are circulating change from year to year, necessitating annual revaccination; CDC and WHO are charged with monitoring flu virus strains and determining which will be used for vaccine development. In addition, these agencies monitor both human and animal influenza cases to identify new strains, watching for the emergence of a ***pandemic*** *virus—one capable of causing a severe, multicontinent outbreak. Uppermost in the minds of these scientists is the desire to prevent a repeat of the 1918 influenza pandemic—the single deadliest infectious disease event in history, infecting half the world's population and killing at least 20 million people in 120 days (***Figure 3.1***; see also References and Supplemental Reading at the end of the chapter).*

Figure 3.1 The rapid spread and severity of the 1918 influenza pandemic placed an enormous burden on healthcare workers and facilities. Depicted here is a demonstration at the Red Cross emergency ambulance station in Washington, D.C. Courtesy of Library of Congress, Prints & Photographs Division [reproduction number LC-USZ62-126995].

In addition to ordinary, seasonal human viruses, WHO and CDC had been keeping tabs for some years on an avian (bird) influenza virus strain known as H_5N_1 that at the time they believed posed the greatest risk of a new pandemic. This "bird flu" virus has caused severe infections in domestic fowl and in humans in direct contact with birds (such as poultry farmers) but has thus far remained incapable of efficient transmission from person to person. In reality, however, the next human pandemic resulted not from H_5N_1 but from a previously unknown strain of H_1N_1 that had escaped detection. When Mexican authorities reported a number of cases of influenza caused by this relative of the 1918 flu strain, public health officials were concerned about the possibility of widespread severe illness. Particularly alarming were reports of severe cases and deaths among the young and middle aged, as virulence for these age groups (seasonal flu has serious health consequences mostly for infants and the elderly) was a hallmark of the 1918 virus. Fortunately, it later became clear that this new H_1N_1 virus was no more dangerous than ordinary seasonal strains. Nonetheless, in the interval between identification of the new strain and development of a vaccine, it caused at least 8,000 deaths and a large number of precautionary school closings.

What exactly is a "new strain" of influenza virus, and how is a new strain identified? What makes one strain a dangerous pandemic virus and another a mild seasonal virus? Why are some strains transmitted easily among humans, whereas others are largely confined to animals?

Bioinformatics Solutions:
Sequence Alignment and Sequence Comparison

Alignment of the sequences of two genes or proteins refers to matching them up in what we hope is a biologically relevant way to determine how similar they are. Sequence alignment is possible when the sequences are evolutionarily related: Similar sequences are similar because they are descended from the same common ancestor, with the differences among them resulting from mutation (for more detail, see BioBackground). **Figure 3.2** shows an example in which many different oxygen-carrying proteins have similar sequences because they all have the same origin.

The problem of alignment was introduced briefly in the last chapter, where sequence comparison was used to detect mutations. Sequence alignment is also used in developing phylogenetic trees based on molecular data, assembling genome sequences, predicting protein structure and function, and numerous other bioinformatics applications. Indeed, it would be fair to say that sequence alignment is *the* key technique in bioinformatics—and also a difficult computational problem because of the complexity of genomic information. This chapter presents an algorithm for identifying the best alignment of two sequences, with projects in which you will use this technique to investigate influenza virus strains and

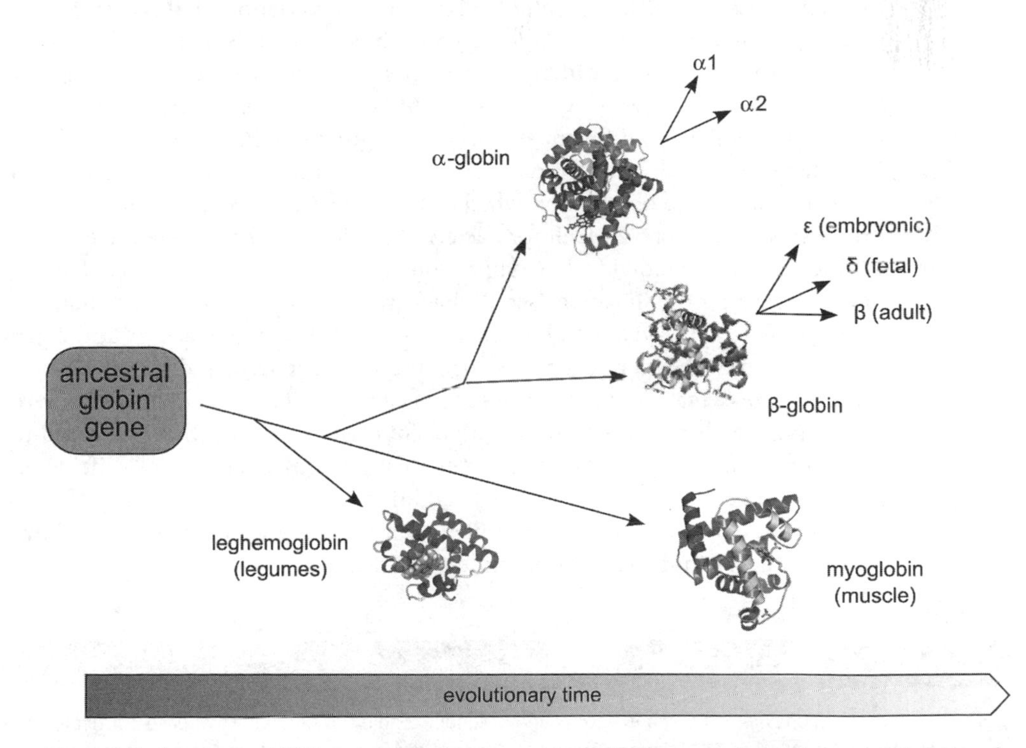

Figure 3.2 Alignment of DNA and protein sequences is possible because of evolutionary relationships. In this example, evolution from an ancestral globin gene is thought to have produced a variety of oxygen-carrying proteins—including the two subunits of hemoglobin found in human blood, myoglobin found in the muscles of mammals, and even leghemoglobin made by leguminous plants. Thus, all these different proteins would be encoded by genes with recognizably similar sequences. Structures from the RCSB PDB (www.pdb.org): leghemoglobin, PDB ID 2GDM (E. H. Harutyunyan et al. (1995) The structure of deoxy- and oxy-leghemoglobin from lupin. *J. Mol. Biol.* 251:104–115); alpha-globin and beta-globin, PDB ID 4HHB (G. Fermi and M. F. Perutz (1894) The crystal structure of human deoxyhaemoglobin at 1.74 A resolution. *J. Mol. Biol.* 175:159–174); myoglobin, PDB ID 1MBO (S. E. V. Phillips (1980) Structure and refinement of oxymyoglobin at 1.6 A resolution. *J. Mol. Biol.* 142:531–554).

Alignment of a gene from two closely related viruses

```
Hemagglutinin gene from virus A:  ATGAACGCAATACTCGTAGTT...
                                  ||||| |||||||| ||||||
Hemagglutinin gene from virus B:  ATGAAGGCAATACTAGTAGTT...
```

Alignment of a gene from two distantly related viruses

```
Hemagglutinin gene from virus A:  ATGAACGCAATACTCGTAGTT...
                                  ||| ||| ||| |||| |   |
Hemagglutinin gene from virus C:  ATGCACGAAATGCTCGGACCT...
```

Figure 3.3 An example showing how sequence alignment can demonstrate similarity, and thus relatedness, of two DNA sequences.

their virulence. Subsequent chapters will explore how variations of this basic algorithm may be extended to apply to many other important biological problems.

Despite their obviously different characteristics, St. Bernards and chihuahuas are members of the same species, *Canis familiaris*. Although we usually use the term "breed," we could think of them as different **strains** of dog: groups within a species that have distinct, inheritable genetic characteristics. Even in animals and plants, it can be very difficult to determine by simple observation whether two organisms belong to the same species; the problem is much more difficult for bacteria and viruses, where there are few if any visual distinctions among individuals. Comparison of DNA or protein sequences has become the new standard for classification (see BioBackground). Bioinformatic techniques provide a means of comparing genes and identifying species or strains. Each year, the genomes of many influenza viruses isolated from patients are sequenced, and it is the comparison of these sequences that allows agencies such as CDC to determine whether new viruses have arisen and whether they are minor variants of existing viruses (this is referred to as antigenic "drift") or are very different from circulating viruses (antigenic "shift" variants) and have pandemic potential (**Figure 3.3**). In addition, comparison of the genes of a new variant with known viruses that are highly virulent or more moderate in their effects allows experts to predict the potential severity of influenza outbreaks.

BioConcept Questions

Computational techniques for gene alignment depend on understanding of the biological basis for gene comparison and the meaning of similarity and variation among the genes of different organisms. Use these questions to test your biological understanding; read the BioBackground box at the end of the chapter if you find that you need a better foundation.

1. How is similarity between genes related to the biological concept of descent from a common ancestor?
2. Given the sequences ACGAT and CGATC, why is the simplest alignment ACGAT
CGATC not a very satisfactory one? What do we have to allow for in order to generate an alignment that appears more biologically relevant?

3. List all the possible ways to align the very short sequences ACC and ACT. Discuss why "brute-force" alignment (trying all the possible combinations to identify the best one) is not a practical method of aligning real genes.
4. Often, it is necessary to introduce gaps into one or both sequences to align them optimally. However, most alignment programs penalize gaps to keep them to a minimum. Why are gaps potentially problematic, particularly for sequences that represent coding regions?
5. The influenza virus mutates so rapidly that you would likely be able to identify at least a couple of mutations over the length of the complete virus genome even if you sequenced two viruses from two different patients within the same influenza outbreak. What might be some considerations in deciding whether two viruses with different genome sequences actually represent two different strains?

Understanding the Algorithm: Global Alignment

Learning Tools

From the *Exploring Bioinformatics* website, you can download a demonstration spreadsheet that shows visually how the Needleman Wunsch algorithm aligns short sequences. Try the examples in the text or make up your own sequences to see how the algorithm deals with mutations, differences in length, and so on. Files are available for Excel and OpenOffice for Windows, Linux, and Mac OS.

The simple algorithms in the previous chapter that in essence align two genes to look for mutations are limited: One algorithm required the genes to be of the same length, whereas the other used an inefficient trial-and-error method. To be able to align *any* two sequences, we need a flexible algorithm that will match them up in a meaningful way, accounting for differences in length due to indels and recognizing that over evolutionary time mutation may have made similar genes look quite different. The algorithm needs a means of discriminating between better and worse alignments and also a scoring system to decide *how* similar the genes are.

Here, we discuss an algorithm for optimal, global alignment of pairs of genes published by Saul Needleman and Christian Wunsch in 1970 (see References and Supplemental Reading). This algorithm and modifications of it (discussed later in this chapter) are still widely used today, and the ideas they are based on are at the root of many other comparison algorithms as well. Indeed, it may interest you to know that when an Internet search engine such as Google asks, "Did you mean . . . ," it is using an algorithm very similar to this one to match what you typed with common search words.

Optimal Alignment and Scoring

To compare two genes, such as the HA genes of two different influenza virus strains, we want to look for matches and mismatches along their entire lengths: a **global alignment**. (Reasons to compare only parts of genes are discussed later.) Global alignment is a technique used to compare sequences in their entirety; the Needleman-Wunsch algorithm is also a **pairwise** alignment algorithm, because it compares a sequence to only one other sequence at a time.

Sequences can be aligned in many different ways. For example, three ways to align the short sequences ACGTACT and ACTACGT are shown below:

```
ACGTACT    ACGTAC-T    ACGTACT----
ACTACGT    AC-TACGT    ----ACTACGT
**    *    ** *** *        ***
```

If we do not allow for insertions or deletions, there is only one way to align these sequences (left), but if we make the biologically reasonable assumption that indels could have occurred, we get many more possibilities. The hyphens used in the center and right alignments represent **gaps**, indicating that insertions in one sequence or deletions in the other occurred at these points.

Which alignment is best (**optimal**)? To decide, we need a scoring system. If we simply count nucleotides that match, then the introduction of one gap in each sequence (center) gives us a much better score (6) than simply aligning the ungapped sequences (3). However, indels pose a biological problem, because they can create frameshifts; thus, we should use them with caution. Intuitively, we recognize that the left alignment is far superior to the rather cumbersome right one, but both have three matches according to our simple scoring system. A more sophisticated scoring system (we could call this a **scoring metric**) would award a **match score** (or **match bonus**) for nucleotides that match, a **mismatch score** (or **mismatch penalty**) for nucleotides that do not, and a **gap penalty** where a gap was introduced. For example, if the match score is 1, the mismatch score is 0, and the gap penalty is −1, then the left alignment still scores 3, the center alignment scores 4, and the right one scores −5 (matching our judgment that this is likely to be a poor choice from a biological standpoint).

An obvious way to do a global alignment is simply to try every possibility and see which one gives the best score. However, even for these two short sequences, permitting gaps gives more than 40,000 possible alignments; that number quickly becomes staggering if we are working with real genes consisting of thousands of nucleotides. This is in fact an intractable problem even for a computer: A programmer would say that it is not bounded by polynomial time, meaning the time required to arrive at a solution increases so rapidly as sequence length increases as to become impractical.

The key element of Needleman and Wunsch's now-famous article was a solution based on **dynamic programming**. A dynamic programming algorithm divides a problem into a series of smaller subproblems, solves them, and then uses these solutions to build the solution to the original problem. Needleman and Wunsch solve the problem of a global, optimal alignment of large sequences by using a matrix of partial alignment scores and then backtracking along a path to the best possible alignment(s). This clever approach allows all optimal alignments to be found quickly, even for long sequences.

Needleman-Wunsch Algorithm

Let's see how the Needleman-Wunsch algorithm works to align two short sequences: `CGCA` and `CACGTAT`. We use a match score of 1, a mismatch score of 0, and a gap penalty of −1. First, construct an $N \times M$ matrix, where N is the length of the first sequence + 1 and M is the length of the second sequence + 1. Each position in the matrix represents a possible way to align part of the sequence. If two identical, equal-length sequences were being aligned, the matching nucleotides would line up right down the diagonal. In our example, however, we will obviously need at least two gaps, because one sequence is two nucleotides shorter than the other. Even when the two sequences are of equal length, gaps could be needed to obtain the optimal alignment in order to account for indels. These gaps move the matching nucleotides off the diagonal. We need to account for this as we initialize the matrix. We start with a zero in the first cell of the matrix and then initialize the first row and first column by adding the gap penalty (−1) to each successive cell, as shown in **Figure 3.4A**. These initial values show what happens if we have to introduce a gap at the beginning of one of the sequences. If a single gap was added to the beginning of the sequence, its maximum score would be reduced by one, two for a double gap, and so on.

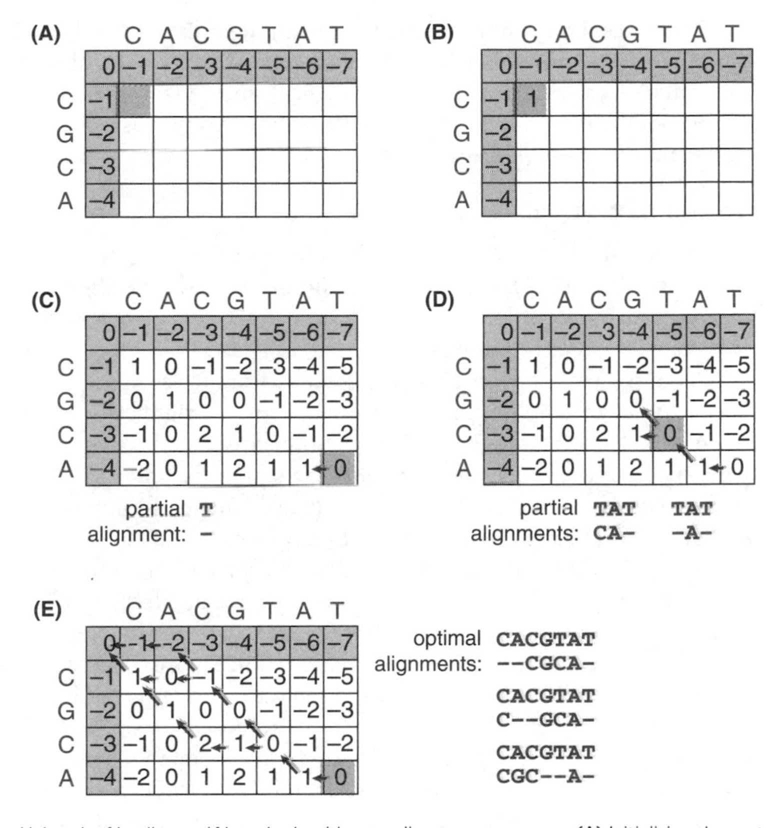

Figure 3.4 Using the Needleman-Wunsch algorithm to align two sequences: (A) Initializing the matrix using gap penalties; (B) Filling in the matrix using the best subscore; (C) The completed matrix with the optimal score (blue cell) and first backtracking step; (D) Backtracking through the matrix, with two possible paths shown; (E) The completed alignments.

Now we are ready to fill out the rest of the matrix, which we do by computing the optimum (maximum) score for each possible partial alignment. Each cell in the matrix represents a partial alignment: For example, the blue cell in Figure 3.4A represents the alignment of the C in the long sequence with the C in the short sequence. At each point, there are three choices:

1. If the two nucleotides match, their score is 1, but if they mismatch, they score zero. Add this match or mismatch score to the score diagonally above and to the left of the cell. This represents aligning nucleotides without leaving a gap. In our example, C matches C, so the score (representing the alignment of C with C) is 0 (from the cell on the diagonal) plus 1 for the match, or 1 total.
2. *Or*, we could introduce a gap in the short sequence, represented by moving horizontally rather than diagonally (moving to the next nucleotide along the top sequence but *not* making a corresponding move to the next nucleotide in the left sequence). The gap penalty is −1, so in our example, we add −1 to the score in the cell to the left of the blue cell: $-1 + -1 = -2$.
3. *Or*, we could introduce a gap in the long sequence, represented by adding the gap penalty to the score in the cell above the blue cell: $-1 + -1 = -2$. We want an optimal alignment in the end, so we should choose the *best* possible score for each partial alignment; in this case, the best of the three options is 1, so we put a 1 in the blue cell (**Figure 3.4B**).

This process now continues for the remaining cells of the matrix. In the cell to the right of the blue cell in Figure 3.4B, our choices are −1 (−1 on the diagonal + 0 for the A vs. C mismatch), 0 (for a gap in the short sequence), and −3 (for a gap in the long sequence), so 0, the best of the three, goes in this cell. Repeating this process for the remaining cells results in the matrix shown in **Figure 3.4C**.

Generating the Alignment

Remember that this is a *global* alignment, so we are comparing the two sequences along their entire lengths. That means the optimum score for the alignment as a whole is always represented by the number in the bottom-right cell of the matrix (at the end of both sequences, the blue cell in Figure 3.4C): in this case, 0.

Of course, we don't want just the score; we want to see *how* the sequences can be aligned optimally. To accomplish this, start from the bottom-right cell and work backward to determine how that subscore was obtained. In this case, the zero resulted from adding the gap penalty to the cell to its left, representing a gap in the short sequence, as indicated by the arrow in Figure 3.4C. So, the T in the long sequence is aligned with a gap in the short sequence (partial alignment at the bottom of Figure 3.4C).

Now, follow the path one cell to the left and consider the 1 there. It must have come from adding the match score to the cell diagonally above and left, so now you know that you can align the two A's and move diagonally (**Figure 3.4D**). Now we have an interesting situation. The zero in the next cell in the path (blue cell in Figure 3.4D) could have been generated *either* by adding the gap penalty to the cell on its left *or* by adding the mismatch score to the cell diagonally left. This means we have *two* possible paths from this point and thus two possible alignments that give equally good scores: one in which we add a gap to the short sequence and one in which we allow C and T to mismatch (arrows in Figure 3.4D). It is entirely possible for there to be more than one way to optimally align two sequences—and this is a great example of why real-world research requires the good judgment of scientists who understand both biology and computational algorithms.

We can now continue this way until we reach the upper-left cell of the matrix. Along the way, another point is reached at which two paths give the same score. Thus, there are three optimal ways to align these sequences, each giving an overall score of zero, as shown in **Figure 3.4E**.

We can change the scoring parameters (match and mismatch scores and gap penalty) based on the problem we are trying to solve. For example, to compare two protein coding genes, it makes sense to penalize gaps significantly because of the frameshift problem. But in genes for noncoding RNAs, a gap may be no worse than a mismatch, and we might set our gap penalty lower. Or, if we only wanted highly similar sequences to give good scores, we might penalize both gaps and mismatches.

The Needleman-Wunsch algorithm provides a straightforward way to find optimal, global alignments, and its use of dynamic programming (each cell in the matrix is the solution to a subproblem that is not computationally intensive to obtain) allows it to run efficiently even when long sequences are being compared. Furthermore, simple modifications of this basic algorithm allow different kinds of alignment that can provide additional information.

Test Your Understanding

1. How would the Needleman-Wunsch algorithm align the sequences ACGTACT and ACTACGT? Try them by hand or use the spreadsheet tool from the text website.
2. For a more challenging problem, find all the possible optimal alignments for the sequences CTAG and CGCTAATC. You should find 10 altogether; the score for each should be −1.

3. Now try aligning CAG with TTTCAGCAGTTT. What do you expect will happen? Are you surprised by what actually happens?

   Question 3 points out a problem with using global alignment to compare two sequences of very dissimilar lengths. There might in fact be a good match for the short sequence within the long sequence (e.g., perhaps the short sequence is one conserved domain of a larger protein), but the introduction of many gaps can prevent a global alignment algorithm from finding it. A solution is to use a **semiglobal** (sometimes called "glocal") alignment technique that does not penalize **terminal gaps**—those that occur at the beginning or end of the alignment.

4. How would you modify the Needleman-Wunsch algorithm to carry out a semiglobal alignment? *Hint: Only two changes in how the matrix is used are required. Consider what parts of the matrix represent the terminal gaps.*

CHAPTER PROJECT:

Investigation of Influenza Virus Strains

When the first known cases of influenza caused by the 2009 H_1N_1 virus appeared, sequencing and analysis of the new virus' genome was a high priority, not only to understand its origin and whether it truly represented a distinct strain but also to understand its potential virulence. In this chapter's projects, you will analyze sequence alignments to examine the relatedness of 2009 H_1N_1 to seasonal H_1N_1 strains and to the 1918 H_1N_1 pandemic virus and investigate the virulence of H_5N_1 human isolates. You will also explore how changing alignment scoring parameters can increase the biological relevance of the results.

Learning Objectives

- Understand the value of aligning genes and some practical applications of this technique
- Gain familiarity with the use of Web-based alignment tools to explore sequence similarity and understand how to modify their parameters
- Know how the Needleman-Wunsch algorithm optimally aligns any two sequences
- Understand how the Needleman-Wunsch algorithm can be modified to yield other alignments

Suggestions for Using the Project

This project is designed to be used in courses that require programming skills as well as those that do not. Below are suggestions for modules of the project that instructors might choose to use in these two types of courses. Instructors should also feel free to ask questions of their own that use these same skills.

Programming courses:

- Web Exploration: Experiment with the Needleman-Wunsch algorithm and the effect of gap penalty parameters as well as the benefits of local alignment (Smith-Waterman algorithm). Parts I, II, and III can be used independently.
- Guided Programming Project: Implement the Needleman-Wunsch algorithm in a programming language of your choice.

- On Your Own Project: Modify the code for the Needleman-Wunsch program to implement a local alignment algorithm.

Nonprogramming courses:

- Web Exploration: Experiment with the Needleman-Wunsch algorithm and the effect of gap penalty parameters as well as the benefits of local alignment (Smith-Waterman algorithm). Parts I, II, and III can be used independently.
- On Your Own Project: Identify modifications to the Needleman-Wunsch algorithm that would convert it to a local alignment algorithm.

Web Exploration

Part I: Pairwise Global Alignment with the Needleman-Wunsch Algorithm

The genomes of influenza viruses are divided into eight segments, each representing essentially the coding information for a single protein. Segment 4 contains the gene for **hemagglutinin (HA)**, the viral surface protein essential for the initial interaction between the virus and its host cell. HA is one key determinant of which host(s) a particular virus can infect, because the virus cannot replicate or cause disease without being able to first bind to a host cell. The HAs of one of the major seasonal human viruses circulating before 2009, the 2009 H_1N_1 pandemic virus, and the 1918 human pandemic virus are all classified as the H_1 type, whereas recent outbreaks of severe avian flu are caused by a virus with HA classified as H_5. These classifications are based on binding of antibodies of known specificity, but sequence alignment provides much more detailed information about similarities and differences and where changes have occurred.

We can use the Needleman-Wunsch algorithm to compare influenza virus HA segments. To start with, let's see how the 2009 H_1N_1 virus—the reference strain is designated A/California/07/2009 (H_1N_1)—compares with the human seasonal H_1N_1 virus that was currently circulating at that time, A/Brisbane/59/2007 (H_1N_1). Download the DNA sequences of segment 4 for both viruses from the *Exploring Bioinformatics* website. We align the sequences using EMBOSS, a suite of alignment tools produced by the European Bioinformatics Institute (somewhat parallel to the U.S. NCBI). At the **EBI-EMBL's EMBOSS Web page** (not the page for the EMBOSS software itself), you should see a list of programs for pairwise sequence alignment. Under the heading `Global Alignment`, the program Needle is an implementation of the Needleman-Wunsch algorithm.

From the EMBOSS site, choose the version of Needle that compares nucleotide sequences, and then paste your two sequences into the designated text boxes. Notice that you can set some parameters for the comparison, most notably the gap penalty. Needle uses an **affine** gap penalty, which means it imposes a larger penalty when a new gap is added and a smaller penalty when that gap is extended (our earlier example used a **linear** gap penalty). Leave the parameters set to the defaults for now.

Run Needle to align your two sequences; your results should look similar to **Figure 3.5**. At the top, you will see parameters such as the gap penalty and two measures of similarity: the number and percentage of matching nucleotides (labeled "Identity") and an alignment score (based on the scoring matrix, in this case awarding a match bonus of 5). In the alignment itself, matching nucleotides are shown by a | character, mismatches by a dot (.), and gaps by a dash (-).

```
#=======================================
# Aligned_sequences: 2
# 1: H5N1_NA
# 2: 2009_H1N1_NA
# Matrix: EDNAFULL
# Gap_penalty: 10.0
# Extend_penalty: 0.5
#
# Length: 1417
# Identity:     1160/1417 (81.9%)
# Similarity:   1160/1417 (81.9%)
# Gaps:           70/1417 ( 4.9%)
# Score: 4944.0
#=======================================

H5N1_NA            1 ATGAATCCAAATCAAAAGATAATAACCATTGGGTCAATCTGTATGGTAAT     50
                     |||||||||||.||||||||||||||||||||.||..||||||||..|||
2009_H1N1_NA       1 ATGAATCCAAACCAAAAGATAATAACCATTGGTTCGGTCTGTATGACAAT     50

H5N1_NA           51 TGGAATAGTTAGCTTAATGTTACAAATTGGGAACATGATCTCAATATGGG    100
                     ||||||.|.||.||||||.|||||||||||.|||||.||||||||||||.
2009_H1N1_NA      51 TGGAATGGCTAACTTAATATTACAAATTGGAAACATAATCTCAATATGGA    100

H5N1_NA          101 TCAGTCATTCAATTCAGAC-AGGGAATCAAAACCAAGTTGAGCCA-----    144
                     |.||.||.|||||||| || .|||||||||||.||..||||..||
2009_H1N1_NA     101 TTAGCCACTCAATTCA-ACTTGGGAATCAAAATCAGATTGAAACATGCAA    149

H5N1_NA          145 --------------------------------------------------    144

2009_H1N1_NA     150 TCAAAGCGTCATTACTTATGAAAACAACACTTGGGTAAATCAGACATATG    199

H5N1_NA          145 -----ATCAGCAATACTAATTTTCTTACTGAGAAAG-CTGTGGCTTCAGT    188
                          ||||||||.||.||.|||..|.||| ||.|| |.||||.|||.||
2009_H1N1_NA     200 TTAACATCAGCAACACCAACTTTGCTGCTG-GACAGTCAGTGGTTTCCGT    248
```

Figure 3.5 Sample output from the EMBOSS Needle program, showing scoring data and part of an alignment for two sequences. Matching nucleotides are represented in the alignment by a vertical line, mismatches by a dot, and gaps by a dash. Generated from: EMBOSS Needle/European Bioinformatics Institute.

Web Exploration Questions

1. How many matching nucleotides are there between your two sequences? What is the alignment score?
2. How many gaps were needed to align these sequences? Is there any particular pattern to where or how the gaps occur?
3. Can you suggest where the coding sequence might occur within this segment? What is your evidence?

Nearly all of segment 4 consists of coding sequence, so we would expect indels—especially one- or two-nucleotide indels—to be mutations with serious consequences for the HA protein. Considering this, perhaps it would be valuable to consider strongly penalizing gaps: Try setting the gap opening penalty to 50, rather than the default 10.

Web Exploration Questions

4. What is the logic behind the affine gap penalty, which imposes a large penalty for opening a new gap but a much smaller penalty for extending the size of an existing gap?

5. When you align the two HA sequences using a higher gap opening penalty, does the percent identity change significantly? How about the number of gaps and their placement or size?

6. Your alignments with higher and lower gap opening penalties are both optimal alignments (the best alignments given the parameters), and they give quite similar scores. Which alignment do you believe is "better," biologically, and what is your justification? (*Hint: What striking observation did you make when looking at the gaps in the second alignment?*)

The origins of the 1918 pandemic virus remain murky, but its H_1 HA gene is thought to be the source of the HA genes found in all modern human and swine H_1 viruses. Download the segment 4 sequence of the 1918 human pandemic virus from the *Exploring Bioinformatics* website and compare it with the others. Consider what gap penalty you would like to use for this alignment.

Web Exploration Questions

7. Discuss how closely the HA segments of the two modern viruses are related to each other and how closely they resemble the 1918 virus. Can you draw any conclusions from your data about the origin of HA in the 2009 pandemic virus?

8. If you were to use a different segment from the same viruses for your sequence comparisons, you might come up with different answers. How is this possible?

Part II: Local Alignment with the Smith-Waterman Algorithm

Another way to use sequence alignment is to find one sequence within another. The influenza virus M2 gene, for example, is another key player in the biology of the virus: Once the virus enters the cell, M2 is involved in the release of the virus genome subunits so they can travel to the nucleus and direct viral replication. Suppose we have sequenced segment 7 from the 2009 H_1N_1 pandemic virus but are uncertain what part of it represents the actual M2 coding region. To find out, we could align the well-characterized M2 coding sequence from the Brisbane strain with the full segment 7 sequence from the newly sequenced virus. Download the DNA sequence for segment 7 from A/California/7/2009 and the coding sequence for M2 from A/Brisbane/59/2007 from the *Exploring Bioinformatics* website and align them using Needle with the default gap opening penalty of 10.

Web Exploration Questions

9. How good are the score and the percentage of sequence identity for this comparison? Why don't these statistics tell the full story in this case?

10. Suppose we only looked at the portion of the 2009 segment that actually aligned with the M2 coding region of the Brisbane strain. How would this change the percent identity? Is this degree of similarity as high as you would expect for these related viruses?

Considering what you know about the Needleman-Wunsch algorithm, you should see why it might not be the best choice for aligning sequences that are so drastically different in length. Because the need to make alignments of this kind arises frequently, in 1981 Smith and Waterman published a modification of the Needleman-Wunsch algorithm that allows for **local** alignments (see References and Supplemental Reading). A local alignment looks for optimal partial (subsequence) matches; how this works is discussed further in the On Your Own Project. EMBOSS includes an implementation of the Smith-Waterman algorithm, called Water. Choose the nucleotide version of the Water method and then set a gap open penalty of 10 and a gap extension penalty of 0.1 and align the sequences.

Web Exploration Questions

11. How does this alignment differ from the previous one? Is the percent identity, either for the whole alignment or just for the regions that actually match, significantly better than before?
12. There is an obvious difference in how the subsequences of the M2 coding region align with the 2009 segment 7 sequence in the local alignment. Can you suggest a hypothesis for *why* the sequences align this way? (*Hint: Remember that the M2 sequence is the protein coding sequence.*) Based on your hypothesis, is the local alignment superior to the global alignment in terms of its ability to help us understand the viruses *biologically*?

This alignment is very sensitive to the parameters used. If you want to demonstrate this, try changing the gap extension penalty (e.g., from 0.1 to 0.5). Although almost all bioinformatic programs come with default settings that are usable for many common purposes, this illustrates the importance of understanding the algorithm and the meaning of the parameters, as well as the value of considering what kind of alignment would be most appropriate for the sequences being aligned.

Part III: Using Alignment to Investigate Virulence

Influenza viruses have received a great deal of study, and the ability to compare many strains has led to significant advances in understanding what allows one strain to cause more severe disease than another. The H_5N_1 "bird flu" virus makes an interesting case in point. This virus causes severe influenza in birds and has become established in populations of domestic chickens and turkeys. Human cases occur sporadically, mostly in individuals heavily exposed to infected birds, such as poultry farmers, and H_5N_1 flu is severe for humans as well. Once a human case occurs, however, spread to another human is exceedingly rare, even among family members in close contact with the infected individual. A 2006 article by van Riel et al. (see References and Supplemental Reading) demonstrated that the avian H_5N_1 virus binds to a form of sialic acid receptor that in humans is found only far down in the lungs and lower respiratory system. Human viruses, in contrast, bind to a form common in the upper respiratory tract. Thus, it is difficult for H_5N_1 to infect humans because our respiratory defenses normally prevent viruses from reaching the lungs. However, a mutant strain in which HA was altered to be able to bind to sialic acid in the upper respiratory tract could be a very dangerous strain indeed.

So far, no such H_5N_1 strains that infect humans efficiently have been observed. However, we might ask whether the strains that do make it into humans tend to have altered HA genes—if so, that would suggest that either adaptive mutations could be occurring within the human host or that the viruses that cause human infections are subpopulations that are already better adapted. There are many avian H_5N_1 sequences available and a number of sequences of H_5N_1 viruses isolated from infected humans, so we can use sequence alignment to see whether these have essentially the same HA or noticeable differences.

Download sequences for segment 4 from two different avian H_5N_1 virus isolates and from a human H_5N_1 isolate from the *Exploring Bioinformatics* website and compare them using the Needleman-Wunsch algorithm.

Web Exploration Questions

13. What are the scores and sequence identities for a comparison of the two avian viruses? Are the differences between the human isolate and the avian isolates greater than the differences among avian isolates?
14. Based on your results (which of course are limited—it would be necessary to do many more comparisons in reality), do you believe there is evidence that human adaptation is occurring in H_5N_1 viruses that might merit concern about human-to-human transmission in the near future?

More to Explore

The sequences for all the influenza virus segments and genes used in this exercise come from the **Influenza Research Database**, which indexes a wealth of sequence information on influenza viruses of all types. If you are interested in exploring influenza virus sequences further, you can retrieve individual genes, segments, or whole genomes from this database using a flexible search interface.

Guided Programming Project: The Needleman-Wunsch Global Alignment Algorithm

In this project, you will gain an understanding of dynamic programming and how it can be used to tackle the difficult problem of sequence alignment by implementing the Needleman-Wunsch algorithm and using it to construct global, optimal alignments. You will then modify your program to implement a semiglobal alignment algorithm. (Local alignments are tackled in the On Your Own Project that follows.)

All the programming examples in this section are written in pseudocode: They are intended to show you the flow of program execution but do not represent the syntax of any particular language. Thus, you can implement them in any language you wish (we recommend Perl or Python). Depending on your programming experience, you may need a syntax guide for your language; some basic syntax related to the chapter projects can be found on the *Exploring Bioinformatics* website. Instructors can find complete programs in Perl or Python and solutions for the Putting Your Skills into Practice exercises and On Your Own Projects in the instructors' section of the *Exploring Bioinformatics* website.

Dynamic Programming and the Needleman-Wunsch Algorithm

The Needleman-Wunsch algorithm was one of the first to implement dynamic programming to solve an alignment problem. Dynamic programming is a problem-solving technique that breaks down a complex problem, such as the global alignment problem, into smaller overlapping subproblems. The solutions of the subproblems are then used to solve the original problem. Problems that can be solved with dynamic programming have a few common characteristics:

- There must be a way to divide the problem into smaller subproblems. (Each subproblem may then be broken down further.)
- The problem-solving process starts by solving these more manageable subproblems.
- Solutions to the smallest subproblems are then used in determining solutions to the next largest problems.
- The process repeats until the original (largest) problem is solved.

You learned earlier (review Understanding the Algorithm before continuing if needed) how the Needleman-Wunsch algorithm works. Building a scoring matrix divides the alignment problem into subproblems: The values in the matrix represent partial alignment scores or partial solutions to the overall problem. The bottom-right cell of the matrix always gives the optimal score, and backtracking through the matrix yields one or more "paths" that are interpreted as a series of aligned nucleotides or gaps that generate the corresponding optimal alignment(s). Figure 3.4E shows the matrix and paths for the sample sequences you have already seen.

Implementing the Needleman-Wunsch Algorithm

To align sequences using the Needleman-Wunsch algorithm, a computer program must (1) build a scoring matrix, (2) find paths through the matrix, and (3) generate alignments from the paths. The scoring matrix should be relatively easy for you to implement. The matrix itself could be implemented as a two-dimensional array. The first row and first column are initialized the same way regardless of the sequences compared. Then, each cell in the matrix is filled using the optimal score from among three choices: match or mismatch, gap in the first sequence, or gap in the second sequence (see Understanding the Algorithm).

The more difficult problem is how to find the path(s) back through the matrix and convert them to actual alignments computationally. Recall that we start at the lower-right cell and then determine the direction to move based on which of the three bordering cells (above, left, or above-left diagonal) could have been used to arrive at the score in the current cell. The directional arrows in Figure 3.4E show how we moved from cell to cell, but computers cannot really deal with these arrows. So, we replace the arrows with directional strings, using "H" for a horizontal move, "V" for a vertical move, and "D" for a diagonal move.

Our example contains three possible paths, so the following three strings are created, following the path from the lower-right corner to the upper-left corner in each case: HDHHDDD, HDDDHHD, and HDDDDHH. Moving from left to right in the directional strings and right to left in the sequences (we start at the ends of the two sequences because the directional strings start with the lower-right cell), we create the alignments as follows:

1. If the directional character is a D, then align the two currently considered nucleotides and obtain new nucleotides to consider by moving to the left one position in each sequence.
2. If the directional character is an H, then align the current nucleotide in the second (top) sequence with a gap character. Obtain a new current nucleotide for sequence 2 (top) by moving to the left one position, but keep the same current nucleotide for sequence 1 (left).
3. If the directional character is a V, then align the nucleotide in the first sequence with a gap character and obtain a new current nucleotide by moving to the left one position in the first sequence but not the second.

This process continues until all nucleotides have been aligned. For our sample sequences, the result is as follows:

```
    Path 1: HDHHDDD        Path 2: HDDDHHD        Path 3: HDDDDHH
Alignment: CACGTAT     Alignment: CACGTAT     Alignment: CACGTAT
           CGC--A-                C--GCA-                --CGCA-
```

The memory usage required by this algorithm is bounded by the size of the two input sequences, because you need to keep an array of size $N \times M$ in memory at all times. The length of the sequences that can be aligned is limited to the memory size of the computer on which the program runs. In the pseudocode that follows, only one directional string is constructed; a function (subroutine) is used for this task to modularize the steps of the algorithm. Finding all possible strings is left as an exercise.

The pseudocode that follows will guide you in writing a Needleman-Wunsch program that prompts the user for sequences to align and for a scoring metric. The Putting Your

Skills into Practice exercises that follow ask you to implement the program in whatever language your course is using and then provide suggestions for further exploration of the algorithm. Alternatively, your instructor may choose to provide the basic code (from the instructor section of the *Exploring Bioinformatics* website) for you to test and modify.

Algorithm

Needleman-Wunsch Algorithm

Goal: Determine the optimal global alignment of two sequences.

Input: Two sequences

Output: Best, global alignment(s) of two input sequences

```
// Initialization
Input the two sequences: s1 and s2
N = length of s1
M = length of s2
matrix = array of size [N+1, M+1]
gap = gap score
mismatch = mismatch score
match = match score

// STEP 1: Build Alignment Matrix
set matrix[0,0] to 0
for each i from 1 to N, inclusive
    matrix[i, 0] = matrix[i-1, 0] + gap
for each j from 1 to M, inclusive
    matrix[0, j] = matrix[0, j-1] + gap
for each i from 1 to N, inclusive
    for each j from 1 to M, inclusive
        if (s1[i-1] equals s2[j-1])
            score1 = matrix[i-1, j-1] + match
        else
            score1 = matrix[i-1, j-1] + mismatch
        score2 = matrix[i,j-1] + gap
        score3 = matrix[i-1, j] + gap
        matrix[i][j] = max(score1, score2, score3)

// STEP 2: Create Directional Strings
dstring = buildDirectionalString(matrix, N, M)

// STEP 3: Build Alignments Using Directional Strings
seq1pos = N-1 // position of last character in seq1
seq2pos = M-1 // position of last character in seq2
dirpos = 0
```

```
while (dirpos < length of directional string)
    if (dstring[dirpos] equals "D")
        align s1[seq1pos] and s2[seq2pos]
        subtract 1 from seq1pos and seq2pos
    else if (dstring[dirpos] equals "V")
        align s1[seq1pos] and a gap
        subtract 1 from seq1pos
    else // must be an H
        align s2[seq2pos] and a gap
        subtract 1 from seq2pos
    increment dirpos

// Function to create directional string
function buildDirectionalString(matrix, N, M)
    dstring = ""
    currentrow = N
    currentcol = M
    while (currentrow != 0 or currentcol != 0)
        if (currentrow is 0)
            add 'H' to dstring
            subtract 1 from currentcol
        else if (currentcol is 0)
            add 'V' to dstring
            subtract 1 from currentrow
        else if (matrix[currentrow][currentcol-1] +
             gap equals matrix[currentrow][currentcol])
            add 'H' to dstring
            subtract 1 from currentcol
        else if (matrix[currentrow-1][currentcol] +
             gap equals matrix[currentrow][currentcol])
            add 'V' to dstring
            subtract 1 from currentrow
        else
            add 'D' to dstring
            subtract 1 from currentcol
            subtract 1 from currentrow
    return dstring
```

Putting Your Skills into Practice

1. Write a program in the language used in your course to implement the above pseudocode. Test your program by using the sample sequences above and the other short sequences you used in the Test Your Understanding exercises and verify that it finds the expected alignment (only *one* alignment,

however: see question 4 for more about finding all possible alignments). Then try it on the influenza virus sequences you compared using Needle in the Web Exploration. (If your class skipped the Web Exploration section, download sequences for HA genes from various influenza virus strains from the *Exploring Bioinformatics* website.)

2. A user-friendly alignment program would format the output for readability, printing a specific number of characters on each line and then leaving a blank line between segments of the alignment. Numbering and a special character to indicate matches are also helpful (similar to the output you saw for EMBOSS). Modify your program to make it a more user-friendly solution.
3. Improve the program further by adding additional information beneficial to users: the alignment score and match percentage. You could also give the user the option to print the matrix and the path string for debugging purposes (which might also help you if your program is not doing exactly what you want it to).
4. The implementation of the Needleman-Wunsch algorithm shown previously finds only a single optimal alignment, but you can modify your program to find *all* possible optimal alignments. If you are familiar with the programming technique of recursion, you may want to consider a recursive solution, but this problem can also be solved without using recursion. Test your modified program to see that it finds all optimal alignments of your short test sequences, then test your program with real influenza HA sequences. Are there multiple optimal alignments for these sequences? In general, would long sequences be more or less likely to lead to multiple optimal alignment paths?

Although global alignment algorithms are useful, they do not solve all alignment problems. An example mentioned earlier is the need to find the coding sequence for a gene within a longer DNA sequence, requiring alignment of a short sequence with a long one. The Needleman-Wunsch algorithm can perform a global alignment, but it will penalize not only internal gaps but also the many terminal gaps—gaps at the beginning and end of the alignment—needed to align the short sequence at its proper position within the large sequence. This idea is illustrated by three sample alignments of a pair of sequences:

```
CGCTATAG    CGCTATAG    CGCTATAG
--CTA---    C--TA---    --C--TA-
```

Using a global alignment, these alignments are all considered "optimal" (three different paths to the same optimal score, –2). However, it is clear that the first alignment would actually be the best, because it includes only terminal gaps used to "position" the short sequence. If you eliminated the gap penalty for terminal gaps, the scores for these three sequences would be 3, 1, and 1, with the best alignment getting the best score. This alignment, where terminal gaps are ignored, is called a semiglobal alignment.

Putting Your Skills into Practice

5. Modify the Needleman-Wunsch program so it implements a semiglobal alignment by eliminating the gap penalty for terminal gaps. (*Hint: This actually requires only a few minor changes in the code. Focus on what the outside rows and columns of the matrix represent and how they are used.*) Try your program on the short sequences above and then on the sequences shown in Test Your Understanding question 3. If it works correctly, try a real-world case by downloading the sequence of 2009 H_1N_1 pandemic influenza virus segment 7 and the coding sequence for the 2009 H_1N_1 virus M1 gene (do not use M2, because that requires a local alignment, discussed later in the chapter) from the *Exploring Bioinformatics* website. Align the sequences and see if your program can successfully pick out the M1 coding sequence within the segment 7 sequence.
6. If you try the M1 coding sequence versus segment 7 alignment just mentioned in the EMBOSS Needle program, you might not expect it to succeed. However, it does. Go back to the parameter page and look closely at how the default parameters are set and see if you can decide why it works.

■ **On Your Own Project**: A Local Alignment Algorithm

Understanding the Problem: Local Alignment

At this point, you should have a good understanding of how the Needleman-Wunsch algorithm constructs optimal, global alignments. You should have considered (in the Testing Your Understanding exercises) how this algorithm could be modified to produce a semiglobal alignment and perhaps actually programmed such a solution (see Putting Your Skills into Practice). Finally, you should have worked with the Water program from EMBOSS and have an idea why a local alignment would be useful.

Local alignments solve the problem of finding and aligning conserved regions in otherwise dissimilar sequences by looking for optimal partial or subsequence matches between the sequences. Consider the sequences `AAAGCTCCGATCTCG` and `TAAAGCAATTTTGGTTTTTTTCCGA`. Two similar regions in these sequences, `AAAGC` and `TCCGA`, are separated by regions that are very different. A global or semiglobal alignment program should find the `AAAGC` alignment but will fail to correctly align the sequences so the `TCCGA` sequences also match up. To find subregions of similarity, large gaps must be expected and should not adversely affect the alignment score; this was the basis for Smith and Waterman's modification of the Needleman-Wunsch algorithm to produce a local alignment (see References and Supplemental Reading). Surprisingly, implementing the Smith-Waterman algorithm requires only a few changes to a semiglobal alignment algorithm.

Solving the Problem

A key element of a local alignment algorithm is the treatment of gaps. As with the semiglobal alignment, we should not penalize terminal gaps. But, for a local alignment, the Smith-Waterman algorithm also needs to consider how internal gaps are handled. For a global or semiglobal alignment, negative values can occur within the matrix, and they are useful because increasing negative values along an alignment path indicate a move away from similarity. However, for a local alignment, negative scores are no longer useful, because we do not necessarily expect the alignment to approximate an "ideal" diagonal path. Indeed, long gaps may be necessary to find optimally aligned subsequences, and these longer gaps should not be penalized so heavily as to negate good partial alignment scores. How might our system for placing a subscore in each cell of the matrix be modified to deal with this issue?

A second important modification involves the alignment score. Both the global and semiglobal alignment algorithms build the alignment path starting with the cell in the lower right of the matrix; this cell contained the optimal alignment score, because both algorithms considered the sequences in their entirety. However, a local alignment must consider subsequence matches, and high subsequence alignment scores could appear anywhere in the matrix, indicating the presence of a similar subsequence somewhere within the longer sequences. There could be many such similar subsequences within the longer sequences, and we want our local alignment algorithm to find all of them.

Finally, once a high score is found, continuing to follow the path until we reach the upper-left cell is not required: A highly conserved subregion may not extend all the way to the beginning of either sequence. Thus, the process of finding the path start and path end also requires modification.

Based on this information, describe a modified algorithm that would find local alignments given two sequences. Be sure to detail how the matrix is initialized, how the subscores are placed into each cell, and where the alignment path(s) should start and end.

Programming the Solution

If your course involves programming, your instructor may ask you to make the necessary modifications to your semiglobal alignment program and actually implement the local alignment algorithm you described. Test your program with the sample sequences shown previously and see if it can find both matches. Then, download the segment 7 sequence for the 2009 H_1N_1 pandemic influenza virus and the coding region of the M2 gene from the Brisbane seasonal strain from the *Exploring Bioinformatics* website and see if your program gives the same result as the EMBOSS implementation of the Smith-Waterman algorithm.

Connections: An Influenza Controversy

In early 2012, two different influenza virus research groups working on the H_5N_1 strain submitted papers to be considered for publication in prestigious scientific journals. Although their methods differed, the goal of both groups was to identify what mutations were necessary for the avian H_5N_1 flu virus to be transmitted readily among humans and whether the resulting virus would be as virulent as the current avian strains. Bioinformatics, including sequence alignment, played a major role in their research, but their work went beyond computational modeling to actually generate new virus strains whose virulence could be tested directly. The aim of this research was to better predict the future pandemic potential of H_5N_1 and thus better prepare medical researchers to deal with a human-transmissible version. Many scientists agreed that their research had significant merit and that the scientific and medical communities would benefit from publication. Others, however, expressed concern about the potential for accidental release of an engineered H_5N_1 virus that could itself become the next pandemic strain. Still others contended that publication of these results would essentially hand the "blueprint" for a bioweapon to any nation or terrorist organization interested in using it. Months of controversy ensued in an attempt to decide whether the work should be published, suppressed, or published with key techniques and details redacted. What do you believe should be done with this research?

BioBackground: The Influenza Virus and Molecular Evolution

Viruses sit at the interface between living and nonliving: Outside a host cell, they are metabolically inert, apparently nothing but nucleic acid in a protein shell, sometimes surrounded by a membrane-like envelope. Yet, every virus has some molecule on its surface capable of interacting with a receptor on the surface of a living cell. When the virus bumps into and attaches to a cell, this interaction results in entry of the virus into the cytoplasm, whereupon the viral genes are expressed and, pirate-like, the virus takes over the host cell machinery and subverts it to the manufacture of more viruses (**Figure 3.6**), ultimately destroying the cell. For the influenza virus, the preferred host cell is an epithelial cell of the upper respiratory system, and the cellular receptor is a sugar called sialic acid that binds the HA protein on the surface of the virus.

An influenza virus can be classified based on the type of HA protein it carries, as well as a second protein, **neuraminidase (NA)** involved in releasing the viral progeny from the host. Several major types of HA (H_1, H_2, H_3) and NA (N_1, N_2, N_3) are known, so a virus can be denoted H_1N_1, H_3N_2, H_5N_1 and so on. However, mutations produce variation even within these types, so subtypes must be defined. For example, in 2009–2010, one major circulating seasonal flu virus was A/Brisbane/59/2007 (H_1N_1), a type A virus first identified in Brisbane in 2007, whereas in 2007–2008, A/Solomon Islands/3/2006 (H_1N_1) was common; both subtypes are different from

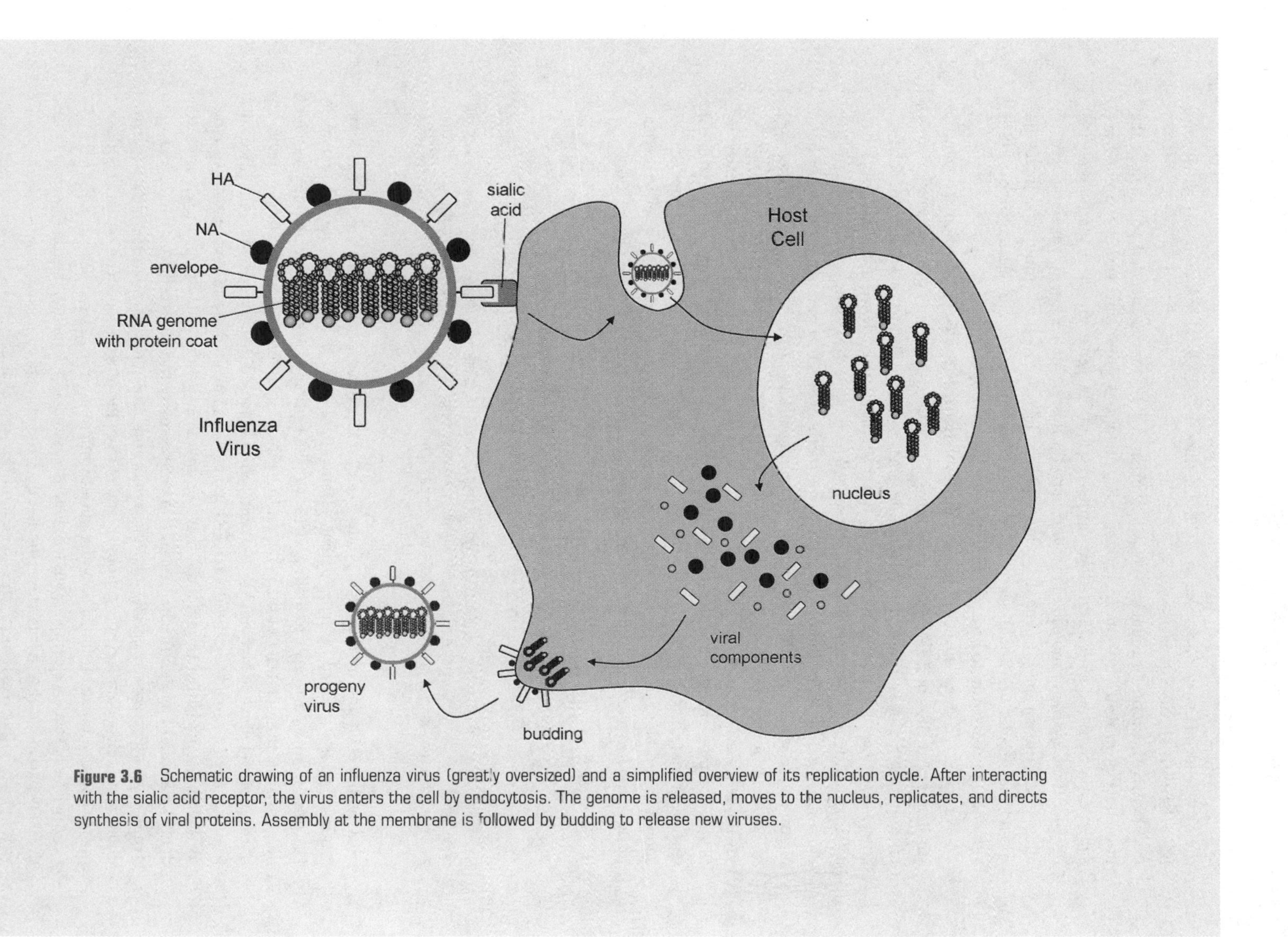

Figure 3.6 Schematic drawing of an influenza virus (greatly oversized) and a simplified overview of its replication cycle. After interacting with the sialic acid receptor, the virus enters the cell by endocytosis. The genome is released, moves to the nucleus, replicates, and directs synthesis of viral proteins. Assembly at the membrane is followed by budding to release new viruses.

the new pandemic virus discovered in 2009, A/California/7/2009 (H_1N_1), even though all three have the same H and N types.

The RNA genome of influenza virus is synthesized by a virus-encoded polymerase that does not "proofread" to remove errors; thus, mutations producing variant strains—new subtypes—occur frequently. Mutations in the HA and NA genes are particularly important because these are major molecules recognized by the host immune system: Variation here can allow a virus to escape immune detection and thus increase its opportunities to infect and spread. Such mutations would clearly be advantageous to the virus and selected for over time, allowing the new strain to become more prevalent in the population.

We would recognize the new strain as being evolutionarily related to the original one by the **similarity** of their genes: Two genes are similar if they have the same DNA sequence to a significant extent. This is determined by aligning genes from two strains (or, more broadly, from any two organisms), and we interpret significant similarity as evidence that these genes have a common origin. Differences between the sequences (**Figure 3.7**) are assumed to result from mutation, including substitutions of one base for another (resulting in mismatched bases in the alignment) as well as insertions or deletions (resulting in gaps in one of the aligned sequences). When a gene in one species or strain is very similar to a gene in a different species or strain, we say the genes are **orthologs** (**Figure 3.8**): Our conclusion is that the two species are descended from a common ancestor and that the genes have become modified by mutation over time in each of the daughter species. In fact, many or most genes in two evolutionarily related species should be orthologs. If we find two similar genes within the *same* species, we refer to these as **paralogs** and conclude that they arose by a gene duplication event followed by mutation.

```
Species 1:  TAAAGACCATAGGAAATAAAGATAA
Species 2:  TAACGACCAT-GGAAACAAAGATAA
```

Figure 3.7 Determining the similarity of two or more genes by aligning them so that their nucleotide sequences match up as well as possible. Differences resulting from mutation are highlighted; dashes represent the locations of insertion or deletion mutations (indels).

Gradual evolution by mutation produces new influenza virus strains that have genome sequences closely related to their "parent" strain; aligning the sequence of, for example, the HA gene from a currently circulating virus with its ortholog from a suspected new variant demonstrates the similarity of the genes and reveals their differences. Differences in regions of the protein known to be bound by host antibodies suggest a new strain of potential medical importance that should be carefully tracked and perhaps included in the next season's vaccine formulation. In addition to mutation, however, influenza viruses can also change more suddenly by a recombination mechanism: If two viruses infect the same cell (this can often happen in pigs, which are susceptible to swine, avian, and human influenza viruses), the progeny of one virus can acquire a whole genome segment from the other. Sequence alignment is again the tool needed to establish that a more radically different virus has evolved.

Analysis of sequence comparisons (see References and Supplemental Reading) revealed that the 2009 pandemic H_1N_1 virus arose through this recombination mechanism: Its parent was a well-known "triple reassortant" strain common in swine that carries an HA gene descended from the 1918 pandemic virus along with other segments from avian and human viruses (**Figure 3.9**). This virus more recently acquired NA and M genes that originated in a Eurasian avian virus, generating a novel virus type that began circulating in the human population probably

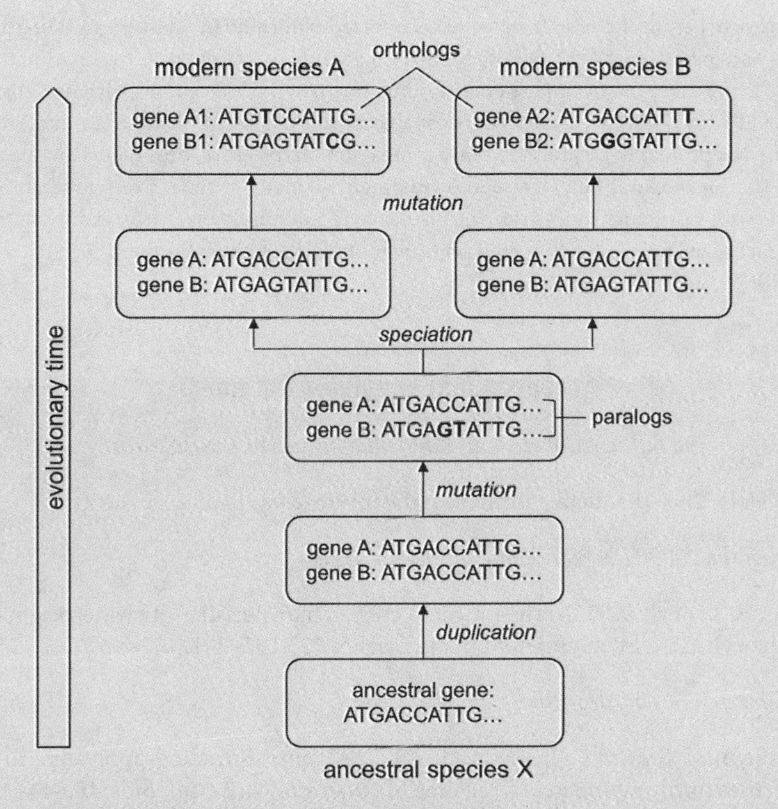

Figure 3.8 Sequences of genes or proteins reflect the pathways of change that have occurred in the evolutionary history of related species or strains.

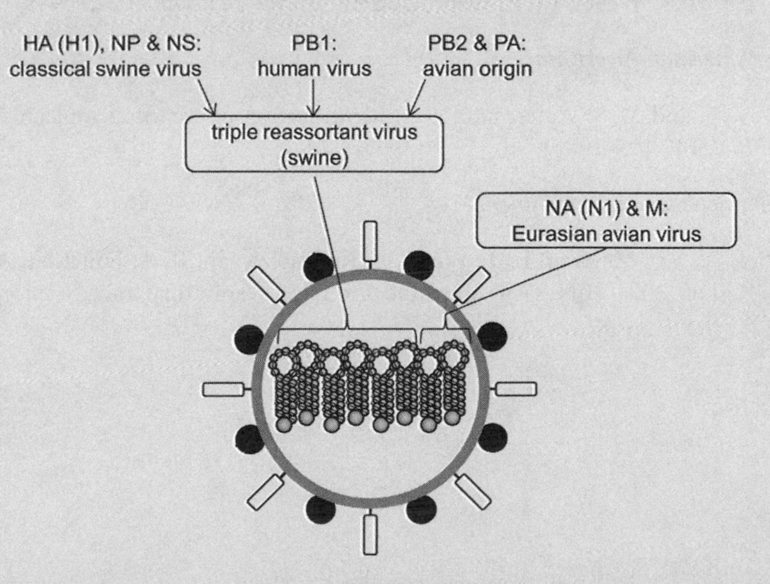

Figure 3.9 Origins of the genome segments of the 2009 pandemic H_1N_1 virus, as determined by sequence alignment.

about a year before the first cases were recognized clinically. In addition to demonstrating origins and pathways of evolution, sequence alignment is a key tool in investigating the functions of genes and proteins. In the case of influenza virus, several specific variations have been associated with highly virulent viruses capable of causing severe disease: a mutation in HA allowing the protein to be processed by a more common protease, thus increasing host range; a mutation in the viral polymerase allowing higher activity at the lower temperature of the human respiratory tract; and so on. The virulence of a new influenza virus strain can thus also be characterized by aligning its genes with their orthologs to look for these specific changes.

References and Supplemental Reading

Variation in the Influenza Virus and Pandemic Influenza Virus Strains

Nicholls, H. 2006. Pandemic influenza: the inside story. *PLoS Biol.* **4**:e50.

Origin of the 2009 H_1N_1 Pandemic Influenza Virus

Garten, R. J., et al. 2009. Antigenic and genetic characteristics of swine-origin 2009 A (H_1N_1) influenza viruses circulating in humans. *Science* **325**:197–201.

Needleman-Wunsch Algorithm

Needleman, S. B., and C. D. Wunsch. 1970. A general method applicable to the search for similarities in the amino acid sequence of two proteins. *J. Mol. Biol.* **48**:443–453.

Dynamic Programming

Eddy, S. R. 2004. What is dynamic programming? *Nat. Biotechnol.* **22**:909–910.

Smith-Waterman Algorithm

Smith, T. F., and M. S. Waterman. 1981. Identification of common molecular subsequences. *J. Mol. Biol.* **147**:195–197.

H_5N_1 Influenza Virus Attachment

van Riel, D., V. J. Munster, E. de Wit, G. F. Rimmelzwaan, R. A. Fouchier, A. D. Osterhaus, and T. Kuiken. 2006. H_5N_1 virus attachment to lower respiratory tract. *Science* **312**:399.

Chapter 4

Database Searching and Multiple Alignment:

Investigating Antibiotic Resistance

Chapter Overview

This chapter develops skills in two very commonly used types of Web-based bioinformatics tools: searching sequence databases for high-scoring matches to a query sequence (using BLAST) and multiple sequence alignment (using ClustalW). No programming project is provided; however, the algorithms and parameters used by these programs, both of which use heuristic methods to speed up complex tasks, are discussed in some detail. This chapter focuses on algorithms for optimal alignment of DNA sequences. This chapter is recommended for both programming and non-programming courses because these techniques and those related to them are used extensively in real-world bioinformatics applications.

Biological problem: Overuse of agricultural antibiotics and development of antibiotic resistance

Bioinformatics skills: One-to-many sequence alignments and multiple sequence alignment

Bioinformatics software: BLAST and ClustalW

Programming skills: Heuristics

Understanding the Problem:

Antibiotic Resistance

Fifty years ago, many people believed the newly discovered antibiotics—drugs that selectively kill bacteria without harming human hosts—would end infectious diseases caused by bacteria. Indeed, these "miracle drugs" have preserved the lives of millions. Today, however, tuberculosis, pneumonia, diarrheal disease, staph infections, and other bacterial diseases remain important—and in some cases increasing—causes of illness and death. One important reason is the dramatic rise of antibiotic-resistant bacteria no longer killed by commonly used antimicrobial drugs.

Resistance results from selection for mutants that can survive antibiotic treatment (see Bio-Background at the end of this chapter). As the use of an antibiotic becomes widespread, bacteria are increasingly exposed to it, escalating selective pressure and resulting in rapid evolution

Figure 4.1 The extensive use of antibiotics in agricultural animals that are not sick has sparked controversy about the role of this practice in speeding the development of antibiotic-resistant bacteria. Courtesy of Scott Bauer/USDA ARS. Inset © AbleStock.

of strains that thrive when antibiotics kill their susceptible cousins. Thus, in an effort to curb resistance, physicians today are much more cautious than in the past, prescribing antibiotics only when the need is clear and holding those least prone to resistance in reserve.

The nontherapeutic use of antibiotics in agricultural animals and even on food crops (**Figure 4.1**) is at the center of a current controversy over resistance. Routine use of antibiotics in animal feed prevents disease and promotes growth, allowing more animals to be raised more cheaply in less space. But many believe these economic benefits come at a high cost: Are the 28 million tons of agricultural antibiotics used annually in the United States and Canada (far outweighing the 3 million tons for all human uses) promoting antibiotic resistance? Most scientists believe that antibiotic overuse is a major contributor to the development and spread of resistance, leading to bans on subtherapeutic agricultural use of antibiotics in Denmark in 1999 and in the European Union in 2006. No such legislation is yet in place in the United States, and those who oppose such laws argue that no causal link has been definitively established between agricultural antibiotics and antibiotic-resistant disease bacteria in humans. We can investigate this link using some more advanced sequence alignment techniques.

Bioinformatics Solutions:
Advanced Sequence Comparison Algorithms

There is no question that intensive use of antibiotics in animals increases the prevalence of antibiotic-resistant bacteria—in animals. But how can a microbiologist determine experimentally whether these bacteria are an important source of resistance genes for bacteria that cause disease in *humans*? In 2001, Abigail Salyers and her colleagues used bioinformatics to look for evidence that bacteria inhabiting the human gut had been the recipients of antibiotic-resistance genes originating in bacteria found in domestic animals (see References and Supplemental

Reading). Taking advantage of the many sequenced bacterial genomes and the huge collection of sequenced genes in public genome databases, they looked for *un*related animal and human bacteria that have closely related resistance genes.

New or altered genes, including those that allow a bacterial cell to resist an antibiotic, arise by random mutation, which is rare. However, once these genes exist in a bacterial community, they can be readily passed from one bacterium to another (usually on plasmids), a phenomenon known as horizontal gene transfer (HGT; see BioBackground), allowing resistance to spread rapidly in a bacterial community. If a "donor" bacterium gives a resistance gene to a "recipient" organism, the two should have the *same* gene—that is, one that encodes a protein with the same amino-acid sequence. Furthermore, if human pathogens have the same antibiotic-resistance genes as bacteria from domestic animals, it would suggest that HGT occurs between them, supporting the conclusion that increased resistance among agricultural bacteria is indeed dangerous to human health. Similarity, of course, can be measured by sequence alignment, so Salyers used alignment first to retrieve genes from GenBank that were similar to a particular resistance gene and then to ask how similar the genes from unrelated species were. Two resistance genes that were ≥95% identical were assumed to have resulted from an interspecies gene transfer event.

The pairwise comparison techniques we have used thus far are of limited value when many sequences must be compared efficiently. In the sections that follow, we explore tools that build on the alignment algorithms we have already seen to allow for the rapid comparison of one sequence to many or the simultaneous alignment of multiple sequences.

BioConcept Questions

To successfully complete this chapter's projects, you need to understand a little about antibiotic resistance, HGT, and how similarity measurement can help us decide whether HGT has occurred. Use these questions to test your biological understanding; read BioBackground at the end of the chapter if you need a better foundation.

1. What is the difference between vertical and horizontal gene transfer? Why are the terms "vertical" and "horizontal" used to describe these processes?
2. Any bacterium could become antibiotic resistant by means of mutation. Why is HGT considered so much more of a threat, at least in terms of medically important resistance?
3. How does the degree of similarity between two genes help us understand whether they descended vertically from a common ancestor (recent or distant) or whether they could have moved from one species to the other by HGT?
4. Suppose you have evidence that two genes in two different bacterial species have a single, common origin. Give two possible explanations for how this might have occurred.

Understanding the Algorithm: Database Searching and Multiple Alignment

BLAST: A Heuristic Approach to Database Searching

The Needleman-Wunsch algorithm is a relatively efficient algorithm for optimal, global pairwise sequence alignment. However, imagine that you wanted to align an antibiotic-resistance gene of interest with every other sequence in GenBank. The computational time required is the time to make one alignment (compute the matrix and alignment paths) times the number of sequences in the database—currently more than 100 million. We would say that the time required to solve this problem is O(NS) or on the order of NS, where *S* is the number of sequences. It gets large quickly: If one alignment took 1 second of computer time, the whole search would take more than 3 years. Yet, **BLAST** (Basic Local Alignment Search Tool; see References and Supplemental Reading) can compare a query sequence to the entire database and return all matching sequences in a matter of seconds.

BLAST and its several variations are perhaps the most widely used of all bioinformatics software. As its name suggests, BLAST implements a local alignment algorithm similar in principle to the Smith-Waterman algorithm. However, it uses a **heuristic** or "shortcut" that makes it a practical and efficient solution to this complex problem.

To understand how BLAST works, we first need to clarify what we mean by a "matching" sequence. The point of comparing an antibiotic-resistance gene to GenBank is to identify similar sequences—generally, orthologs. Thus, a sequence matches the query if it shows statistically significant and/or biologically relevant similarity when aligned to the query. But how does BLAST make 100 million alignments so quickly? In fact, it does not make a full alignment for every sequence. Its first step is to break the query sequence into short "words" called *k*-tuples: subsequences *k* characters long. The default setting for *k* is 11 for DNA alignments and 3 for protein alignments. BLAST then scores close matches between these short sequences and each database sequence; this process is known as "seeding." Where it finds a good match, a local alignment algorithm finally comes into play, and the program tries to extend the alignment in both directions, comparing the resulting score with a threshold value. An alignment that can be extended to score above the threshold is referred to as a **high-scoring pair (HSP)**.

Figure 4.2 shows an example of the BLAST algorithm using a famous quotation for which several variations can be found instead of a sequence (remember, alignment programs can compare any two strings). The alignments in the figure are scored with a simple match = 1, mismatch = 0, gap = −1 system. If the query sequence is broken down into three-letter words

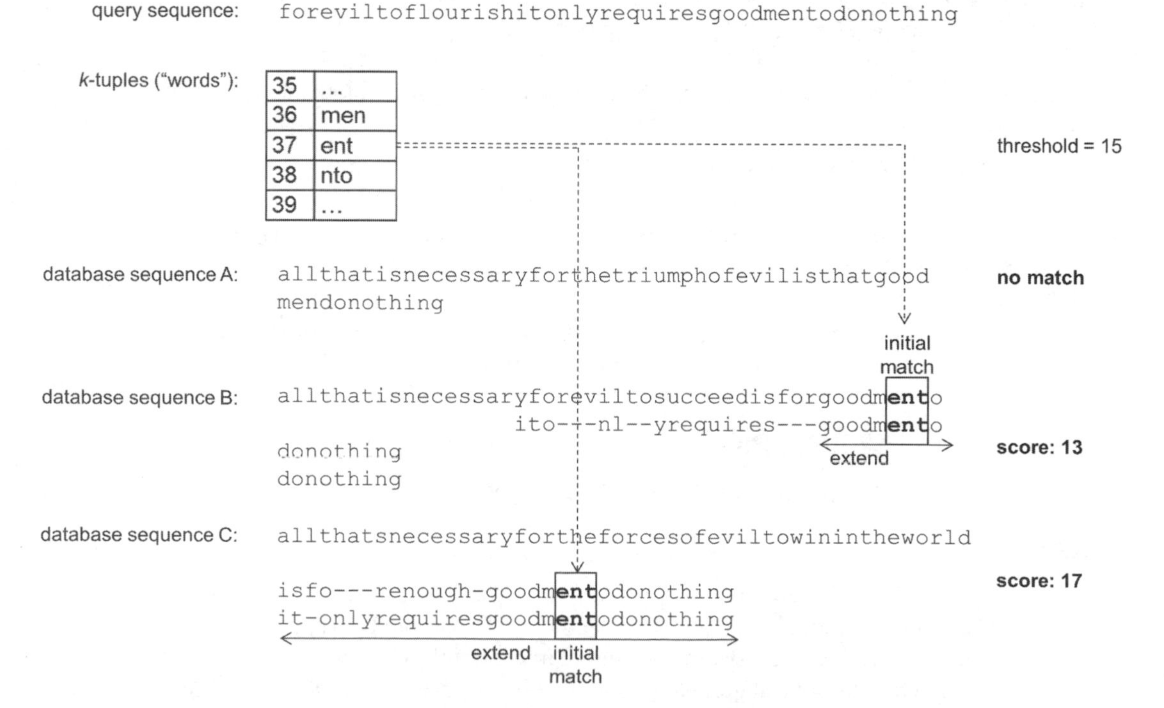

Figure 4.2 An example of how the BLAST algorithm finds an initial match between a short subsequence (*k*-tuple or "word") of the query and the target sequence, then extends the match to find a local alignment with scoring above a threshold value.

(*k*-tuples) and we focus on the 37th *k*-tuple, "ent," there is no match for any of the words in database sequence A, so this sequence can be discarded. Sequence B has an initial match, but attempting to extend the alignment does not increase the score above the threshold, so this sequence would not be reported as a significant alignment. Sequence C, however, has an alignment that exceeds the threshold score and would be reported as a match.

BLAST then calculates the statistical likelihood that a given score would occur based on mere chance alignment of unrelated sequences (the *e*-value) and orders the matching sequences according to this measure of statistical significance. As we will see in the next section, BLAST reports back to the user the name of the matching sequence, the score, the *e*-value, and the alignment itself. In addition to changing scoring parameters such as the gap penalty, BLAST allows the user to adjust the *k*-tuple value if desired. Although the default value typically works well, decreasing the word size allows the identification of sequences that match less well (useful when similarity of the query to other sequenced genes is weak) and is also needed if the sequence to be compared is very short (current implementations of BLAST do this automatically when a short query is entered).

You use heuristics all the time without realizing it. Consider, for example, how you decide which route to take when you have several alternatives. It is extremely difficult to calculate a truly optimal solution (accounting for traffic, construction, traffic lights, speed limits, school zones, and many more variables), so you apply a heuristic: You decide to take the route that is shortest in mileage or the one you believe has the least traffic. This allows you to choose rapidly but does not guarantee that you will in fact choose the fastest option. Similarly, BLAST's heuristic approach allows it to quickly discriminate possible matches from unrelated sequences. Although it may not find optimal alignments, it deals with large volumes of data extremely rapidly while finding solutions that are acceptably close to optimal.

ClustalW: Multiple Sequence Alignment

Although BLAST can quickly identify a large number of sequences similar to a query, it displays only individual alignments of the query with each matching sequence. However, we might instead want to see an alignment of a whole group of similar sequences at once (**Figure 4.3A**). For example, perhaps the sequences of genes similar to our query resistance gene fall into two or three distinctly identifiable groups. Or, we might want to identify a **consensus sequence**: the nucleotides or amino acids that appear the most frequently at each position in a given region of the sequence. Rather than a pairwise alignment, this requires a **multiple sequence alignment** algorithm.

The computational complexity problem for multiple sequence alignment is even greater than for database searching. Here, the order of adding sequences to the alignment matters. Suppose, for example, we have optimally aligned two sequences, GTCT and GGT as in **Figure 4.4A**. If we now want to align the sequence CT with the other two, we might get the alignment in **Figure 4.4B**. However, if we aligned GTCT with CT first, we might find the optimal alignment to be the one in **Figure 4.4C** instead. The dynamic programming approach of Needleman and Wunsch could deal with this problem by building a matrix of size $L \times M \times N$, each dimension one character longer than the length of one sequence. However, as more sequences are added, the matrix becomes four-, five-, six-dimensional, and so on and the computational time required becomes $O(N^S)$: the time required for one alignment raised to the *power* of the number of sequences, which obviously becomes impractical very fast.

Thus, multiple sequence alignment algorithms again use heuristics to manage the complexity of the problem. ClustalW (see References and Supplemental Reading) is one of the most popular multiple sequence alignment algorithms; it uses a **progressive alignment** algorithm in which the order of adding new sequences to the alignment is determined by first calculating a rough phylogenetic tree called a **guide tree** (**Figure 4.3B**). The guide tree

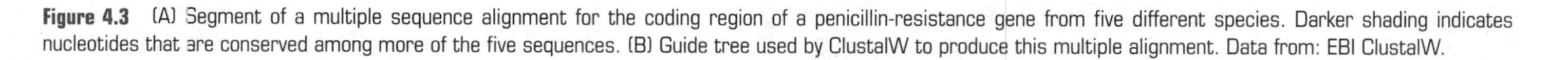

Figure 4.3 (A) Segment of a multiple sequence alignment for the coding region of a penicillin-resistance gene from five different species. Darker shading indicates nucleotides that are conserved among more of the five sequences. (B) Guide tree used by ClustalW to produce this multiple alignment. Data from: EBI ClustalW.

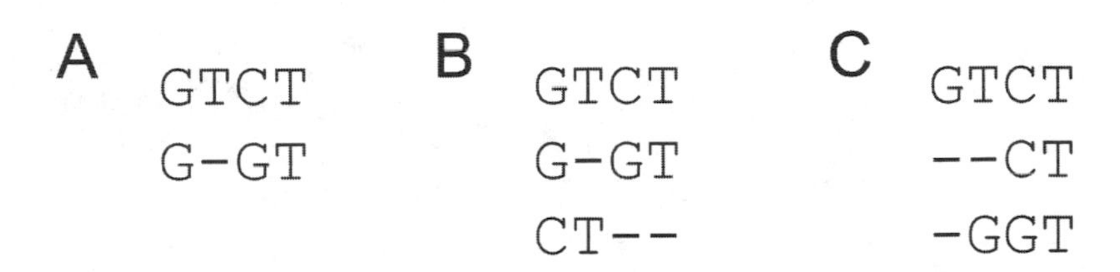

Figure 4.4 Multiple sequence alignment is complex because the order of adding sequences to the alignment can affect the alignment results.

is generated by first doing pairwise alignments and then using the score or percent similarity from those alignments to draw a tree showing which sequences are more and less closely related (we will have much more to say about the mechanics of generating a phylogenetic tree in subsequent chapters). Starting with the two most closely related sequences (in the example in Figure 4.3B, these are *Bacteroides xylanisolvens* and *B. fragilis*), ClustalW then does global, pairwise alignments to align each new sequence with those already aligned, in order of decreasing relatedness. Note that although this is an efficient way to produce a multiple alignment, the fact that it is based on global alignment means ClustalW may not correctly align sequences that share regions of similarity if the sequences are not very similar overall.

Test Your Understanding

1. Describe two features of the BLAST algorithm that enable it to complete a database search much faster than the Needleman-Wunsch algorithm would.
2. For the BLAST example in Figure 4.2, are there *k*-tuples within the query sequence that give a very different result? What might be an example of a query sequence that would yield an HSP for all three database sequences?
3. Describe briefly how the sequence differences you can see in Figure 4.3A relate to the lengths of the branches in Figure 4.3B.
4. In Figure 4.3, the sequence labeled CA_F7SDb01 is from an organism that has not yet been characterized sufficiently to give it a species name; all other sequences are from species within the genus *Bacteroides*. Based on the region of the multiple alignment shown in this figure, would you characterize CA_F7SDb01 as likely to belong to some *Bacteroides* species or likely to come from a different genus?
5. Write out a set of six short (seven or eight nucleotides) DNA sequences in which all six are related but there are two sets of three that are more closely related to each other than to the other set. Show how the guide tree might look for your sequences and then what the multiple sequence alignment might look like.

CHAPTER PROJECT:

Horizontal Gene Transfer of Antibiotic Resistance

Salyers and her colleagues (see References and Supplemental Reading) used bioinformatics methods to look for evidence of horizontal transfer of antibiotic genes between bacteria found in animals routinely fed antibiotics and bacteria that might affect human health. Because of the enormous number of bacteria residing normally in the human large intestine, they hypothesized that these bacteria serve as a reservoir for HGT (**Figure 4.5**) and could easily

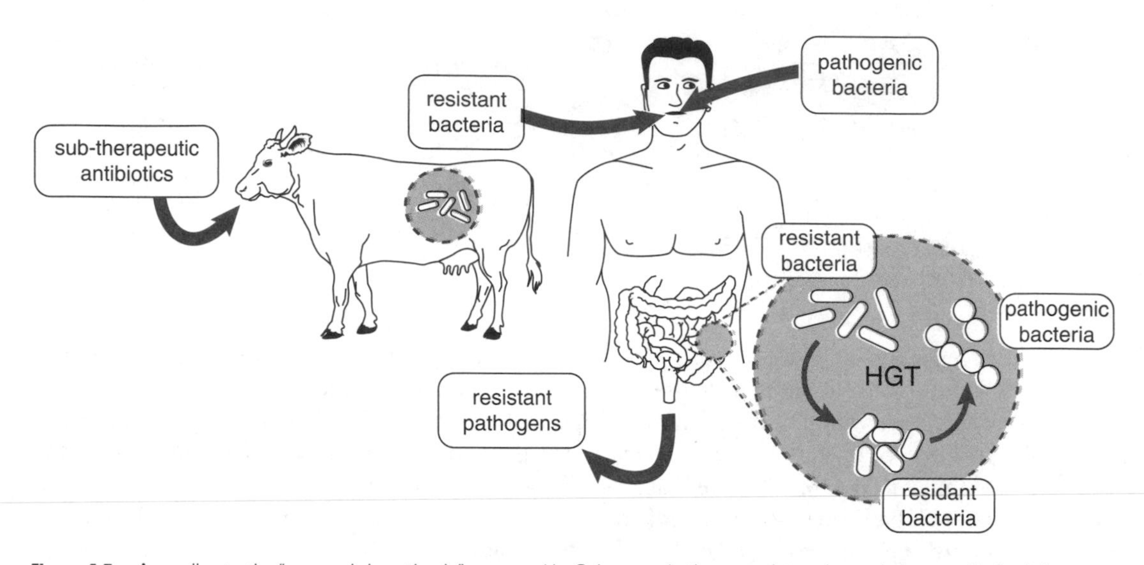

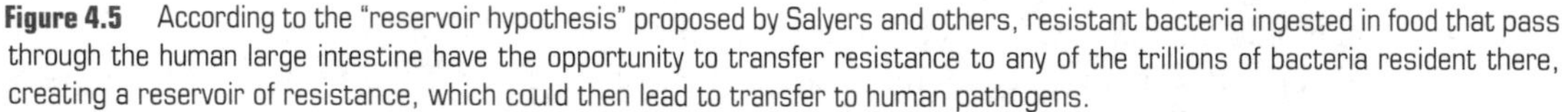

Figure 4.5 According to the "reservoir hypothesis" proposed by Salyers and others, resistant bacteria ingested in food that pass through the human large intestine have the opportunity to transfer resistance to any of the trillions of bacteria resident there, creating a reservoir of resistance, which could then lead to transfer to human pathogens.

exchange genes with ingested bacteria, including antibiotic-resistant bacteria originating in agricultural animals. Thus, using alignment methods, Salyers focused on determining whether common intestinal bacteria might carry the *same* genes for antibiotic resistance as unrelated species that are not gut residents. Related species, of course, are likely to have highly similar genes, but a high degree of similarity between genes of otherwise *dissimilar* organisms strongly suggests horizontal transfer. Salyers used the criterion of ≥95% similarity to decide whether sequences from two organisms in fact represented the same gene. In this project, we will use BLAST to identify a set of resistance genes of interest and ClustalW to examine the similarity among them, enabling us to draw some conclusions about the impacts of subtherapeutic agricultural antibiotic use. We will focus on genes enabling bacteria to resist the antibiotic erythromycin, a drug commonly used in both therapeutic and agricultural applications.

Learning Objectives

- Understand the value of searching a database for sequences matching a query
- Gain experience with the use of BLAST in database searching and understand its parameters
- Appreciate the importance of a heuristic in processing large amounts of data rapidly
- Understand the use of multiple sequence alignment and know how to use ClustalW for this purpose

Suggestions for Using the Project

This project is designed to build skills in using two very important pieces of bioinformatics software: BLAST and ClustalW. Because of their wide use, familiarity with these tools is highly recommended for students in both programming and nonprogramming courses. The BLAST and ClustalW sections that follow can be used independently; instructors can download a set of *ermB* sequences from the *Exploring Bioinformatics* website if they would like their students to do the multiple alignment without first using BLAST to identify sequences of interest. Instructors could also ask students in programming courses to implement a BLAST-like algorithm based on the earlier discussion.

Searching for Erythromycin Resistance Genes with BLAST

Obtaining the ermB *Sequence*

Erythromycin is an antibiotic that halts bacterial growth by binding to the bacterial ribosome and blocking translation. Two different mechanisms of erythromycin resistance have been observed: Some resistant bacteria have acquired a gene whose product modifies the ribosome so erythromycin can no longer bind, whereas others have acquired a gene encoding a transport protein (called an efflux pump) that rapidly removes erythromycin from the cell. You already know how to find sequences in GenBank via a text search; however, a key word such as "erythromycin" will retrieve both kinds of genes and will fail to retrieve any resistance genes that were not annotated as such. Instead, using BLAST, we can search using a *sequence* as our query and retrieve all similar sequences, regardless of how they are annotated.

As our query sequence, we use an erythromycin-resistance gene called *ermB* from *Streptococcus agalactiae*, a Gram-positive bacterial species commonly associated with the udder of cows, where it can cause mastitis. This gene produces one of several known resistance proteins of the ribosome-modification type. Erythromycin resistance due to *ermB* has commonly been seen in the human pathogen *Streptococcus pneumonia*, the most common cause of bacterial pneumonia, so it will be interesting to determine whether HGT of this gene has occurred among diverse bacteria. Start by obtaining the DNA sequence for the *S. agalactiae ermB* coding region from GenBank in FASTA format by using a text search, by searching for the accession number DQ355148.1, or by downloading the file from the *Exploring Bioinformatics* website.

Understanding BLAST Results

BLAST results are shown in three sections. The top section is a graphical view (see sample of some representative BLAST results in **Figure 4.6A**), with a bar for each sequence that matches the query. The length of the bar shows the length(s) of the matching region(s), and its color represents the score for each segment. The middle section (**Figure 4.6B**) gives details about each match: the accession number and description for the gene matched and five parameters related to the quality of the match:

- **Max score**: the score of the best matching segment (remember, this is a local alignment, not a global one).
- **Total score**: the total scores of all matching segments found (same as max score if there is only one matching segment).
- **Query coverage**: the percentage of the query sequence that aligned to some part of the match.
- ***e*-Value**: a statistical measure evaluating how likely it is that a match this good would occur by chance. The lower the *e*-value, the more likely it is that the two sequences are truly similar and not just chance matches. Two identical sequences would have an *e*-value of zero.
- **Max ident**: the percentage of nucleotides that are identical between the query and target sequences within the matching regions.

The third section (**Figure 4.6C**) shows the actual pairwise alignments between the query sequence and the top matching database sequences. Links in each section provide direct access to a variety of additional information about the matching sequences.

Identifying ermB *Orthologs with BLAST*

From the **NCBI BLAST home page**, you can see several ways to run BLAST, including both nucleotide and protein comparisons. For this exercise, we compare DNA sequences, so you should choose the nucleotide option. This should take you to a search form where you can either paste or upload your *S. agalactiae ermB* sequence.

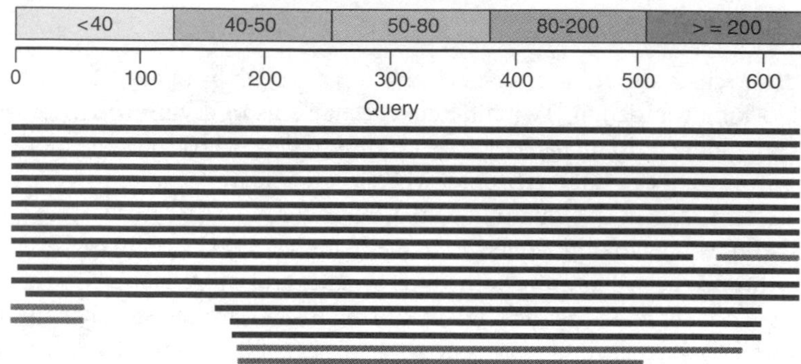

(A)

Sequences producing significant alignments:
(Click headers to sort columns)

Accession	Description	Max score	Total score	Query coverage	△ E-value	Max ident
CP000948.1	Escherichia coli str. K12 substr. DH10B, complete genome	1132	1132	100%	0.0	100%
AP009048.1	Escherichia coli W3110 DNA, complete genome	1132	1132	100%	0.0	100%
U00096.2	Escherichia coli str. K-12 substr. MG1655, complete genome	1132	1128	100%	0.0	99%
CP000946.1	Escherichia coli ATCC 8739, complete genome	1126	1126	100%	0.0	99%
CP000802.1	Escherichia coli HS, complete genome	1126	1126	100%	0.0	99%
CP000970.1	Escherichia coli SMS-3-5, complete genome	1122	1122	100%	0.0	99%
CP000266.1	Shigella flexneri 5 str. 8401, complete genome	1113	1113	100%	0.0	99%
AE014073.1	Shigella flexneri 2a str. 2457T, complete genome	1113	1113	100%	0.0	99%
BA000007.2	Escherichia coli O127:H7 str. Sakai DNA, complete genome	1113	1113	100%	0.0	99%
AE005174.2	Escherichia coli O157:H7 EDL933, complete genome	1113	1113	100%	0.0	99%
AE005674.1	Shigella flexneri 2a str. 301, complete genome	1108	1108	100%	0.0	99%
CP000034.1	Shigella dysenteriae Sd197, complete genome	1108	1108	100%	0.0	99%
CP000038.1	Shigella sonnei Ss046, complete genome	1108	1108	100%	0.0	99%
CP000800.1	Escherichia coli E24377A, complete genome	1099	1099	100%	0.0	98%
CP000036.1	Shigella boydii Sb227, complete genome	1095	1095	100%	0.0	98%
CP001063.1	Shigella boydii CDC 3083-94, complete genome	1090	1090	100%	0.0	98%
CP000468.1	Escherichia coli APEC 01, complete genome	722	722	99%	0.0	85%

(B)

```
Score = 1132 bits (1254), Expect = 0.0
Identities = 627/627 (100%), Gaps = 0/627 (0%)
Strand = Plus/Plus

Query  1        TTAAGCCAGCTCACCCTTCACTAAAGGGACAAAGCGCACGGCCTCC
                ||||||||||||||||||||||||||||||||||||||||||||||
Sbjct  2959457  TTAAGCCAGCTCACCCTTCACTAAAGGGACAAAGCGCACGGCCTCC

Query  61       AAATTCGCCTCCCCGACGACGCACCCGTTTCAAATACTGGTGCTCC
                ||||||||||||||||||||||||||||||||||||||||||||||
Sbjct  2959517  AAATTCGCCTCCCCGACGACGCACCCGTTTCAAATACTGGTGCTCC

Query  121      GACGAGAATCCCGCCTTCGTCCAGCTGCGTCATTAGCGCAGTTGGA
                ||||||||||||||||||||||||||||||||||||||||||||||
Sbjct  2959577  GACGAGAATCCCGCCTTCGTCCAGCTGCGTCATTAGCGCAGTTGGA

Query  181      CGCCGTAACAATGATAGCGTCAAACGGCGCACGTGCCTGCCAACCT
                ||||||||||||||||||||||||||||||||||||||||||||||
Sbjct  2959637  CGCCGTAACAATGATAGCGTCAAACGGCGCACGTGCCTGCCAACCT

Query  241      ATGACGGGTTGAAACATTATGTAAATCAAGATTTTTCAGGCGGCGA
                ||||||||||||||||||||||||||||||||||||||||||||||
Sbjct  2959697  ATGACGGGTTGAAACATTATGTAAATCAAGATTTTTCAGGCGGCGA

Query  301      CAAGCCTTTAATCCGTTCAACCGAGCAAACATGCTGGACAAGATGC
                ||||||||||||||||||||||||||||||||||||||||||||||
Sbjct  2959757  CAAGCCTTTAATCCGTTCAACCGAGCAAACATGCTGGACAAGATGC
```

(C)

Figure 4.6 Sample results of a BLAST search for database sequences matching a nucleotide query sequence: (A) graphical summary of results, (B) table of scores, and (C) alignments.

Many options and parameters are available on this page. Notice the section labeled `Choose Search Set`, where you can specify the sequences to be searched. Importantly, the default set of sequences is the subset of GenBank containing human DNA sequences. This obviously will not work in our case, where we want to retrieve bacterial sequences. Change the database to `nucleotide collection (nr/nt)`, which will search all the unique ("nonredundant" or nr) sequences in GenBank. Furthermore, many sequences in GenBank are from bacteria that have been sequenced (using DNA harvested from an environmental sample) but never cultured; these are not useful to us because we do not know what species they come from, so check the box to exclude sequences from uncultured samples. To further refine the results, there is also an input box where you can limit your search to a particular organism or group of organisms; you could type `bacteria` here to exclude any nonbacterial sequences that might happen to match. Finally, there is a box where you can type an Entrez query to include or exclude specific kinds of sequences.

If you click `Algorithm parameters` near the bottom, you can set the parameters that BLAST uses for its comparison. These options should be starting to look familiar to you: For example, you can set a linear or affine gap penalty, change the match and mismatch scores, and alter the word size (*k*-tuple) for the initial match. Some of these parameters are set automatically when you make a choice from the `Program selection` section, where you choose the specific algorithm that will be used by selecting options such as `Highly similar sequences (megablast)` or `Somewhat similar sequences (blastn)`. With the parameters visible, try clicking each of these options and notice how the parameters change. For example, megablast has a default word size of 28, whereas blastn has a default of 11; how would this change the results? When you have finished exploring, choose `blastn` for now to see both very similar and less-similar sequences the program might identify. Click the `BLAST` button to start the search and compare your *ermB* sequence with the selected sequences. In a short time, you should get a page of results (see Figure 4.6 for an example of what this page would look like).

Web Exploration Questions

1. In their original survey, Salyers and colleagues used a cutoff of 95% identity for sequences considered similar enough to have been shared by HGT. You can get a quick measure of identity by using the `max ident` score in the BLAST results—however, you can also get a high max ident for a very small matched region, so also consider the query coverage. Looking at these parameters, are the matches that BLAST retrieved highly similar to your query or less similar? Do your data suggest that all or most of them represent the same gene, transferred from organism to organism by HGT?
2. You may notice in your list that a number of the sequence matches come from cloning vectors—engineered DNA molecules used for laboratory manipulations. Construct an Entrez query to exclude these from your results and run your search again—but be careful not to exclude too much. Remember that unless you limit the field, the entire text of each entry will be searched for a match. What query did you use?
3. What evidence can you find among your BLAST results to support or refute the hypothesis that resistance genes are being shared between unrelated species—especially between agricultural species and human pathogens or human gut bacteria that might come into contact with pathogens? You will have to do some detective work to answer this question: For example, find a bacterial phylogenetic tree online to help you decide how closely related the different species in your list are, and then try to find out which ones might be found in domestic animals, which are residents of the human gut, and which are human pathogens.
4. There are so many sequences in GenBank today, including many whole genome sequences, that BLAST often fills up its list of top matching sequences without ever getting down to less related but potentially more interesting matches. In your initial BLAST results, for example, it is likely that most

if not all sequences come from Gram-positive organisms, one major division of the bacteria. HGT to the more distantly related Gram-negative organisms would be very interesting but is hard to assess from this list. Construct a BLAST search that excludes Gram-positive matches. Or, another way to get interesting results might be to require matches to specific groups of Gram-negative organisms that you know live in the human gut, such as *Bacteroides* (the most common genus among human gut bacteria) or *Escherichia*. Be careful to exclude from consideration sequences that come from cloning vectors in this case—you only want sequences naturally found in these bacteria. Describe how you searched, the similarity of your results to the query, and whether the percent identity suggests that your results represent horizontally transferred genes or genes arising by mutation.

5. Based on your results thus far, would you say that you have evidence for (a) extensive HGT, (b) a mix of HGT and evolution by mutation, (c) evolution mostly by mutation with occasional HGT, or (d) a number of unrelated resistance genes? Support your answer with evidence.

Retrieving Sequences

In the next section, we will carry out a multiple alignment of some *ermB* genes from different species, which requires retrieving their sequences in FASTA format. The NCBI implementation of BLAST includes a number of useful tools for working with the sequences it finds, including a means of quickly retrieving the ones in which you are interested. Checkboxes next to the sequences BLAST aligned allow you to select interesting matches; chose some that are from different genera, from human pathogens or gut organisms, from Gram-negative organisms, and so on. Then, you should see a download link allowing you to retrieve the sequences in FASTA format. You can combine the results of several searches simply by downloading each set and then cutting and pasting in the resulting text files. Compile a file with several interesting sequences that you can go on to align with ClustalW.

Before leaving BLAST, take a look at the sequences you retrieved. In some cases, BLAST will have retrieved an entire plasmid or even genome sequence, even though only a short region of this sequence is actually of interest. You can use the accession numbers of these sequences to retrieve the GenBank entry and then obtain just the coding sequence (see Chapter 1). Or, even though BLAST aligned your query with a correctly oriented nontemplate strand of the gene from the database, it might retrieve the template strand if that is how the matching sequence was entered into GenBank; you can get the reverse complement using Sequence Manipulation Suite (Chapter 2) if this is the case. Your text file should ultimately contain correctly oriented coding sequences for all the *ermB* orthologs you intend to align. Finally, the comment lines may be long and not terribly helpful. Because the ClustalW implementation we will use does not like spaces and will truncate the comments, replace the comment lines with something more useful, such as simply the name of the species with no spaces (e.g., >Streptococcus_agalactiae).

Multiple Sequence Alignment with ClustalW

Although you were able to get some information about the similarity of many sequences to your query sequence from your BLAST results, you undoubtedly noticed that BLAST still only made pairwise comparisons: It showed alignments between your query and one other sequence at a time. When comparing many sequences, it can be much easier to analyze the results when all alignments can be visualized at once. Furthermore, some questions might be better answered by aligning a group of sequences: for example, to ask if there are particular regions of the sequences that are more or less conserved. ClustalW is an example of a multiple sequence alignment program designed for this purpose; sample output is shown in Figure 4.3A.

For this part of the project, you will need a text file containing the sequences of at least six to eight sequences similar to *ermB* in FASTA format. You should have all your sequences

in a single file, separated by their comment lines; be sure you have the coding regions only. If your class did not do the BLAST part of the project, your instructor can download a file with some interesting sequences from the *Exploring Bioinformatics* instructor website and make it available to you.

A good Web implementation of **ClustalW** is maintained by the EBI. Once you have loaded ClustalW, paste your entire list of sequences into the input box or upload your text file. Notice that two sets of parameters can be set: one for the initial pairwise alignments used to generate the guide tree and another for the subsequent multiple alignment itself. You will notice familiar ideas such as gap opening and extension penalties. Run your alignment initially with the default parameters.

When the results are returned, you will see the alignment in simple text format, with asterisks below the alignment wherever a particular nucleotide is found in all sequences. You can view the guide tree by clicking the appropriate tab, and the `Result Summary` tab shows the results of the individual pairwise alignments that were done. A more sophisticated presentation can be obtained by using Jalview, a Java-based viewer: click the `Result Summary` tab and then click `Start Jalview`. Here, you can see a consensus sequence representing the most conserved nucleotides at each position, and you can format and color the alignment in various ways. A convenient way to visualize differences among the sequences is by selecting `Percentage Identity` from the `Colour` menu; this gives a dark background for nucleotides conserved in all sequences and lighter colors for nucleotides conserved in fewer sequences.

Web Exploration Questions

6. Which *ermB*-like genes are the most similar? Which are less similar? Are there particular regions of the gene that are highly conserved or less conserved?
7. What kinds of differences can you see among these genes? Do substitutions outnumber indels or vice versa? What do you notice about the indels that occur in the alignment?
8. Try running ClustalW again with a very low gap penalty. Do the alignments change significantly? Which alignment is more biologically relevant, and what is your evidence for this view?
9. Based on the criterion of closely related genes from unrelated organisms, do your results support the HGT hypothesis?

How would you summarize your findings and conclusions regarding the likelihood that agricultural use of antibiotics can result in resistant human gut residents and/or resistant human pathogens? Your instructor may ask you to write up your findings in the form of a short report.

BioBackground: Antibiotic Resistance and Gene Transfer

Bacteria can have **natural (intrinsic) resistance** to some antibiotics because of their cell structure. For example, Gram negative bacteria (such as the common intestinal organism *Escherichia coli*) are resistant to penicillin simply because the cell wall that penicillin attacks is protected by an outer membrane that other bacteria lack. But the resistance that is really important medically is **acquired resistance**: when bacteria that were previously sensitive to (killed by) an antibiotic become resistant to it, making that antibiotic useless for treatment. Acquired resistance requires genetic change: Either a new gene or new variant of a gene arises by mutation or a cell acquires a preexisting gene by horizontal transfer.

Many people have the idea that using an antibiotic "makes" bacteria resistant. This is not true, however: Antibiotics do not *cause* resistance to occur (nor is it true that antibiotic use makes the *person* resistant to the antibiotic). However, antibiotic use can **select** for bacteria that have already become resistant, allowing them to become more prevalent in a population. As shown in **Figure 4.7**, if some bacteria in a population are more resistant to an antibiotic than others (due to mutation or to genes they have acquired), they will not be killed as easily when they encounter it. Thus, the antibiotic kills the most sensitive cells first and leaves the more resistant ones to pass their genes on. This can happen in your own body if you do not finish your antibiotic prescription: The most resistant cells remain alive and can then multiply and cause a relapse. The more we expose bacteria to antibiotics—whether in the body, in animals, or in the environment—the more we select for resistant organisms and thus the more prevalent the resistant bacteria become.

If a mutation gives a bacterial cell some advantage—and antibiotic resistance is just one of many possible examples—that cell's descendants inherit the altered gene. This is sometimes called **vertical gene transfer** (**Figure 4.8**, left panel) and could lead to increased resistance by selection if the population is challenged by an antibiotic. However, mutations are relatively rare, and resistance would develop slowly if bacteria had to rely on inheriting a rare mutation from their parents. A major reason for the rapid spread of resistance is that bacteria can also acquire genes by HGT. This refers to genetic material being transferred from one cell to another that is not its descendant (Figure 4.8, right panel). For example, many antibiotic resistance genes are carried on plasmids: small, circular, independent DNA molecules. A cell with a resistance plasmid can often transfer that plasmid to nonresistant cells around it, so that the

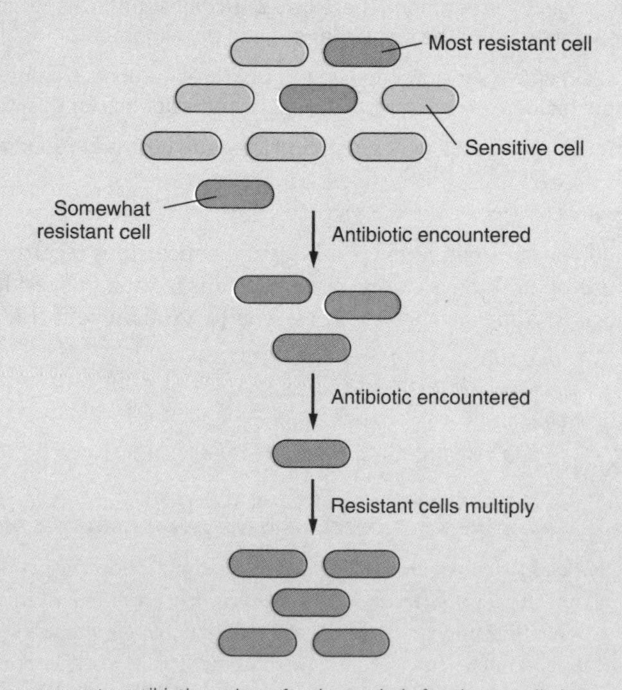

Figure 4.7 How exposure to antibiotics selects for the survival of resistant cells in a population of bacteria.

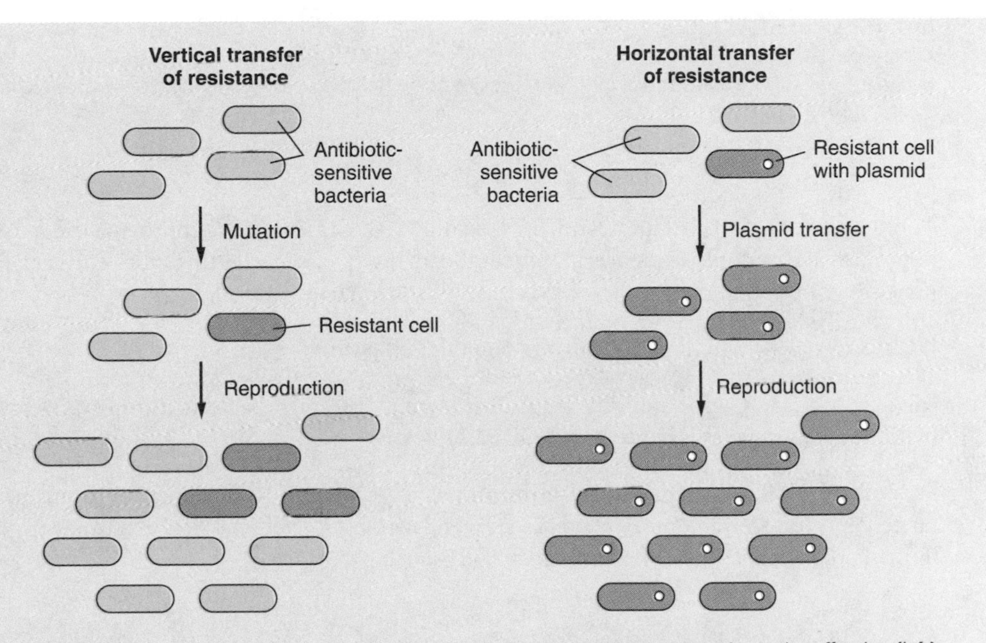

Figure 4.8 Vertical gene transfer occurs when a cell passes a resistance mutation to its offspring (left); horizontal transfer from cell to cell (right) allows much faster spread of resistance.

resistance gene is passed not only to a cell's descendants but to its peers and to their descendants. Depending on the circumstances, this transfer could occur by cell-to-cell contact (conjugation), by means of a bacterial virus (transduction), or by direct uptake of DNA released into the environment (transformation). Antibiotic resistance genes are also often found within transposons, semi-independent DNA sequences that can move within a genome, further promoting their mobility.

As discussed in the preceding chapter, when the sequences of two genes are similar, we conclude that they have a common origin; furthermore, we assume that highly similar genes diverged from that common origin only recently and have not had much time to evolve independently. Two very similar sequences found in dissimilar organisms—those that do not have a recent common ancestor—suggest that HGT has occurred: The gene evolved in one species but was then transferred intact to another relatively recently, so there has been limited opportunity for mutation.

References and Supplemental Reading

Original BLAST Algorithm

Altschul, S. F., W. Gish, W. Miller, E. W. Myers, and D. J. Lipman. 1990. Basic local alignment search tool. *J. Mol. Biol.* **215**:403–410.

Modified BLAST Algorithms

Altschul, S. F., T. L. Madden, A. A. Schaeffer, J. Zhang, Z. Zhang, W. Miller, and D. J. Lipman. 1997. Gapped BLAST and PSI-BLAST: a new generation of protein database search programs. *Nucleic Acids Res.* **25**:3389–3402.

Importance of BLAST

Harding, A. 2005. BLAST: how 90,000 lines of code helped spark the bioinformatics explosion. *The Scientist* **19**(16):21–25.

ClustalW

Thompson, J. D., D. G. Higgins, and T. J. Gibson. 1994. CLUSTAL W: improving the sensitivity of progressive multiple sequence alignment through sequence weighting, position-specific gap penalties and weight matrix choice. *Nucleic Acids Res.* **22**:4673–4680.

HGT Between Agricultural Bacteria and Human Pathogens

Salyers, A. A., A. Gupta, and Y. Wang. 2004. Human intestinal bacteria as reservoirs for antibiotic resistance genes. *Trends Microbiol.* **12**:412–416.

Shoemaker, N. B., H. Vlamakis, K. Hayes, and A. A. Salyers. 2001. Evidence for extensive resistance gene transfer among *Bacteroides* spp. and among *Bacteroides* and other genera in the human colon. *Appl. Environ. Microbiol.* **67**:561–568.

Chapter 5

Substitution Matrices and Protein Alignments:

Virulence Factors in *E. coli*

Chapter Overview

The preceding chapters discussed dynamic programming to find optimal alignments of nucleotide sequences (global, semiglobal, or local) and related algorithms for multiple alignments and database searches. This chapter extends our understanding of alignment to the amino-acid sequences of proteins, introducing the idea of scoring matrices based on amino-acid similarity. Students will learn how substitution matrices are developed, understand the advantages of protein alignment, and appreciate why different substitution matrices might be of value in solving particular biological problems. Students in both programming and nonprogramming courses will gain experience using alignments to investigate protein function "*in silico*," an essential part of genomic research. Students in programming courses will also learn how to implement substitution matrices within dynamic programming alignment programs discussed earlier and how iterations of a substitution matrix are computed.

Biological problem: Determining possible functions of a newly discovered virulence factor

Bioinformatics skills: Alignment of protein sequences, selection of substitution matrices, identification of putative functions based on database searching and alignment

Bioinformatics software: BLAST, substitution matrices, PSI-BLAST, Pfam, MOTIF, DAS, MicrobesOnline

Programming skills: Scoring matrices incorporating degrees of similarity, deriving substitution matrices, nested hash tables

Understanding the Problem:

Virulence Factors in *E. coli* Outbreaks

If it seems to take a long time to get a hamburger at your favorite restaurant, perhaps Escherichia coli *strain O157:H7* (**Figure 5.1**) *is to blame. Since its recognition as an important cause of human disease as a result of an outbreak of gastrointestinal illness in 1982, this bacterium*

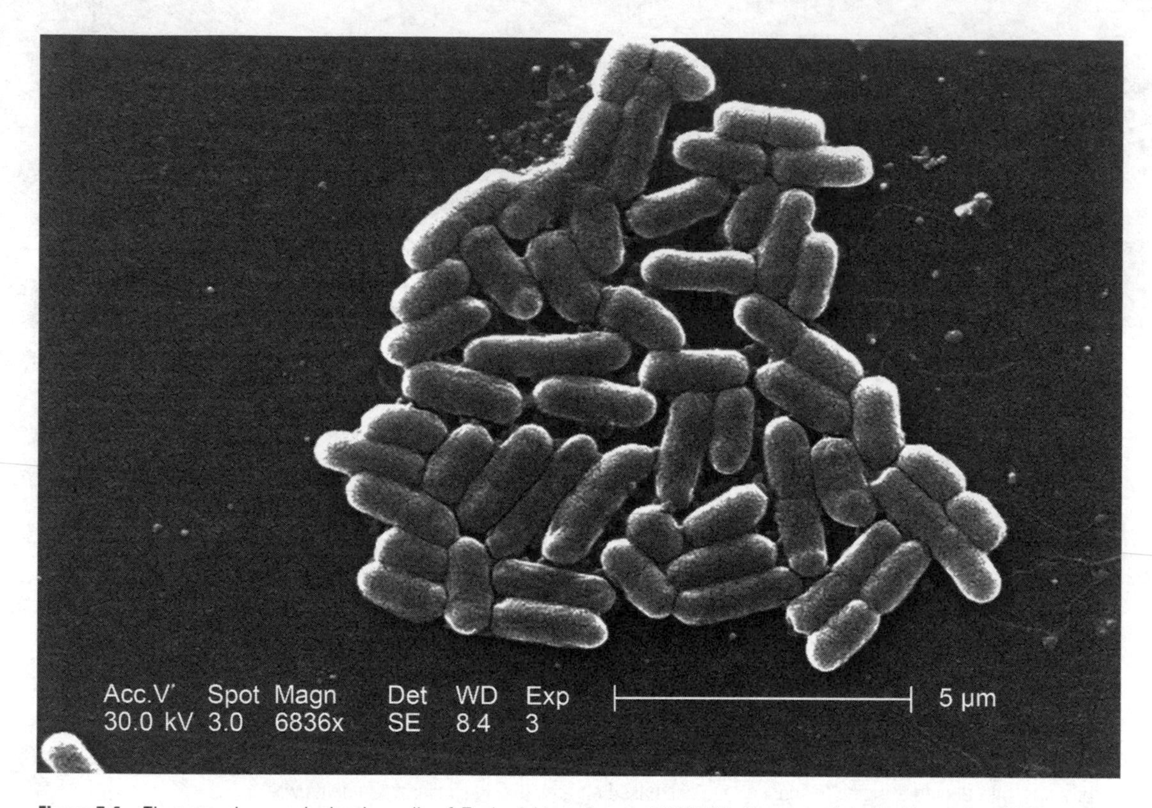

Figure 5.1 Electron micrograph showing cells of *Escherichia coli* strain O157:H7. Courtesy of Janice Haney Carr/CDC.

has gained considerable public attention and is the prime reason for labels urging cooking of ground beef to 160°F and even irradiation of raw ground beef prior to its sale. E. coli *O157:H7 now causes hundreds of infections annually, dozens of which lead to hospitalization and several of which are ultimately fatal, usually due to the bacterium's ability to spread from the intestine to infect the kidneys (hemolytic uremic syndrome). Highly publicized outbreaks have been associated with contaminated meat, water, spinach, lettuce, and other products, giving* E. coli *a very negative public image.*

Despite its association with recent outbreaks of foodborne disease, most strains of *E. coli* are harmless (even beneficial) residents of the large intestines of humans and other mammals. At worst, they usually cause only minor diarrheal disease (e.g., "travelers' diarrhea"). Strain O157:H7, in contrast, is a highly virulent pathogen, and serious or even potentially fatal disease can be the result of swallowing as few as 10 cells. What makes strain O157:H7 so different? One key factor is the acquisition of the gene for a toxin called Shiga toxin (or Stx) not present in other *E. coli* strains. Stx binds receptors found in human kidney tissue but, importantly, is not found in cattle, enabling these animals to be symptom-free carriers of the bacteria. However, genome sequencing reveals many other differences between the O157:H7 genome and the genome of "tame" *E. coli* inhabiting the human gut. At least some of the genes specific to O157:H7 are likely to encode **virulence factors**: proteins that contribute to the ability of the organism to cause disease (Stx itself, for example, is one major virulence factor).

What do these pathogen-specific genes encode? How do their products contribute to the virulence of *E. coli* O157:H7, and how might we exploit that knowledge to better prevent and treat disease? The existence of genome data opens up the opportunity to investigate the functions of these virulence-factor genes. Protein alignment is a key technique that can be used to develop hypotheses about protein function.

Bioinformatics Solutions:
Protein Alignment and Clues to Function

You already know how to align two DNA sequences using either global or local alignment algorithms, how to do a multiple alignment of several sequences at once, and how to make rapid alignments between a query sequence and a large database to identify similar sequences. However, the problem with DNA alignments is that they are "blind" to the genetic code. As illustrated in **Figure 5.2**, an optimal alignment between two nucleotide sequences could be meaningless in the context of the reading frame of a coding region, suggesting a relationship between the sequences that is not genuine. And, even if codons align correctly, a look at the genetic code (see Figure 2.4) demonstrates that mismatches can have very different biological consequences. For example, if the sequence AGG is aligned with AGA, we score two matches and one mismatch, and the same is true for the alignment of AGG with AGC. However, if AGG is a codon encoding arginine within a protein coding sequence, then a mutation to AGA would still encode arginine and have no effect on the protein. A mutation to AGC, however, encodes serine, a rather different amino acid that might have a significant effect on the protein. From an evolutionary standpoint, a change from AGG to AGA has no selective disadvantage, and thus the alignment of AGG with AGA is more likely to represent genuine evolutionary relatedness than an alignment of AGG with AGC.

The Needleman-Wunsch algorithm has no mechanism for deciding which nucleotide changes are more or less likely to have occurred over evolutionary time—or even which nucleotides represent codons or parts of coding sequences. One can imagine an algorithm that would require locations of coding sequences as input and that would take into account the genetic code, but we can get the same effect more easily by dealing with protein sequences. If we align the amino-acid sequences of two *proteins* rather than the nucleotide sequences of their genes, then it becomes obvious that arginine aligned with arginine is better than arginine aligned with serine—nature has done the work for us.

Furthermore, when a mutation occurs that does change a codon, some changes would be less damaging to the protein than others. For example, arginine and lysine are chemically similar: Both are positively charged amino acids of similar size. Therefore, a change from

coding sequence 1:

ATGGAACTAACTCCG
MetGluLeuThrPro

coding sequence 2:

ATGGCTACCCCGAGT
MetAlaThrProSer

optimal nucleotide alignment:

ATGGAACTAAC-TCCG---
ATGG--CTA-CCCC-GAGT

Figure 5.2 Ignoring the coding sequence can result in a "good" DNA alignment between sequences that actually encode proteins with very little similarity.

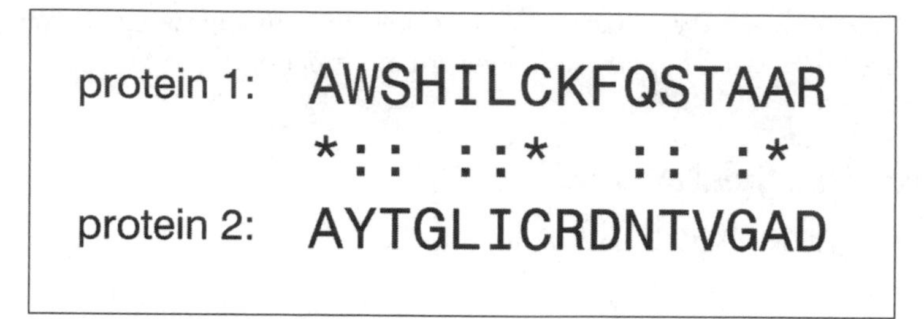

Figure 5.3 Alignment of two amino-acid sequences (using the one-letter code to represent amino acids). The alignment program reports both identical (*) and similar (:) amino acids, so similarity that might not have been apparent from the sequences can now be observed.

arginine to lysine might make less difference to the structure or function of the protein than a change from arginine to the smaller, uncharged serine. Thus, whereas nucleotides are either the same or different, amino acids can be identical, similar, or dissimilar. Accounting for the degree of similarity can allow us to discover that nucleotide or amino-acid sequences that appear dissimilar are actually orthologous (**Figure 5.3**). Therefore, we need an algorithm that can (1) align amino acids rather than nucleotides and (2) take into account some measure of how similar two amino acids are. With this algorithm in hand, we can ask in-depth questions about the relatedness of two proteins and use the answers to draw conclusions about evolutionary pathways and about the functions of the proteins. This latter idea is a very important part of genomic analysis: Once a gene is identified and the amino-acid sequence of its protein product inferred, we can use alignment to find similar proteins—orthologs or paralogs. If those related proteins have known functions, we can hypothesize that our new protein has the same or a similar function. It is important to realize, however, that bioinformatics can only provide *hypotheses* about protein function. Genetic and biochemical experiments in the laboratory are then required to test those hypotheses.

BioConcept Questions

This chapter's projects require you to understand the basics of amino-acid structure and the chemical nature of these building blocks, as well as how proteins are put together. Use these questions to test your biological understanding; read BioBackground at the end of the chapter if you need a better foundation. You may also want to review the introduction to the genetic code in Chapter 2.

1. Suppose a mutation changes a codon in a gene from GTA (or GUA in mRNA) to GAA. What is the corresponding amino-acid change? What are two ways in which this small change in DNA could produce a drastic change in the function of the protein encoded by this gene?
2. Even though this particular mutation (GTA to GAA) changes only a single nucleotide and in turn only a single amino acid, it should be quite rare to observe this change when comparing genes that are genuinely related. Why would this be an uncommon change?
3. A mutation that is even more rarely observed within a coding region is a change from a TGG codon to a TGA codon. Explain why this mutation would be selected against so strongly.
4. The enzyme lactase is found in your small intestine and converts lactose from dairy products into two simple sugars. Some people are lactose intolerant because they carry mutations in the lactase gene that result in production of a less-active enzyme. As you would expect, some of these mutations change amino acids that are found at the active site of the enzyme, the region where the lactose actually binds and is split. However, some mutations that affect lactase activity

change amino acids located far from the active site. What could explain the effect of these mutations?

5. Suppose a gene's coding region begins with the sequence ATGCTCCGGCAAAGG. A gene in another organism begins with the sequence ATGTTAAGAAACCGT, so there does not seem to be much sequence similarity. Is it likely that these two genes encode proteins of similar functions? (*Hint: Translate the sequences before answering the question.*)
6. The coding sequence for the Stx toxin in a particular strain of *E. coli* O157:H7 begins with ATGAAGTGTATATTATTTAAA. Suggest one possible single-nucleotide substitution mutation that *would* change an amino acid (missense mutation) but would *not* have a drastic effect on protein function. Then suggest one missense mutation that *would* have a drastic effect on protein function.

Understanding the Algorithm: Protein Alignment and Substitution Matrices

Consider the Needleman-Wunsch algorithm for aligning protein sequences you used in Chapter 3: Is there any reason the two sequences could not be amino-acid sequences (using the one-letter code)? Try it! Using your Needleman-Wunsch program, you should be able to make a perfectly reasonable global alignment between, for example, GTIKENIIFGVSYDEYR and GTIKENIIGVSYDEYR, the amino acid sequences from two *CFTR* variants you looked at in Chapter 2. (Indeed, you could equally well use the Needleman-Wunsch algorithm, which has no way to know what the sequences mean, to generate an alignment between "UNIVERSAL" and "UNIVERSITY.") Most algorithms for local alignment, multiple alignment, and database searching are similarly flexible and can align any two strings. However, a protein alignment becomes powerful only when we can consider both those amino acids that match exactly and those that are similar. The key to this lies in the scoring. The Needleman-Wunsch algorithm as we implemented it in Chapter 3 could assign only match, mismatch, and gap scores, but it can easily be modified to assign a score based on the degree of similarity between two amino acids.

Substitution Matrices

To accomplish this, a protein alignment algorithm needs a **substitution matrix**—essentially just a table of scores for different pairs of amino acids. Aligning two similar amino acids should get a higher score than aligning two dissimilar ones. The substitution matrix gives the scores that should be assigned for each possible pair based on some set of criteria.

There is no one absolute standard to describe amino-acid similarity. Many different substitution matrices can be developed, and one may fit a particular situation better than another (discussed later in the chapter). As a straightforward example, a substitution matrix could be based on the hydrophobicity of the amino acids: Lower scores would be given for replacement of a more hydrophilic amino acid with a more hydrophobic one, whereas higher scores would be given for pairs more similar in their hydrophobicity. **Table 5.1** shows simple hydrophobicity values for the 20 amino acids, with larger numbers representing more hydrophobic amino acids. We might then decide to give a score of 1 if two amino acids match exactly, a score of 0.5 if they are both hydrophilic (hydrophobicity < 0, such as aspartate and asparagine) or both hydrophobic (hydrophobicity > 0, such as alanine and cysteine), and zero if one is hydrophilic and the other hydrophobic (such as alanine and aspartate). (As we will discuss, we could make the scoring much more complex than this simple example.) Now we could develop the substitution matrix to score any possible pair of amino acids; part of such a matrix is shown in **Figure 5.4**.

Table 5.1 Hydrophobicity values for the 20 amino acids. A more positive value represents a more hydrophobic amino acid.

Amino Acid	Hydrophobicity	Amino Acid	Hydrophobicity	Amino Acid	Hydrophobicity
D	−3.5	Y	−1.3	I	4.5
K	−3.9	N	−3.5	C	2.5
H	−3.2	L	3.8	A	1.8
T	−0.7	E	−3.5	S	−0.8
V	4.2	R	−4.5	G	−0.4
F	2.8	W	−0.9	P	−1.6
M	1.9	Q	−3.5		

The problem with a substitution matrix based on chemical properties is that even if we try to account for our best understanding of which amino acids are similar, we are not always right about what substitutions would replace an amino acid with a similar one (**conservative** substitutions). Luckily, we have evidence from billions of years of evolution to help us understand which substitutions are actually likely. Most substitution matrices today are based on observed conservation of amino acids: Two very common ones are the PAM (*p*oint *a*ccepted *m*utation) and BLOSUM (*blo*cks *su*bstitution *m*atrix) matrices.

The PAM matrix was developed by Dayhoff et al. in 1978 (see References and Supplemental Reading) by examining very closely related sequences for amino-acid changes. The use of closely related sequences minimizes the likelihood that an observed difference really represents a sequence of two or more individual mutations. The observed frequencies of each possible amino-acid substitution were determined and converted into scores using the equation $10 \log_{10} (M_{ij}/f_j)$, where M_{ij} is the probability of a mutation replacing amino acid i with j and f_j is the frequency of occurrence of amino acid j in a large set of sequences. In essence, we are comparing the hypothesis that amino acid j occurs due to evolutionary conservation of a substitution with the hypothesis that its occurrence is random. If this **odds ratio** is 1, then the substitution is no more likely than the chance of finding j randomly. If a substitution is

	A	C	D	F	S	N	W
A	1						
C	0.5	1					
D	0	0	1				
F	0.5	0.5	0	1			
S	0	0	0.5	0	1		
N	0	0	0.5	0	0.5	1	
W	0	0	0.5	0	0.5	0.5	1

Figure 5.4 Partial substitution matrix based on amino-acid hydrophobicity (Table 5.1). Scores for each pair of amino acids are based on having the same (1), similar (0.5) or different (0) hydrophobicity.

likely to be evolutionarily conserved, this ratio increases, and if it is selected against, the ratio decreases. The $\log_{10}$ is taken for convenience in scoring: This **log-odds ratio** is more positive for likely (conservative) substitutions and more negative for unlikely (nonconservative) substitutions. For further convenience, it is multiplied by 10 to allow rounding to the nearest integer. **Figure 5.5** shows an example of the resulting scoring matrix.

The initial PAM matrix, PAM 1, was built from 71 known protein families (later updated to include much larger sets of sequences) and normalized to represent an average change of 1 amino acid in 100. This means the PAM matrix effectively represents a unit of evolutionary time: the time required for change to occur in 1% of amino acids. This is relatively short on the evolutionary time scale, so to effectively compare less closely related sequences, we can recompute the matrix for an evolutionary time span of 2 PAMs or 10 PAMs or 100 PAMs—indeed, the matrix in Figure 5.5 is PAM 250, very commonly used to compare protein sequences thought to be distantly related. This computation is simply 10 $\log_{10}(M^k_{ij}/f_j)$, where k is the desired PAM distance.

		A	R	N	D	C	Q	E	G	H	I	L	K	M	F	P	S	T	W	Y	V
Ala	A	2																			
Arg	R	–1	5																		
Asn	N	0	0	3																	
Asp	D	0	–1	2	5																
Cys	C	–1	–1	–1	–3	11															
Gln	Q	–1	2	0	1	–3	5														
Glu	E	–1	0	1	4	–4	2	5													
Gly	G	1	0	0	1	–1	–1	0	5												
His	H	–2	2	1	0	0	2	0	–2	6											
Ile	I	0	–3	–2	–3	–2	–3	–3	–3	–3	4										
Leu	L	–1	–3	–3	–4	–3	–2	–4	–4	–2	2	5									
Lys	K	–1	4	1	0	–3	2	1	–1	1	–3	–3	5								
Met	M	–1	–2	–2	–3	–2	–2	3	3	–2	3	3	–2	6							
Phe	F	–3	–4	–3	–5	0	–4	–5	–5	0	0	2	–5	0	8						
Pro	P	1	–1	–1	–2	–2	0	–2	–1	0	–2	0	–2	–2	–3	6					
Ser	S	1	–1	1	0	1	–1	–1	1	–1	–1	–2	–1	–1	–2	1	2				
Thr	T	2	–1	1	–1	–1	–1	–1	–1	–1	1	–1	–1	0	–2	1	1	2			
Trp	W	–4	0	–5	–5	1	–3	–5	–2	–3	–4	–2	–3	–3	–1	–4	–3	–4	15		
Tyr	Y	–3	–2	–1	–2	2	–2	–4	–4	4	–2	–1	–3	–2	5	–3	–1	–3	0	9	
Val	V	1	–3	–2	–2	–2	–3	–2	–2	–3	4	2	–3	2	0	–1	–1	0	–3	–3	4

Figure 5.5 The PAM 250 substitution matrix.

Notice how different the matrix in Figure 5.5 is from what we might have expected based only on our knowledge of chemical properties. For example, cysteine, a smallish hydrophobic amino acid, would seem similar to several others based on our hydrophobicity table (see Table 5.1), but the PAM matrix shows that in fact it has a much higher probability of remaining cysteine than being replaced by any other amino acid. This strong bias is likely due in this case to the unique role of cysteine in forming structurally important disulfide bonds, but it illustrates the value of a matrix based on observed evolutionary change.

The BLOSUM matrix is conceptually similar. The starting point for this matrix was some 500 groups of local alignments in which ungapped blocks of amino-acid sequence had been conserved. Scoring is again based on a log-odds ratio, the log of the frequency with which two amino acids align in a multiple alignment divided by the expected frequency for a random alignment. As with PAM, the more positive the score, the more likely a particular substitution is to occur. The default matrix for a BLAST search is BLOSUM 62, which means a BLOSUM matrix computed by choosing blocks more than 62% identical.

The choice of a substitution matrix depends on the proteins being aligned. PAM and BLOSUM generally give similar results; some researchers prefer PAM because it is based on global alignment and the generation of an initial phylogenetic tree, whereas BLOSUM is based on local alignment and an implicit evolutionary model. More important is the choice of which PAM or BLOSUM model to use: When aligning distantly related proteins, a higher-numbered PAM matrix (e.g., PAM 250) or *lower*-numbered BLOSUM matrix (e.g., BLOSUM 45) is used, whereas PAM 1 or BLOSUM 85 might be used for closely related proteins. PAM 120 and BLOSUM 62 are commonly used for initial database searches when no specific set of sequences has been selected for comparison. Low-numbered PAM matrices such as PAM 30 and PAM 70 are often recommended for searches involving short sequences.

Protein Alignment Algorithm: Scoring with the Substitution Matrix

The substitution matrix values represent the likelihood that one amino acid will substitute for another. This can also be seen as representing the desirability of aligning a particular pair of amino acids (a biologically better alignment would have fewer nonconservative substitutions), and thus the substitution matrix can act as a scoring matrix. Instead of a single match score, the match scores become the values along the diagonal (likelihood of not changing the amino acid), and the mismatch scores are looked up in the matrix based on the particular amino acids being compared at each point (see Figure 5.4). Only the gap penalty is not contained within the matrix; it is important to decide on a score that is comparable with the scale used in the substitution matrix. Looking at the PAM 250 matrix (Figure 5.5) for example, scores range from –5 to 15, so one might choose a gap penalty of 0 (essentially the same as a random amino acid), –5 (as bad as the worst mismatch), or maybe even –6 (worse than the worst mismatch). Note, however, that the effect of a single amino-acid deletion is likely to be much less extreme than a single nucleotide deletion, so you might not want to over-penalize gaps.

Alignment of two protein sequences, just like two nucleotide sequences, begins with an alignment matrix (not to be confused with the substitution matrix, which is simply used as a lookup table to determine scores). Suppose we wish to align the amino-acid sequences `ANFNNASWF` and `ANFNCFWS` using our hydrophobicity scoring matrix (Figure 5.4) and that we have decided on a gap penalty of –1. Recall that the first row and column are initialized using the gap penalty (**Figure 5.6A**). The rest of the matrix can be populated using the Needleman-Wunsch algorithm from Chapter 3, except that at the step where three scores are computed (cell to the left plus gap penalty, cell above plus gap penalty, or cell diagonally above left plus match/mismatch value), the match/mismatch value is looked up from the

A

		A	N	F	N	N	A	S	W	F
	0	-1	-2	-3	-4	-5	-6	-7	-8	-9
A	-1									
N	-2									
F	-3									
N	-4									
C	-5									
F	-6									
W	-7									
S	-8									

B

		A	N	F	N	N	A	S	W	F
	0	-1	-2	-3	-4	-5	-6	-7	-8	-9
A	-1	1	0	-1	-2	-3	-4	-5	-6	-7
N	-2	0	2	1	0	-1	-2	-3	-4	-5
F	-3	-1	1	3	2	1	0	-1	-2	-3
N	-4	-2	0	1	4	3	2	1	0	-1
C	-5	-3	-1	0.5	3	4	3.5	2.5	1.5	0.5
F	-6	-4	-2	0	2	3	4.5	3.5	2.5	2.5
W	-7	-5	-3	-1	1	2.5	3.5	5	4.5	3.5
S	-8	-6	-4	-2	0	1.5	2.5	4.5	5.5	4.5

Figure 5.6 Alignment matrix for a sample protein alignment using a scoring matrix based on hydrophobicity. (A) The matrix initialized with a gap penalty of −1. (B) The completed alignment matrix.

substitution table. **Figure 5.6B** shows the result. At this point, optimal paths through the matrix from lower right to upper left are computed and alignments built just as for nucleotide alignments. We could equally well have used the PAM matrix or some other substitution matrix, selecting an appropriate gap penalty and simply substituting one table for the other when looking up scores. Indeed, most protein alignment programs allow the user a choice of substitution matrix.

Test Your Understanding

1. Examine the PAM 250 matrix shown in Figure 5.5. Which amino acids are the least likely to be replaced by another amino acid in a protein? Are any of these surprising based on our naïve assumption that conservative substitutions are based on shared chemical properties? What might you infer about the roles of these amino acids in proteins?
2. Often, a researcher studying a particular protein will test the hypothesis that a specific amino acid is important in the protein's function by deliberately making a mutation that substitutes another amino acid. Very often, alanine is chosen for the replacement, because it is a small amino acid thought to be unlikely to grossly disrupt protein structure. Based on the PAM substitution matrix, what arguments could you make for or against the use of alanine?
3. Complete the alignment begun in Figure 5.6 by traversing the matrix to determine the optimal global alignment(s) for the two short sequences shown. Compare this alignment and its score with the results of a simple match/mismatch/gap scoring scheme. Can you see evidence that even this very simplistic scoring matrix led to better alignment?
4. Complete the alignment matrix for the two sequences using the PAM 250 substitution matrix with a gap score of −14 (default gap penalty for BLAST when PAM 250 is used). What is the best alignment for the two sequences using this substitution matrix? Can you see evidence that this results in a biologically better alignment than our very basic hydrophobicity scoring system?
5. How would you have to modify the protein alignment algorithm to do a local alignment incorporating a scoring matrix rather than a global alignment?

CHAPTER PROJECT:

Using Protein Alignment to Investigate Functions of Virulence Factors

Novel virulence genes evolved in or acquired by highly pathogenic strains of *E. coli* such as O157:H7 could be very important in dealing with this important foodborne disease. Understanding how these bacteria cause disease and why they have more severe effects than typical *E. coli* strains may lead us to new and better ways to treat and prevent disease. In this chapter's projects, you will work with genes found in *E. coli* strain O157:H7 but not in the nonpathogenic strain K-12. Using protein alignment tools, you will look for clues to the functions of these potential virulence factors. You will examine the effect of changing substitution matrices and in the On Your Own Project explore the advantages of creating a substitution matrix specific to a particular problem.

Learning Objectives

- Understand the use of a substitution matrix to score amino acid similarity in a protein sequence alignment
- Gain experience using protein alignment to develop hypotheses about protein function based on sequence similarity
- Know how protein alignment differs algorithmically from DNA alignment
- Know how a substitution matrix is developed and how different matrices might be used to produce better alignments in particular situations

Suggestions for Using the Project

This project is designed to develop skills in aligning proteins and evaluating potential functions using existing Web-based tools for all students. It also provides an opportunity for students in programming-based courses to take a more nuts-and-bolts look at how protein alignment programs use scoring matrices and for all students to understand how scoring matrices are developed and used. Of course, an enormous variety of questions can be explored using protein alignment tools, and instructors with interests in particular areas should not feel limited to the specific project suggestions here but could adapt the same methods to other problems.

Programming courses:

- Web Exploration: Use BLAST and other alignment tools to find potential orthologs of novel *E. coli* O157:H7 proteins and explore their functions. Part I is recommended for all students; Part II can be used optionally to add breadth (instructors may wish to specify which tools students use in Part II).
- Guided Programming Project: Modify the Needleman-Wunsch algorithm to perform protein alignments using a substitution matrix for scoring, using the programming language of your choice. Instructors that choose to skip the Web Exploration can download files from the *Exploring Bioinformatics* website to provide students with orthologs to test in this section.
- On Your Own Project: Develop a program to generate a project-specific substitution matrix from a given set of protein sequences.

Nonprogramming courses:

- Web Exploration: Use BLAST and other alignment tools to find potential orthologs of novel *E. coli* O157:H7 proteins and explore their functions. Parts I and II are recommended for a nonprogramming course, but Part II could be omitted to meet time constraints or instructors could specify which of the several tools students should try for this part.

- On Your Own Project: Compile a set of protein sequences suitable for generation of a project-specific substitution matrix and complete exercises to understand how this is accomplished.

Web Exploration

One of the first completely sequenced genomes was that of *E. coli* strain K-12, specifically a substrain commonly used in molecular biology labs called MG1655. This strain is a descendant of benign intestinal *E. coli* isolates. Subsequently, a number of different *E. coli* genomes have been sequenced, including O157:H7 strains. The first O157:H7 genome sequence, published in 2001 (see References and Supplemental Reading), came from strain EDL933, isolated from contaminated ground beef from a McDonald's restaurant in Michigan. With these sequences in hand, a key question was to find out how the genomes differed (see More to Explore). Genes unique to O157:H7 might contribute to the ability of this bacterium to cause disease; these candidate virulence factor genes can then be investigated further to identify potential functions.

The degree of difference between the genomes of MG1655 and EDL933 is surprising: MG1655 has more than 500,000 bp of sequence not found in EDL933, whereas more than 1.3 million bp of sequence unique to EDL933 were identified, including about one-fourth of its 5,416 total genes. Thus, hundreds of distinct genes could be virulence factors for EDL933. A short list of candidates is given in **Table 5.2**, along with their NCBI gene id numbers. (The entire EDL933 genome sequence is published under accession number AE005174; individual genes are given gene ID numbers to identify them within the sequence.)

Bioinformatics allows us to develop hypotheses about the functions of proteins. Simply being present in EDL933 but not MG1655 suggests that a gene could be a virulence factor. Evidence to strengthen the hypothesis can be acquired by using protein alignment to look for orthologs of putative virulence proteins that have been studied in other organisms. Sequence similarity to a protein with a known virulence function or identification of protein domains suggestive of a virulence function are examples of such evidence. Much is known about bacterial virulence, and based on that background knowledge, we would expect virulence factors to function in roles such as toxins, systems for delivering toxins to host cells (e.g., Type III or Type IV secretion systems), components of pili and other bacterial surface features allowing attachment to host cells, enzymes that break down host proteins, and proteins that sequester iron or other nutrients. However, it is important to bear in mind that even strong bioinformatics-based hypotheses require experimental testing—even minor sequence variations might result in altered functions or characterization of a gene with no obvious disease function might lead to the discovery of a new type of virulence factor.

Table 5.2 Candidate virulence genes from *E. coli* O157:H7 strain EDL933.

Gene Name[1]	NCBI ID	Gene Name[1]	NCBI ID
yadK	12512854	*ydgE*	12515577
yagW	12513076	*yeeJ*	12516151
ybbK	12513379	*yehC*	12516323
ybgP	12513628	*yhiF*	12518204
ycjZ	12515432	*ysaS*	12517366

[1]In bacterial genomes, gene designations beginning with *y* indicate genes whose identity is not yet sufficiently certain to merit a specific name.

Part I: Using Protein Alignment to Explore Protein Function

BLAST. You already know how to use BLAST to compare a nucleotide sequence with every known sequence in the NCBI database. We can do the same for a protein sequence, increasing the quality of the search results by adding to the already powerful BLAST search the benefit of a substitution matrix to score amino-acid similarity. From the **BLAST home page**, choose `protein blast` to align an amino-acid sequence query with database sequences. Select one or more candidate virulence factors to investigate from Table 5.2 and obtain their amino-acid sequences in FASTA format using the **NCBI Protein database**. Paste a candidate sequence into the BLAST query sequence box.

It would make sense to search for orthologs of your candidate virulence factor among the bacteria, and you could also choose to limit the search to the Gram-negative bacteria or even to the Enterobacteria (the large family of intestinal bacteria to which *E. coli* belongs). Use the `Organism` field to limit your search appropriately. Then add an additional Organism field and use it to exclude *E. coli* from the search results—this prevents your results from being cluttered with high-scoring matches from EDL933 itself or other sequenced pathogenic *E. coli* strains. At the bottom of the window, click `Algorithm parameters` to choose an appropriate substitution matrix: BLOSUM 62 is the default, but because the search is limited to relatively closely related organisms, perhaps it makes sense to try a matrix optimized to more closely related sequences, such as BLOSUM 80 or PAM 70 (remember higher BLOSUM numbers and lower PAM numbers represent more similar sequences used to generate the matrix). Run your BLAST search.

Now comes the important work of analyzing the results. Obviously, a high-scoring match (indicating a high degree of similarity between your query and some other protein) provides stronger evidence for a conserved function than a lower-scoring match, and good alignment along the whole length of the protein similarly better supports functional conservation than a partial match. Review Chapter 4 if necessary to refresh your memory of what the score and *e*-value mean. If you find a good match, investigate the function of the putative ortholog: Is it found in a pathogenic bacterium? What is known about its function? Is there evidence that it is a virulence factor?

Conserved domains. While your BLAST search was running, you may have seen a page informing you that "conserved domains" had been detected in your query protein. If so, you should see a box at the top of your BLAST results page headed `Putative conserved domains have been detected`. A **domain** is a functional region of a protein. For example, an energy-requiring enzyme might have an ATP-binding domain as well as a substrate-binding domain where its catalytic function is carried out. A transcription factor would likely have a DNA-binding domain as well as a domain that interacts with RNA polymerase. Even if two proteins are not terribly similar overall, they might have a particular domain in common: Two DNA-binding proteins that have different functions might have similarity in their DNA-binding domains but be very different in a domain used for interactions with their distinct molecular partners.

BLAST looks for patterns in the query protein that resemble known functional domains and reports these results. The conserved domains box shows the regions of your protein that are similar to well-characterized functional domains; clicking on this display takes you to more information about the conserved domains and the other proteins that contain them. You can also run a conserved domain search directly without a BLAST search by searching **NCBI's Conserved Domains database**.

Substitution matrices. What would happen if you changed the substitution matrix used in your search? You initially optimized it to give higher scores to substitutions likely to occur in closely related sequences, but what if you used a matrix like PAM 250 or BLOSUM 45 that is based on more distantly related sequences? Try it and see. Although it is likely that BLAST will still pick up the same high-scoring matches, there could be some less closely related proteins in the list, or you may notice changes in the score or *e*-value resulting from scoring with this matrix. What would happen if you searched for matches to a really distantly related organism? Because the goal of this exercise is to identify potential virulence factors in *E. coli*, it is appropriate to limit the matches to related bacteria, but perhaps you are curious whether your virulence factor might have a human ortholog. Use BLAST to find out—with a substitution matrix chosen to score such distant relationships appropriately. Some bacteria-specific proteins have no identifiable human orthologs, whereas others have been conserved across this long span of evolutionary time. Still others are surprisingly similar to human proteins, leading to speculation about recent horizontal transfer between species.

Web Exploration Questions

For each candidate virulence factor you selected, write a one-page summary discussing its likely functions based on conserved domains and orthologous proteins. Discuss the organisms in which similar proteins were found, the known functions of those proteins, and the quality of the your BLAST matches. Based on the evidence you have accumulated thus far, is it reasonable to identify your protein as a virulence factor? How would the function you have hypothesized for the protein contribute to the ability of EDL933 to cause disease? Comment on the strength of your evidence: How confident are you in assigning this function to your protein or in characterizing its role in virulence?

Part II: More Tools for Exploring Protein Function

Inferring putative functions for proteins based on amino-acid sequence is a major role of bioinformatics in modern biology. Not only do researchers using genetic methods continue to discover new genes involved in important aspects of physiology and disease, but new genome sequences are being completed at an ever-increasing pace. In both cases, one winds up with genes of unknown function, and various alignment-based tools are the major means of developing hypotheses. Thus, many bioinformatics tools have been developed that are directed toward understanding protein function. Below are a few such tools that might be used to learn more about candidate *E. coli* O157:H7 virulence factors. You can choose to try any or all of the programs listed, or your instructor may assign you to use specific tools.

PSI-BLAST. PSI-BLAST is a variation of BLAST in which initial matches are used to refine the substitution matrix to identify even more distant matches. This is a good tool when you want to identify meaningful alignments to distantly related proteins, such as when a simple BLAST search reveals no good orthologs. To use PSI-BLAST, start at the **BLAST home page**, choose `protein blast` as before, and then on the next page click on the PSI-BLAST radio button before starting the search.

Pfam. **Pfam** is a database of protein families—groups of proteins already shown to be similar in structure and function. Particularly when a protein sequence of interest does not have a strong ortholog identifiable by a BLAST search or when the closest matches are partial or relatively low scoring, aligning the sequence with the Pfam protein families may yield information about specific domains or regions of the protein. When matches are found, the Pfam database provides considerable information about the known functions, sequences, and structures of the matching families, including links to still more sources of information.

MOTIF. Like Pfam, **MOTIF** looks for alignments between a query amino-acid sequence and functional domains and motifs (short sequence segments associated with some function). The difference here is that MOTIF is a "meta site" that allows you to search up to six databases at once.

DAS. The localization of a protein within the cell can also provide clues to possible functions. **DAS** (dense alignment surface) deals with one aspect of protein localization: whether the protein contains potential transmembrane domains that would suggest it is an integral membrane protein. Although there are various ways to approach this question (e.g., mapping hydrophobic amino acids), DAS uses an alignment-based approach in which it essentially looks for meaningful local alignments between the query protein and a set of *un*related known membrane proteins.

Web Exploration Questions

Try some additional tools to extend your understanding of the candidate virulence factors you have chosen and add any new information you gain to your summary. If your original analysis yielded few clues to the function of your protein, discuss whether these additional analyses helped identify a putative function. If you already had a likely function for your protein, discuss whether the further analysis you did strengthened your hypotheses.

More to Explore

Did you happen to wonder how one would go about developing a list of candidate virulence factors that are found in *E. coli* O157:H7 but not in the ordinary *E. coli* found in the human gut? Building on the alignment techniques we have discussed in this and previous chapters, programs have been developed that are specialized for aligning entire genomes and highlighting differences, as well as for conducting automated BLAST-like comparisons. An instructor with an interest in genome alignment could incorporate genome alignment into the course by asking students to first build their own list of virulence factor candidates, or an interested student could explore the genomes of these *E. coli* strains further. A number of Web-based tools for bacterial genome comparison can be found at **MicrobesOnline**; the On Your Own Project at the end of this chapter uses this site. Another good program is **MAUVE**, a visually oriented program for multiple genome alignment that can be downloaded freely from the University of Wisconsin.

Guided Programming Project: Using a Substitution Matrix in a Protein Alignment Program

As you saw in Understanding the Algorithm, a global protein alignment using a substitution matrix requires only a few changes to the Needleman-Wunsch global alignment program you developed in Chapter 3. The method of constructing the alignment, using dynamic programming and an alignment matrix, is in fact no different; the key change is to obtain scores for each pair of amino acids using a substitution matrix.

Substitution Matrices and the Protein Alignment Algorithm

A substitution matrix could be coded into the alignment program, but that would give limited flexibility. Given that the user may wish to use a particular scoring matrix, it would be desirable to read the scoring matrix from a file. The *Exploring Bioinformatics* website includes several substitution matrix files, including the simple hydrophobicity matrix shown in Figure 5.4 and the PAM 250 and BLOSUM 62 matrices. You could use a two-dimensional array to hold the scores, but a better approach is a hash table or equivalent data structure, so you can look up scores by using amino-acid letters directly rather than converting them to numerical equivalents. In the pseudocode presented next, we assume that a hash table is used to hold the substitution matrix and that the substitution values for amino-acid pairs are read from a data file.

The first four lines of a data file representing the PAM 250 matrix may look like the following:

```
>PAM250
 A, R, N, D, C, Q, E, G, H, I, L, K, M, F, P, S, T, W, Y, V
 2,-1, 0, 0,-1,-1,-1, 1,-2, 0,-1,-1,-1,-3, 1, 1, 2,-4,-3,-1
-1, 5, 0,-1,-1, 2, 0, 0, 2,-3,-3, 4,-2,-4,-1,-1,-1, 0,-2,-3
```

The file contains a FASTA-style comment to identify the matrix followed by the amino acids and then the substitution values. Each line of data values corresponds to a particular amino acid's substitution values. In the example above, the first line after the amino acids represents the substitution values for A with all other amino acids. Therefore, A substituted for A has a value of 2, R substituted for A has a value of −1, N substituted for A has a value of 0, and so on. The second line of data values represents the substitution values for R and all other amino acids. Therefore, A substituted for R has a value of −1, R substituted for R has a value of 5, N substituted for R has a value of 0, and so on.

Remember that the one score that cannot be looked up in the scoring matrix is the gap score. Because the gap score needs to be on the same scale as the match/mismatch scores (for example, our hydrophobicity scoring matrix has scores ranging from 0 to 1, whereas the PAM 250 matrix has scores ranging from −5 to 15), the program could prompt the user to input a gap score. For convenience, the prompt could display the high and low score values from the selected matrix as a guide. Or, an even more sophisticated approach is to add a default gap penalty to each scoring matrix file and then prompt the user to input a different value if desired.

Below is pseudocode for the protein alignment algorithm. This example uses a nested hash table. A **nested hash table**, essentially a hash of a hash table, allows us to store more than one value per element. We can think of a nested hash table as a two-dimensional structure, where the outer key represents the row and the inner key represents the column. For this problem, the outer key represents the first amino acid, the inner key represents the second amino acid, and the value is their score. Modifications from the Needleman-Wunsch alignment program appear in bold. When you implement this program in your desired programming language, you need to add the user input options discussed previously. Consider also the output format. For an amino-acid alignment, output similar to Figure 5.3 is typical, using characters between the two sequences to indicate identical or similar sequences (* and :, respectively, in our example).

Algorithm

Protein Alignment Algorithm

Goal: Determine the optimal global alignment of two amino-acid sequences

Input: Two sequences

Output: Best, global alignment(s) of two input sequences

```
// Initialization
Input the two sequences: s1 and s2
N = length of s1
M = length of s2
matrix = array of size [N+1, M+1]
gap = gap score
// note - there is no mismatch or match score assigned
```

```
// STEP 1: Initialize Substitution Matrix
// create substitution matrix file from data file
open input file containing substitution matrix: infile
read and discard FASTA comment line
read 2nd line of infile and store as array - aminoacids
i = 0
for each remaining line of data in infile
    j = 0
    for each value in line
        hashTable[aminoacids[i], aminoacids[j]] = value
        j++
    i++

// STEP 2: Build Alignment Matrix
set matrix[0,0] to 0
for each i from 1 to N
    matrix[i, 0] = matrix[i-1, 0] + gap
for each j from 1 to M
    matrix[0, j] = matrix[0, j-1] + gap
for each i from 1 to N
    for each j from 1 to M
        score1 = matrix[i-1, j-1] + hashTable{s1[i-1], s2[j-1]}
        score2 = matrix[i,j-1] + gap
        score3 = matrix[i-1, j] + gap
        matrix[i,j] = max(score1, score2, score3)

// STEP 3: Create Directional Strings
dstring = getDirectionalString(matrix, N, M);

// STEP 4: Build Alignments Using Directional Strings
seq1pos = N-1
seq2pos = M-1
dirpos = 0
while (dirpos < length of directional string)
    if (dstring[dirpos] equals "D")
        align s1[seq1pos] and s2[seq2pos]
        subtract 1 from seq1pos and seq2pos
    else if (dstring[dirpos] equals "V")
        align s1[seq1pos] and a gap
        subtract 1 from seq1pos
    else // must be an H
        align s2[seq2pos] and a gap
        subtract 1 from seq2pos
    increment dirpos

// Function to create directional string
function getDirectionalString(matrix, N, M)
    dstring = ""
    currentrow = N
    currentcol = M
```

```
        while (currentrow != 0 or currentcol != 0)
            if (currentrow is 0)
                add 'H' to dstring
                subtract 1 from currentcol
            else if (currentcol is 0)
                add 'V' to dstring
                subtract 1 from currentrow
            else if (matrix[currentrow][currentcol-1] + gap equals
                     matrix[currentrow][currentcol])
                add 'H' to dstring
                subtract 1 from currentcol
            else if (matrix[currentrow-1][currentcol] + gap equals
                     matrix[currentrow][currentcol])
                add 'V' to dstring
                subtract 1 from currentrow
            else
                add 'D' to dstring
                subtract 1 from currentcol
                subtract 1 from currentrow
    return dstring
```

Putting Your Skills into Practice

1. Write a program in the language used in your course to implement the previously given pseudocode. Create a substitution matrix file that represents a hydrophobicity substitution matrix (see Figure 5.4 for a partial example of the matrix). Align the amino acid sequences `EREHSISIVLE` and `QNHKTLGFICN` using your solution and using your global alignment solution from Chapter 3. Note the alignments found and their scores. Does the alignment change? Does the score? What happens if you use the PAM 250 or BLOSUM 62 matrix?
2. Now, try comparing two real genes. Use the results from BLAST search you did in the Web Exploration Project to select a protein nonidentical but orthologous to the EDL933 protein you investigated (if your class skipped the Web Exploration, your instructor can make a file containing pairs of orthologs from the instructor section of the *Exploring Bioinformatics* website available to use as test sequences). Compare these genes with substitution matrices appropriate for the relatively short evolutionary distances that separate them.
3. The partial hydrophobicity scoring matrix shown in Figure 5.4 is oversimplified and does not take advantage of all the hydrophobicity data in Table 5.1. For a more sophisticated comparison, we could assign each amino acid a value based on whether it is very hydrophobic (>1), somewhat hydrophobic (>0 but <1), somewhat hydrophilic (<0 but >−1), or very hydrophilic (<−1). We would then have four possible scores for each comparison: match, good hydrophobicity match (both in same category), weak hydrophobicity match (both hydrophilic or hydrophobic but in different categories), or mismatch. Develop a hydrophobicity substitution matrix for this improved scoring matrix and run your program using this new matrix. Test with the sequences you used for question 1.
4. For an even more sophisticated hydrophobicity-based scoring system, can you come up with a means of converting the hydrophobicity data in Table 5.1 to a log-odds score in which each possible pair of amino acids would have a specific score derived from a comparison of their hydrophobicities? Try this matrix with your test sequences and note the differences.
5. Use your BLAST search results from the Web Exploration Project (or download sequences provided by your instructor from the instructor section of the *Exploring Bioinformatics* website) to obtain amino-acid sequences of several orthologs of one of the virulence factor candidates you investigated. Try

to choose sequences that have a range of scores. Then, compare pairs of these sequences with your alignment algorithm, using a comparable scoring matrix, and compare your scores to your BLAST results. Remember that BLAST is using a very different algorithm, but do you get a similar ranking of sequences with your program?

6. You can use your protein alignment program for semiglobal or local (see Chapter 3) protein alignments with only small modifications. Modify your program for local alignments, and then test some orthologs of a candidate virulence gene. Again compare your results with the BLAST results. This comparison would be particularly interesting if you found some orthologs that share conserved domains with your candidate gene but are not highly similar along their full lengths.

On Your Own Project: Building Your Own Substitution Matrix

This project asks you to build a substitution matrix from scratch. The Solving the Problem section discusses an algorithm to accomplish this. If you are in a nonprogramming course, an appropriate exercise to help you understand the algorithm and how substitution matrices are built is to calculate substitution values for two or three amino acid pairs by hand and to answer the questions in the Solving the Problem section. If you are in a programming course, your instructor may ask you to go on to the Programming the Solution section and implement this algorithm in a programming language.

Understanding the Problem: Scoring Matrices Based on Observed Substitutions

In the earlier Understanding the Algorithm, you saw how a scoring matrix could be based on chemical properties of amino acids or—letting nature do the work—observed substitutions over evolutionary time. Building your own substitution matrix based on observed substitutions is the focus of this project. Although PAM and BLOSUM scoring matrices are widely used and work well for searching databases for orthologs, they may not be the best choice for every situation. In fact, at least 80 substitution matrices are available in various alignment programs, many of which are built for specific purposes such as aligning proteins with transmembrane segments. Furthermore, recognizing that different kinds of proteins or different organisms have biases in the amino acids that are used, a bioinformatician studying a specific problem might well create a scoring matrix applicable to that particular situation.

If comparisons of protein sequences among *E. coli* strains were a major focus of your research, perhaps you would want to develop a scoring matrix that would give higher scores for the amino-acid substitutions that are most strongly conserved within this specific bacterial group. Algorithms that build substitution matrices use an initial set of aligned sequences as the basis for the matrix, so in this case you would want a set of aligned sequences from a variety of *E. coli* strains. From there, the algorithm should identify the substitutions in these aligned sequences and develop a substitution model based on these observations.

At first glance, this may appear somewhat troubling. How can we obtain the initial set of alignments if we don't yet have a substitution matrix? However, think about the PAM 1 matrix: It was built from sequences with no more than 1% variation—sequences that could be aligned quite accurately with simple match, mismatch, and gap scores. This initial set of data is often called the **training set**: Although these are not perfect alignments, they are useful to "train" the model to produce a substitution matrix that will lead to better alignments. This process is illustrated in **Figure 5.7**. Ideally, the training set should contain a significant number of sequences that are relevant to the problem and have a known relationship, and of course it should be as error-free as possible. We start with a modest number of sequences to keep the project manageable, but in reality the most accurate results will be

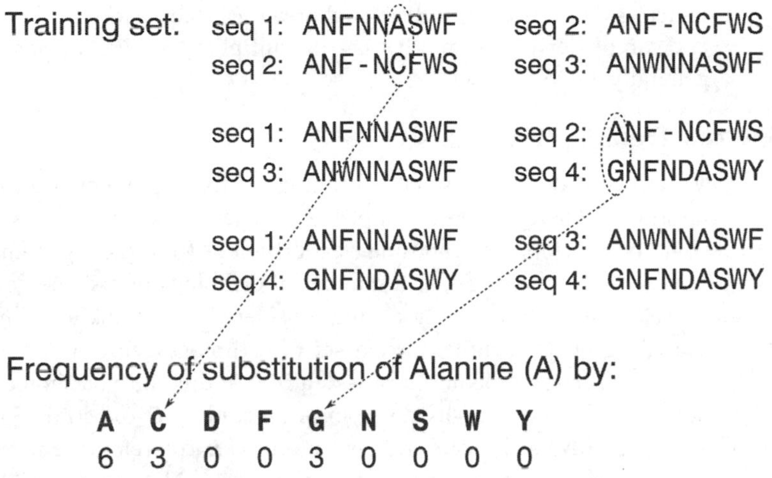

Training set:

seq 1: ANFNNASWF
seq 2: ANF-NCFWS

seq 2: ANF-NCFWS
seq 3: ANWNNASWF

seq 1: ANFNNASWF
seq 3: ANWNNASWF

seq 2: ANF-NCFWS
seq 4: GNFNDASWY

seq 1: ANFNNASWF
seq 4: GNFNDASWY

seq 3: ANWNNASWF
seq 4: GNFNDASWY

Frequency of substitution of Alanine (A) by:

A	C	D	F	G	N	S	W	Y
6	3	0	0	3	0	0	0	0

Substitution matrix:

		Amino acid								
		A	C	D	F	G	N	S	W	Y
Ala	A	2.0	2.4	0.4	-1.8	2.4	-2.4	-1.3	-1.6	0.4

Figure 5.7 Development of a substitution matrix from a training set of aligned sequences. The frequency of observed substitutions is shown for clarity, but in calculating the matrix, a pseudocount is used to account for substitutions that don't appear in the training set.

obtained by having as many good-quality training sequences as possible. The substitution model can also be further refined as new sequences become available.

Your program should build a substitution matrix by starting with pairwise alignments of closely related sequences. The values for each pair of amino acids should reflect the substitutions observed in the training set. For example, if the training set shows that alanine is frequently replaced by lysine, the calculated substitution rate should be relatively high. If you think carefully about how this might be implemented, you will probably recognize some important considerations:

1. Gap handling: If an amino acid aligns with a gap, identification of a substitution becomes ambiguous. Algorithms generally ignore ambiguous decisions; in this case, we simply discard these positions from the dataset.
2. Unrepresented amino acids: It is likely that not all 400 possible substitutions will have actually occurred within the training set. To overcome this issue, the substitution matrix algorithm uses a **pseudocount** of 1 as a default starting value when counting substitutions (instead of zero). This effectively establishes a baseline substitution rate, and observed substitutions raise this rate. In Figure 5.7, for example, the alignment of A with D never occurs in the training set, but in calculating the substitution matrix, q_{ij} is given a value of 1 divided by the number of aligned positions (and the count of each observed alignment is similarly incremented by 1).
3. Limitations on using the matrix: If our substitution matrix is built using closely related sequences, it will have limited use for alignment of more distantly related proteins. Sometimes this is no problem: Perhaps it is only intended for a specific purpose. However, it is also possible to use an initial substitution matrix to build additional

matrices: For example, as discussed previously, all PAM matrices are based on PAM 1 and can be derived from it by matrix multiplication: PAM 250 = PAM 1^{250} and, more generally, PAM x = PAM 1^x.

Obtaining the Training Set

How would we develop a suitable training set for alignment of proteins from different *E. coli* strains? There are many complete *E. coli* genomes available today, so a good training set could be developed by choosing clear orthologs from these genomes. For the purposes of this exercise, we will gather a number of orthologs of a single gene; however, a more sophisticated substitution matrix could consider substitutions across multiple genes.

We can quickly identify a good set of orthologs using the Web-based prokaryotic genome comparison tools at **MicrobesOnline**. At this site, you should see in the upper left corner a list of some often-used sequences, including *E. coli* strain K-12 substr. MG1655, which we already used as our ordinary intestinal *E. coli* reference strain. Add this strain to the list of strains to be considered by clicking `Add`. Now, type "coli" in the search box to list all the *E. coli* strains currently in the genome database. Select several strains (include EDL933 but do not bias the set toward O157:H7 strains: try to get a variety) and add them to your list. Then click `Set selected`. Now we can do some analysis of the genomes by clicking `Info`.

At this point, you should see a list of the genomes you selected. Click the `Find Shared Genes` link at the top to do just that: identify genes that have recognizable orthologs (the program's default is 75% identity) in all of the genomes. Then choose a shared gene you would like to use for your training set and click its link under each genome to obtain its amino-acid sequence in FASTA format (use the `Sequences` tab when you get to the gene information display). (If your instructor does not want you to go through this process, both raw amino-acid sequences and formatted alignments may be found in the instructor section of the *Exploring Bioinformatics* website and may be made available to you.)

Now you will need pairwise protein alignments of the orthologous sequences (see Figure 5.7). The alignments should be in a text file where each sequence appears on one line and gaps are represented with dashes (-). Therefore, if there are five different proteins to align, your input file would contain 20 rows of data. The first two rows represent the alignment between the first two sequences, the second two rows represent the alignment between sequences one and three, and so on. You could obtain these alignments by using your Needleman-Wunsch program with a simple match/mismatch/gap scoring system or by using your protein alignment program with a PAM or BLOSUM matrix, or you could use a Web-based pairwise alignment tool such as **Emboss**.

Solving the Problem

Our algorithm will use a log-odds scoring system. Remember (see Understanding the Algorithm) that the goal is to compare the observed frequency of finding one amino acid (j) substituted for another (i) with the likelihood of finding amino acid j by chance. For each pairwise alignment, we need to count how many times each amino acid occurs and the number of times each pair aligns. We then apply the formulas that follow to determine substitution values. Gap positions are ignored in these calculations, and the count of each alignment begins with a pseudocount of 1. The value s_{ij} is the final substitution value for these two amino acids.

q_{ij} = number of times i and j are aligned / total number of ungapped aligned positions
p_i = total number of times i appears / total number of ungapped amino acid positions

p_j = total number of times j appears / total number of ungapped amino acid positions
e_{ij} = frequency at which we expect j to be substituted for i (by chance) = $2p_ip_j$ (if i and j are the same value, then this becomes p_i^2)
s_{ij} = log-odds for the actual substitution of j for i = $\log_2(q_{ij} / e_{ij})$

To see how the algorithm should work, let's use a very small training set in which three short proteins are aligned as an example:

```
YFRFR  FRFRFR  YFYFR-F
YF-FR  ARFRFR  YFRFRYF
```

Notice that not all amino acids are present. Therefore, we will generate a matrix including only the amino acids present in the training set. (This is not an issue when dealing with real proteins of any normal size: All amino acids will be represented, although not all possible alignments will be present.) Let's apply the formulas to determine the substitution value for *Y* and *R*. First, q_{YR} is the observed substitution frequency, so we count the number of times we see *Y* aligned with *R*, starting with an initial value (pseudocount) of 1. We don't know whether *Y* substituted for *R* or *R* substituted for *Y*, so we will count *Y-R* alignments regardless of which is on the "top" or "bottom" sequence. In our example, we see only one occurrence of *Y* aligned with *R*, so q_{YR} is 2. There are 16 total aligned positions in our three pairs of sequences, ignoring gaps, so $q_{YR} = 2/16 = 0.125$.

The values p_Y and p_R represent the estimated "background" frequencies, or a measure of the overall likelihood of having *Y* or *R* in our sequences. Of the 32 total ungapped positions, 6 are *Y* and 12 are *R* (remember the pseudocount!), so $p_Y = 6/32 = 0.19$ and $p_R = 12/32 = 0.38$. Again, we added one to each total for the pseudocount, and here we use 32 positions rather than 16 because we are looking at the frequency of occurrence of *one* amino acid rather than an aligned pair. Now, e_{YR} represents the expected frequency of aligning these two amino acids by chance, which is the product of their individual frequencies (if you have one chance in four that a certain position will have *Y* and one chance in four that the opposite position will have *R*, the total probability is $1/4 \times 1/4 = 1/16$). Therefore, the frequency of a *Y-R* alignment is $0.19 \times 0.38 = 0.070$, which we multiply by two because an *R-Y* alignment is equivalent: $e_{YR} = 0.14$.

Now we can determine our odds ratio of observed substitutions to expected chance occurrences: $q_{YR} / e_{YR} = 0.125 / 0.14 = 0.89$. This value is less than 1, so it indicates that the *Y-R* alignment occurred *less* frequently than would have been predicted by chance: an unlikely alignment in these sequences and thus a presumably nonconservative substitution. Finally, we take the log of the odds ratio to get a number that's more useful for scoring. The PAM matrix, as we saw earlier, uses base 10, which means a substitution that occurs 10 times more often than expected by chance would give a log-odds ratio of 1 and a substitution that occurred 10 times less often than expected gives a log-odds ratio of −1. PAM then multiplies this number by 10 to get an integer score, because most values fall between 1 and −1. For this exercise, we will use base 2, so that a substitution gets a score of 1 if it is twice as common as expected by chance. Therefore, our final substitution value becomes $s_{YR} = \log_2(0.89) = -0.17$.

Calculate the remaining substitution values for our small example set by hand or with a spreadsheet to be sure you understand how the algorithm works. Notice there are alignments that do not occur in this small training set, such as *F* with *Y*. You also need to calculate a value for these substitutions; use the pseudocount to do this calculation. You can download the calculated matrix from the *Exploring Bioinformatics* website to check your calculations.

Questions for Nonprogramming Courses

1. Using the training set you developed from the *E. coli* orthologs, calculate the substitution values for a few pairs of amino acids by hand or using a spreadsheet. (Note: You can easily automate the counting of the amino acids and aligned pairs using spreadsheet formulas; your instructor could then ask you to generate a complete substitution matrix.) Compare your substitution values with those found in a BLOSUM or PAM matrix; how different are the scores based on your specific set of proteins?
2. Your instructor can download and make available a complete substitution matrix program (and, if desired, aligned training sequences) from the *Exploring Bioinformatics* website. Use this program to generate a substitution matrix. Download test sequences (these are amino-acid sequences of proteins found in diverse *E. coli* strains and in other Gram-negative bacteria) and then use your matrix to align pairs of the test sequences. How do the alignments compare with those made using other substitution matrices? Does your substitution matrix give better scores for more closely related proteins than for distantly related ones? Do you notice any reduction in reliability of alignments when the proteins do not come from closely related organisms?
3. Exercise 2 under Programming the Solution discusses matrix multiplication to derive substitution matrices applicable to larger units of evolutionary time. If you wish to play with this, you could use a spreadsheet to produce the COLI2 (or any other number) matrix from the COLI1 matrix and then test this matrix for aligning the test sequences.

Programming the Solution

Using the algorithm as discussed, write a program in the language of your choice to build a substitution matrix based on these formulas and output to a text file in the format we have used previously. Test your program on the small sample training set provided in the previous section and make sure your program gives the same results that you got by hand. Once you have a working solution, use the training set you developed to build the *E. coli*–specific substitution matrix.

Exercises for Programming Courses

1. Using your *E. coli* substitution matrix in your protein alignment program, download test sequences representing conserved genes from *E. coli* strains and less-related Gram-negative bacteria from the *Exploring Bioinformatics* website. Try various alignments and compare the results to the alignments and scores you get with other matrices. Can you see advantages to using this matrix for *E. coli* comparisons? What happens when you try it on distantly related sequences?
2. The *E. coli* substitution matrix used very similar sequences as its training set; its utility to look even at related organisms is probably limited. However, if we consider this matrix to be COLI1, we could generate COLI2, a matrix that like PAM2 would represent a greater evolutionary distance, using the rule $COLIx = COLI1^x$: $COLI2 = COLI1^2 = COLI1 \times COLI1$. This requires matrix multiplication. When multiplying two matrices, the row/column value of the resulting matrix is the sum of the products of the particular row and particular column of the two multiplying matrices. Assuming we wish to multiply matrix A with B to obtain matrix C, we would use the following formula for each iteration of the derived matrix:

$$C_{ij} = \sum_{x=0}^{n} A_{ix} \times B_{ix}$$

For example:

$$\begin{bmatrix} 2 & -1 \\ 3 & 0 \end{bmatrix} \times \begin{bmatrix} 1 & -2 \\ -3 & 4 \end{bmatrix} = \begin{bmatrix} (2\times1)+(-1\times-3) & (2\times-2)+(-1\times4) \\ (3\times1)+(0\times-3) & (3\times-2)+(0\times4) \end{bmatrix} = \begin{bmatrix} 5 & -8 \\ 3 & -6 \end{bmatrix}$$

In our COLI1 substitution matrix, the cell value at row *i* and column *j* represents the likelihood of amino acid *i* being substituted by *j* in a given span of evolutionary time. Therefore, to get COLI2, we calculate the likelihood of substituting *j* for *i* in two time units—allowing for amino acid *i* to potentially change first to an amino acid *k* and then in a second step to the observed amino acid, *j*. This is done by applying the above formula; in other words, by multiplying COLI1 by itself one time, we obtain COLI2. Note that to accomplish this, we cannot multiply the log-odds scores; the occurrence of aligned positions would change over time, but the expected value would not. So, we would need to create a matrix containing the q_{ij} values and multiply that matrix by itself the desired number of times and then calculate the log-odds scores from the resulting matrix. Modify your program so the user can choose the COLI matrix to build: for example, COLI1, COLI15, COLI30, and so on. Try a couple of these matrices to see if they produce better scores for the less closely related orthologs in your test data set.

Connections: Annotating Genomes

Advances in technology have made it much easier to obtain and assemble the actual DNA sequence (discussed in Chapter 8) of a complete genome, but it remains a daunting task to identify which sequences represent genes (see Chapters 9 and 10) and then try to assign functions to each gene. The process of assigning functional information to genome sequences is called **annotating** the genome, and no genome sequence is useful unless it has been well annotated. This is an ongoing process even for genomes sequenced years ago, because new information is constantly becoming available. Automated annotation is necessary because of the enormous complexity of a genome sequence (more than 20,000 genes in over 3 billion bases of DNA sequence, in the case of the human genome). This means that essentially all initial work of assigning functions to genes is done by bioinformatics programs that align predicted proteins from the genome with known protein sequences and make determinations about functional regions. Unfortunately, this process is not foolproof, and often genes are assigned functions that must later be corrected by researchers doing actual laboratory experiments.

As our alignment algorithms and substitution matrices improve, the initial annotation of genomes becomes more accurate. New methods are published frequently that can yield new insights into the potential functions of sequenced genes. For example, specialized substitution matrices are commonly used to look for membrane proteins: These proteins have additional constraints, because many substitutions would interfere with membrane interaction. Substitution matrices based on predicted protein structures instead of amino acid sequences have also been developed; this kind of analysis could allow for identification of structural similarities (discussed further in Chapter 11) that would not be picked up by ordinary amino acid comparison.

BioBackground: Amino Acids, Protein Sequences, and Protein Function

The coding sequence of a gene consists of three-nucleotide codons that specify the amino acids of a protein and the order in which they are assembled. Amino acids are small molecules consisting of a carboxyl group (COOH), an amino group (NH_3), and a hydrogen atom all attached to a central carbon (**Figure 5.8A**). What distinguishes different amino acids is a fourth group, called the side chain (designated "R" in Figure 5.8A). There are 20 different amino acids used in proteins, and each has a distinct side chain with distinct chemical properties (some examples are shown in **Figure 5.8A–H**). For convenience in working with protein sequences, there

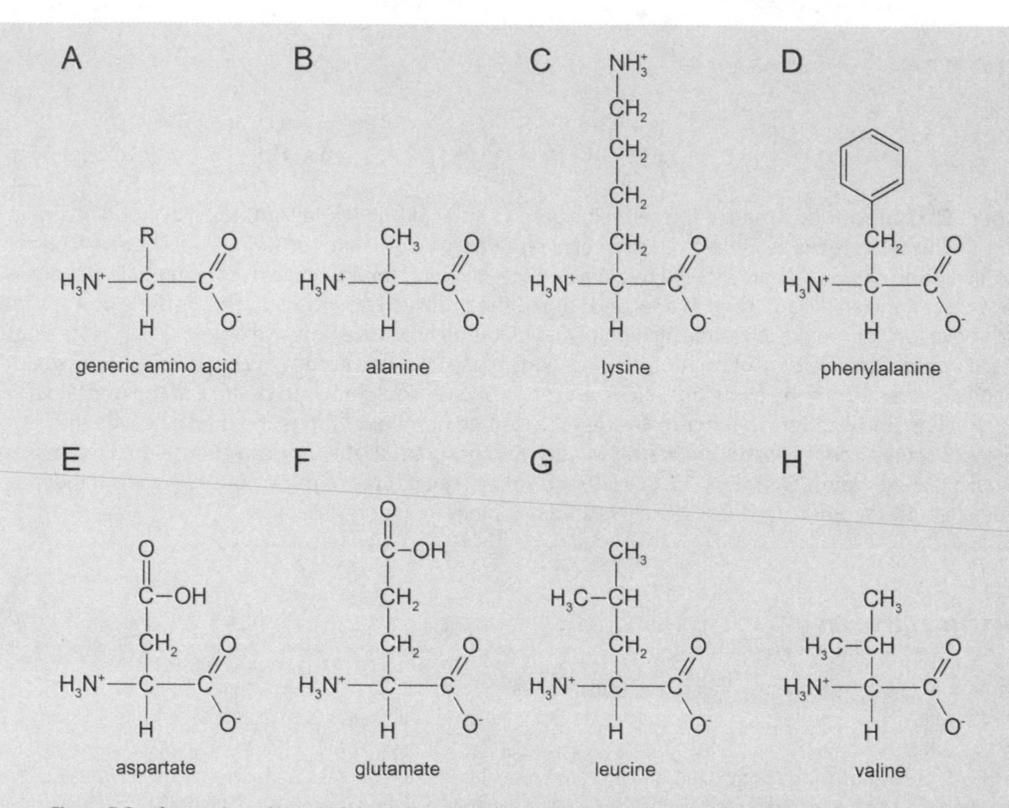

Figure 5.8 A generic amino acid (A) and 7 of the 20 amino acids used in proteins (B–H).

are standard abbreviations for amino acids. Each of the 20 amino acids can be represented by a three-letter abbreviation (Gly = glycine, Glu = glutamic acid, Gln = glutamine, etc.) or by a one-letter code (G = glycine, E = glutamic acid, Q = glutamine, etc.). Bioinformatics programs typically work with the one-letter code, so that each amino acid is just one letter in the protein string. Some, however, convert their output to three-letter abbreviations, which are easier for the reader. You can see all the amino-acid abbreviations in the genetic code table (Figure 2.4).

The nucleotide sequence of a gene directly determines the linear sequence of amino acids that makes up a protein, which we call the protein's **primary structure**. The primary structure, in turn, constrains the interactions of the amino-acids and thus (along with its local environment and sometimes interactions with other proteins) how the protein will fold into a functional three-dimensional structure (see Chapter 11). The protein can only fold correctly if it has appropriate amino acids with appropriate side chains and properties in particular locations; once folded, its interactions with other molecules (such as the substrate for an enzyme) are similarly dependent on having amino acids with the correct chemical properties at the correct sites. Imagine, for example, a protein that needs to bind DNA: It must fold into a shape that can fit around the DNA molecule, build a pocket lined with positively charged amino acids to bind the negatively charged DNA, and include in the binding site amino acids positioned to interact with a specific sequence of nucleotides.

Two proteins that have similar functions, such as two different DNA-binding proteins, are likely to have similar structures. For example, a common feature of many DNA-binding proteins is a stretch of amino acids that curve into a relatively rigid helix that fits into the major groove of the DNA molecule (**Figure 5.9**) When we see this structure in some part of the amino-acid sequence of an unknown protein, we have reason to suspect the unknown protein binds DNA and therefore might function in controlling gene expression, DNA replication, or some related process.

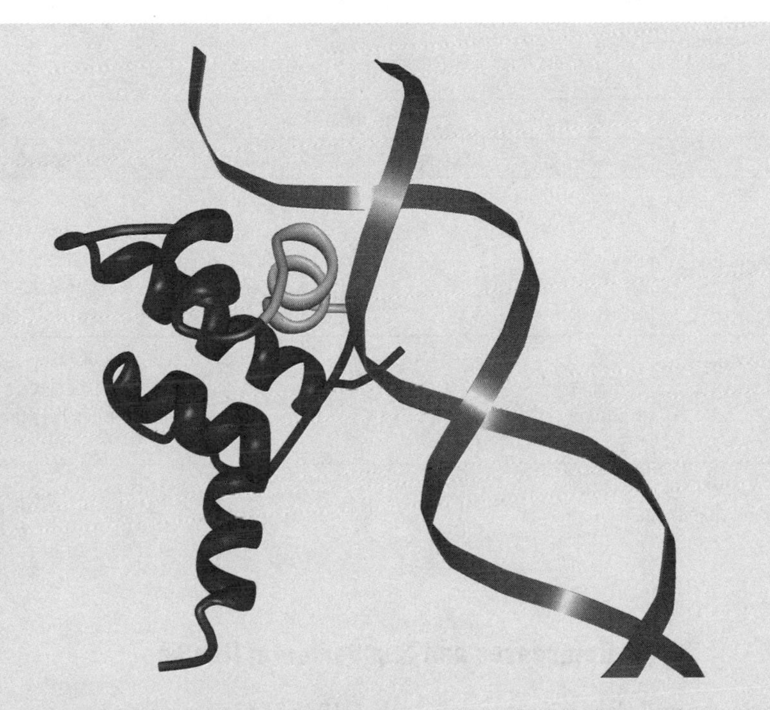

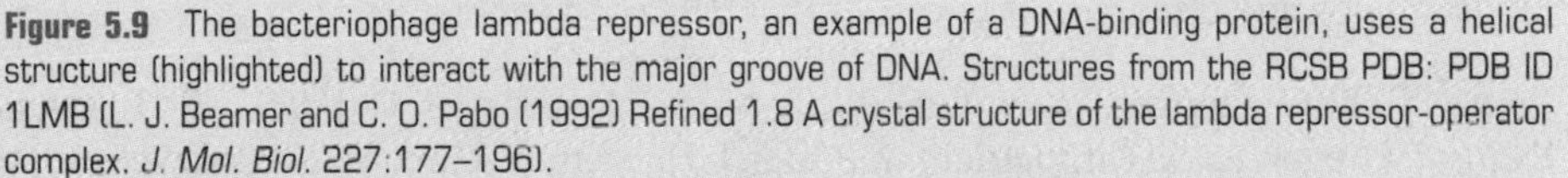

Figure 5.9 The bacteriophage lambda repressor, an example of a DNA-binding protein, uses a helical structure (highlighted) to interact with the major groove of DNA. Structures from the RCSB PDB: PDB ID 1LMB (L. J. Beamer and C. O. Pabo (1992) Refined 1.8 A crystal structure of the lambda repressor-operator complex. *J. Mol. Biol.* 227:177–196).

The challenge, then, is how to determine how similar two amino-acid sequences are. The power of a protein alignment is that we can score not only *exact* matches but also what we would call **conservative substitutions**: mutations that result in the substitution of one amino acid for a similar one (see Figure 5.2). Alignment of two similar amino acids makes good biological sense because these conservative substitutions would not be selected against in evolution (see Chapters 6 and 7). However, which amino acids could be considered "similar?" One approach is to look at their chemical properties: their size, whether they are **hydrophilic** (able to interact with water and charged molecules) or **hydrophobic** (oily, unable to interact with water), their charge, and so on. **Table 5.3** shows some of the chemical properties of the 20 common amino acids. You can see from the structures in Figure 5.8, for example, that valine (Figure 5.8H) might substitute effectively for the similarly sized and similarly hydrophobic leucine (G) and that their carboxyl groups might make aspartate (E) a good substitute for glutamate (F). A bulky amino acid like phenylalanine (D), however, is unlikely to successfully replace a small amino acid like alanine (B), and hydrophobic valine (H) seems likely to affect protein function if it replaced hydrophilic glutamate (F)—indeed, this latter hydrophilic-to-hydrophobic change is exactly the result of the most common mutation that causes sickle-cell anemia.

Initially, our understanding of the chemical properties of amino acids was our only real means of deciding on similarity. However, once the sequences of many proteins were known (first through direct sequencing of proteins but later by inferring amino-acid sequence from the much easier process of DNA sequencing), it became possible to use an evolutionary measurement. The likelihood of a particular substitution can be determined by seeing how frequently it has *actually* occurred over evolutionary time, and this in turn is done by aligning proteins already shown to be similar. This method is the basis for the commonly used PAM and BLOSUM substitution matrices discussed in detail in this chapter.

Table 5.3 Some chemical properties of amino acids.

	Hydrophilic			
Size	Positive	Negative	Uncharged	Hydrophobic
Small		aspartate glutamate	glycine serine threonine asparagine glutamine	alanine valine cysteine[1] methionine[1] proline
Large	lysine arginine histidine		tyrosine[2]	leucine isoleucine phenylalanine[2] tryptophan[2]

[1]Sulfur-containing side chains
[2]Aromatic side chains

References and Supplemental Reading

Epidemiology and Genomics of E. coli *strain O157:H7*

Perna, N. T., G. Plunkett III, V. Burland, B. Mau, J. D. Glasner, D. J. Rose, G. F. Mayhew, P. S. Evans, J. Gregor, H. A. Kirkpatrick, G. Pósfai, J. Hackett, S. Klink, A. Boutin, Y. Shao, L. Miller, E. J. Grotbeck, N. W. Davis, A. Limk, E. T. Dimalantak, K. D. Potamousis, J. Apodaca, T. S. Anantharaman, J. Lin, G. Yen, D. C. Schwartz, R. A. Welch, and F. R. Blattner. 2001. Genome sequence of enterohaemorrhagic *Escherichia coli* O157:H7. *Nature* **409**:529–533.

Rangel, J. M., P. H. Sparling, C. Crowe, P. M. Griffin, and D. L. Swerdlow. 2005. Epidemiology of *Escherichia coli* O157:H7 outbreaks, United States, 1982–2002. *Emerg. Infect. Dis.* **11**:603–609.

Substitution Matrices

Dayhoff, M. O., R. M. Schwartz, and B. C. Orcutt. 1978. A model of evolutionary change in proteins. *Atlas Protein Seq. Struct.* **5**:345–352.

Eddy, S. R. 2004. Where did the BLOSUM62 alignment score matrix come from? *Nature Biotechnol.* **22**:1035–1036.

Henikoff, S., and J. Henikoff. Amino acid substitution matrices from protein blocks. *Proc. Natl. Acad. Sci.* USA **89**:10915–10919.

Jones, D. T., W. R. Taylor, and J. M. Thornton. 1992. The rapid generation of mutation data matrices from protein sequences. *Comp. Appl. Biosci.* **8**:275–282.

MicrobesOnline

Dehal, P. S., M. P. Joachimiak, M. N. Price, J. T. Bates, J. K. Baumohl, D. Chivian, G. D. Friedland, K. H. Huang, K. Keller, P. S. Novichkov, I. L. Dubchak, E. J. Alm, and A. P. Arkin. 2010. MicrobesOnline: an integrated portal for comparative and functional genomics. *Nucleic Acids Res.* **38**:D396–D400.

Chapter 6

Distance Measurement in Molecular Phylogenetics:

Evolution of Mammals

Chapter Overview

Investigating evolutionary relationships is one of the most important current applications of bioinformatics. Building on the alignment techniques discussed in the preceding chapters, similarity of genes and proteins can be used to infer evolutionary relatedness and to build phylogenetic trees. Methods of reconstructing evolutionary history are the focus of this chapter and the next, dealing first with measurement of evolutionary distance and then with tree-building. Students will understand key concepts of molecular phylogenetics and how various distance metrics use sequence evidence and biological understanding to estimate change over evolutionary time. Students in programming courses will develop their own solutions that implement these algorithms.

Biological problem: Evolutionary relationships of marine and terrestrial mammals

Bioinformatics skills: Distance metrics (Jukes-Cantor, Kimura, and Tamura models), curating sequence alignments, introduction to phylogenetic trees

Bioinformatics software: phylogeny.fr, MUSCLE, Gblocks, distance calculator tool

Programming skills: Applying distance metrics

Understanding the Problem:

Mammalian Evolution

*Whales and porpoises are mammals adapted to life in the sea. All mammals are evolutionarily related, so these marine mammals must have at least a distant ancestor in common with land mammals. But what was the evolutionary pathway that gave rise to them? Who are their closest relatives among the land mammals and at what point did their lines diverge? This has been a point of contention among evolutionary biologists: The fossil record suggests a relationship to meat-eating mammals, whereas other evidence points to common ancestry with hippos, cows, and deer. As DNA sequencing opened up the field of molecular evolution, new studies using genes as molecular clocks enabled us to resolve the issue: Bioinformatic analysis shows that hippos (***Figure 6.1***) are the closest living relatives of whales—a conclusion substantiated by subsequent fossil finds. How can molecules tell us what anatomy and paleontology cannot?*

Figure 6.1 Killer whale and hippo: Are they related?

The first systematic classification of living things was proposed by Linnaeus in the mid-1700s. Linnaeus grouped life into plant and animal kingdoms and then into phyla, classes, orders, families, genera, and species—all terms still familiar to us today. However, the only criteria he had to form his groups were the observable morphological characteristics of the different species. Darwin's 1859 proposal that new species arise from common ancestors as natural selection acts on individual variation provided a new basis for classification: evolutionary relatedness. The systematics of Darwin's time has gradually morphed into phylogenetics: the grouping of living things according to their evolutionary relationships, especially by means of genetic and molecular genetic data. Modern phylogenetics recognizes that humans are far more closely related to oak trees than to *E. coli*, and many apparently similar or dissimilar species have had to be regrouped as their evolutionary histories have been uncovered.

Because we cannot directly observe the evolutionary history of life, relationships among species—the "tree of life"—have to be inferred from the fossil record and from similarities and differences among modern-day species. Gene and genome sequencing allow us to expand the scope of evolutionary evidence to include variation in genes—variation that arises from mutations that are conserved over evolutionary time. Careful analysis of this genetic variation can succeed in establishing genuine evolutionary relationships even where the fossil record is incomplete or morphological clues may be misleading. Modern phylogenetics is inseparable from the bioinformatic techniques that let us analyze complex genetic data, allowing us to find solutions to such thorny problems as the origins of whales.

Bioinformatics Solutions:
Molecular Phylogenetics and Distance Measurement

Evolution is the great unifying principle of modern biology. We can study genes in a model organism such as yeast and expect that similar genes in humans encode similar proteins with similar functions because underlying molecular genetics is the evolutionary foundation that humans and yeast are descended from a (distant) common ancestor. In fact, all DNA and protein alignment techniques are based on the idea of evolutionary relatedness. Molecular phylogenetics is the direct application of these same kinds of bioinformatic techniques to reconstructing the evolutionary pathways that led to the species we see today.

Evolution can be defined as descent with modification from a common ancestor. Consider, for example, the salamanders of the genus *Ensatina*, which are found in forests from Baja, California to British Columbia. In California, western populations in the coastal mountains are separated from eastern populations in the Sierra Nevada by the wide Central Valley (**Figure 6.2A**); these two populations have become distinct subspecies that cannot successfully interbreed. It is hypothesized that if the separation persists, they will eventually become two distinct species. A **phylogenetic tree** (**Figure 6.2C**; if you are unfamiliar with reading trees, see the BioBackground section) shows the evolutionary history of these subspecies and shows that the western and eastern subspecies have evolved some distance from their common ancestor, with the western population then branching again to produce a distinct southern subspecies. However, we cannot directly observe evolutionary changes or speciation events that occurred in the past but can only infer them based on the study of modern species and the limited clues ancient species have left behind. The phylogenetic tree is therefore always a *model* of evolution: a hypothesis about the evolutionary path leading to modern species. It may be based on anatomy (e.g., humans, cats, and bats all have the same number of bones in their wrists), on comparison of modern species with the fossil record, on biochemistry and metabolism, or on other kinds of evidence, and it is often revised as new evidence comes to light.

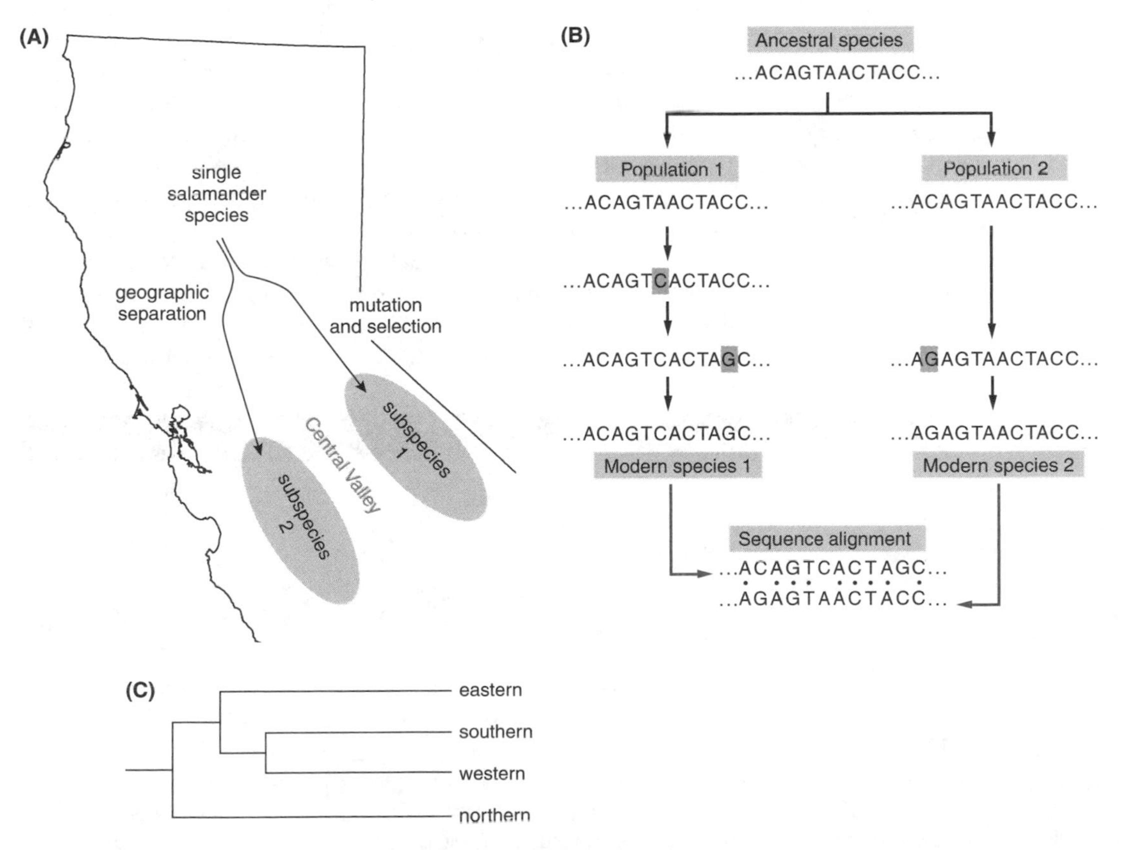

Figure 6.2 (A) Evolution of a salamander population into two distinct subspecies, driven by geographic separation followed by mutation and selection over time. (B) Example of change in a gene sequence due to independent mutation and selection in two distinct subpopulations. (C) Phylogenetic tree for California salamander subpopulations based on molecular data. Data from: "The molecules support the morphology" *Understanding Evolution*. University of California Museum of Paleontology. 20 March 2013.

The modification that occurs over evolutionary time is really genetic change, resulting from mutation. Once our salamander populations became physically separated (Figure 6.2A), for example, mutations occurred independently in each population, creating new alleles that could not be shared with the other population because the geographical barrier prevented interpopulation mating. The populations changed over time as favorable new alleles allowed individuals to have more offspring and deleterious alleles reduced others' contribution to the next generation: This process is **natural selection**. Eventually, the two populations became so genetically different (**Figure 6.2B**) that although their genes were orthologous, they could not successfully interbreed.

Before the advent of DNA sequencing, we could not directly measure the differences between orthologous genes. Today, **molecular evolution**—which is really just a specific application of bioinformatics—has revolutionized phylogenetics. We can use a gene that is well conserved among the species of interest as a **molecular clock** and directly measure the substitutions that have occurred over evolutionary time. The more differences we see among a set of sequences, the longer we conclude the molecular clock has been ticking since they last had an ancestor in common. Quantifying differences among sequences, as you have surely guessed by now, is merely an application of sequence alignment. Indeed, we used this idea implicitly in the preceding chapters, and the BioBackground section in Chapter 3 explains it in more detail.

Because molecular evolution is not dependent on physical characteristics, physiology, or fossils, it can determine relationships in the absence of these traditional clues and can often resolve difficult questions, as we will see when we investigate whale ancestry. However, molecular techniques also have their complications and limitations. For example, Figure 6.2B shows a total of three differences between two orthologous modern sequences. The pathway shown in Figure 6.2B clarifies that one change occurred in one species, whereas two occurred in the other. In reality, however, we cannot directly observe the path of mutational change, so we could equally well hypothesize that all three changes occurred in just one species. Worse, what if an A mutated to a T and then the T mutated back to A? From our perspective, this would look like no change had occurred, suggesting close relatedness when actually enough time had passed for two distinct mutational events to occur. Molecular evolution can never establish with certainty what happened over evolutionary time, but the bioinformatics algorithms discussed in this chapter allow us to model some of the complexity of evolutionary change and more accurately determine distances between sequences.

BioConcept Questions

To understand bioinformatic methods of measuring evolutionary distance, it is important to understand the relationship between mutations and sequences in the genome, how a molecular clock works, and the issues that add complexity to this process. Use these questions to test your biological understanding; read BioBackground: Measuring Evolution at the end of the chapter if you need a better foundation.

1. In your own words, what is a molecular clock?
2. Measuring the rate of substitution between two species is not the same as measuring the rate of mutation. What is the difference between mutation and what an evolutionary biologist understands by substitution?
3. What advantages does the use of gene comparisons have over traditional phylogenetic methods such as comparing the bone structure or other observable features of various species? Can you think of any potential disadvantages of these molecular methods?
4. If you wanted to draw a phylogenetic tree for all living things (bacteria, fungi, plants, animals, etc.), what sorts of genes might serve as reasonable molecular clocks? What genes would *not* be good clocks? How might your choice change if you wanted to look at evolutionary relationships only among primates?

5. Suppose there are four differences in the sequence of a gene between species A and species B and also four between the sequences for species B and species C. You would initially believe that A is just as closely related to B as B is to C. But what might have happened that would make this conclusion false?

Understanding the Algorithm:
Measuring Distance

Learning Tools

The *Exploring Bioinformatics* website has a short exercise available for download that uses English sentences to model gene sequences and asks you to compare them and generate a "phylogenetic tree." For some experience with how sequence comparison relates to evolution, try this exercise.

Constructing a phylogenetic tree based on molecular evidence requires estimating the evolutionary distance between each pair of species based on sequence differences within an orthologous gene. This is not as easy as it sounds, and simply counting mismatches may not accurately model evolution (see BioBackground). Here we deal with **distance metrics**: quantitative measures of the evolutionary time required to account for observed sequence divergence. These metrics incorporate our biological understanding of the evolution of genes, and numerous different algorithms can be used (much as you might choose to use the straight-line distance, freeway distance, transit-line distance, length of the most scenic route, or even travel time to measure the distance between your home and where you work). Once the distances are determined, we can group related species (using clustering algorithms) and generate the actual tree; that process is discussed in the next chapter.

To compare two species, a gene must be chosen whose sequence can represent the species (the "molecular clock"). Then, we can align the sequences for the two species, observe the substitutions, and consider the **substitution rate**. The substitution rate, r, measures the rate of change of two DNA sequences over the time since they last shared a common ancestor (**Figure 6.3**). If we know the number of changes that have actually occurred (K) and the amount of time that has passed (T), we can calculate r by $r = \frac{K}{2T}$. As shown in Figure 6.3,

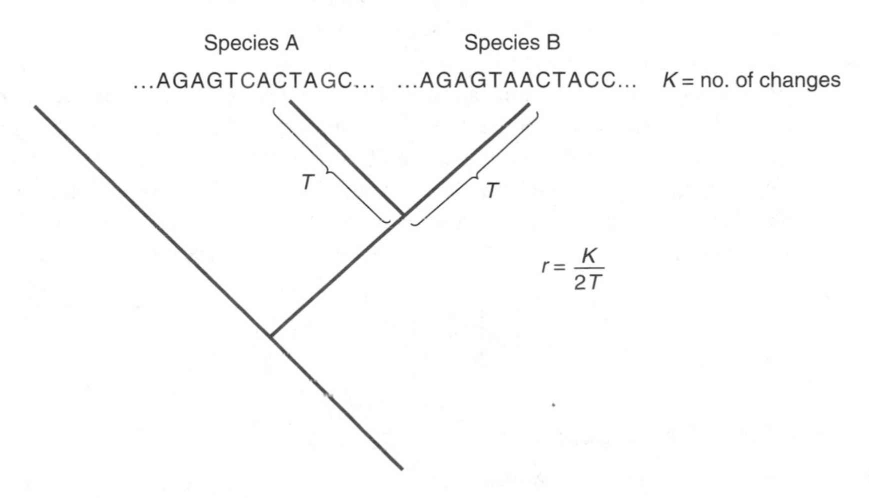

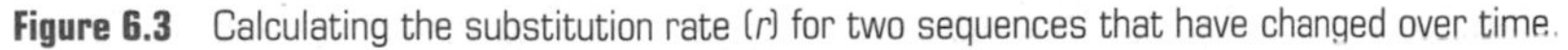
Figure 6.3 Calculating the substitution rate (r) for two sequences that have changed over time.

we need $2T$ because *each* species has been evolving independently for time T. Sometimes we can "calibrate" a particular molecular clock by using external data (such as radiometric dating and the fossil record) to estimate T. However, a more common application of molecular data is to assume a constant r (an assumption that requires some caution) and then use the observed substitutions themselves to determine T. This requires an accurate value for K, which is not trivial because there is usually no way to observe or directly determine the actual pathway of mutation. We discuss three distance metrics that model K based on what we know about mutation.

Jukes-Cantor Model

One key issue in looking at substitutions is how many "hidden" mutations might have occurred in addition to the changes we can see (see BioBackground). A model for estimating K developed by Thomas Jukes and Charles Cantor (see References and Supplemental Reading) attempts to correct for this problem. The Jukes-Cantor model makes the assumption that a set of sequences with few overall variations is less likely to have undergone multiple substitutions at any particular site over evolutionary time than a set of sequences with a large number of variations. If α represents the rate of change per unit time and t represents a small unit of time, then the probability of one of the three possible substitutions occurring at a given nucleotide (e.g., for an A nucleotide, A→C, A→G, or A→T could occur) is $3\alpha\Delta t$. This model assumes all changes are equally likely: A mutation is just as likely to replace A with T as with G or C. After some manipulation of these variables, we can express K, our estimate of the number of substitutions (visible and hidden) likely to have taken place between sequences a and b, as

$$K_{ab} = -\tfrac{3}{4}\ln\left(1 - \tfrac{4}{3}D_{ab}\right)$$

In this equation, D is the fraction of observed substitutions (total substitutions/total nucleotides). We can then directly examine a sequence alignment, calculate K under the model, and obtain an estimate of evolutionary distance that can be compared with the distance for some other pair of aligned sequences.

Kimura's Two-Parameter Substitution Model

Motoo Kimura's two-parameter model, published in 1980 (see References and Supplemental Reading) is an extension of the Jukes-Cantor model, so it again accounts for "hidden" substitutions in estimating K. However, where the Jukes-Cantor model treats all substitutions as equally likely, Kimura's model recognizes that transitions occur more frequently than transversions (see BioBackground) and introduces additional parameters to account for this difference. Given an observed fraction of tranSitions, S (transitions/total substitutions), and an observed fraction of transVersions, V (transversions/total substitutions), Kimura's formula for the number of substitutions that occurred between two aligned sequences is then based on observed transitions and transversions as follows:

$$K_{ab} = \tfrac{1}{2}\ln\left(\frac{1}{1-2S-V}\right) + \tfrac{1}{4}\ln\left(\frac{1}{1-2V}\right)$$

Tamura's Three-Parameter Model

The frequencies of nucleotides are not uniform across the tree of life. The G+C nucleotide content can vary widely (the human genome, for example, is about 40% G+C, whereas some bacterial genomes are more than 60% G+C); this also affects the outcome of mutation and selection. In 1992, Koichiro Tamura (see References and Supplemental Reading) extended Kimura's model by accounting for G+C content bias as well as transition/transversion bias:

$$K_{ab} = -C\ln\left(1 - \frac{S}{C} - V\right) - \tfrac{1}{2}(1 - C)\ln(1 - 2V)$$

where S represents the fraction of transitions, V represents the fraction of transversions, and C is calculated using the following:

$$C = GC_{s1} + GC_{s2} - 2 * GC_{s1} * GC_{s2}$$

$$GC_{s1} = \text{fraction } G + C \text{ in sequence 1}$$

$$GC_{s2} = \text{fraction } G + C \text{ in sequence 2}$$

Other Models

Although the three models discussed above are some of the more popular ones, a number of other models exist that take additional factors into consideration. These models include the Tamura-Nei model, which recognizes that not all transitions occur with the same frequency (a transition from A to G or G to A is counted differently from a transition from C to T or T to C); the Felsenstein model, which allows for differences in the frequency of mutation at different sites; and the Hasegawa-Kishino-Yano (HKY85) model, which allows variable base frequencies and a separate transition and transversion rate.

Is there a "best" model? The choice of distance metric may depend on the sequence data used. For example, a strong G+C content bias might suggest the use of the Tamura model. However, all these models can only estimate evolutionary change, so there is no one clear choice. Any good distance metric can enhance the estimation of evolutionary distance based on nucleotide or amino-acid substitutions between two sequences.

Test Your Understanding

1. Two aligned sequences for the same gene differ by four substitutions, all transitions. Two more sequences for the same gene also differ by four substitutions, two transitions, and two transversions. Which pair of sequences has probably been evolving independently longer, and how do you know?
2. Using the Jukes-Cantor equation and a spreadsheet, determine *K* for a range of values of *D*. (Note that the Jukes-Cantor formula can only be used for values of $D < 0.75$, because one cannot take the log of zero or a negative number. Remember, these distance metrics are intended for comparing related sequences, so the fraction of substitutions is not expected to get this high.) Graph *K* versus *D*. What happens to the rate of change of *K* as the number of observed substitutions increases over a small range? What happens when the total number of substitutions starts to get larger? Why did Jukes and Cantor believe this was a desirable result?
3. Using the Kimura equation and a spreadsheet, determine *K* for some values of *S* and *V*. Remember that the sum of *S* and *V* is 1, and note that in this formula, *V* must be less than 0.5. Graph *K* versus *V*. What happens to *K* as *V* increases? Is the pattern the same for a larger or smaller total number of substitutions? Why did Kimura believe this was a biologically relevant pattern?

CHAPTER PROJECT:

Evolution of the Whale

Learning Objectives

- Understand how sequence similarity reflects evolutionary change
- Understand the strengths and limitations of using sequence data to determine evolutionary relationships
- Gain experience with Web-based phylogenetic tools and how they are used to measure evolutionary relatedness of sequences
- See how different distance metrics can be used to model evolutionary change and use them to measure relatedness among DNA sequences

Suggestions for Using the Project

This project is designed to be used either in courses that require programming skills or in nonprogramming courses. Below are suggestions for modules of the project that instructors might choose to use in these two types of courses. Instructors should also feel free to ask questions of their own that use these same skills.

Programming courses:

- Web Exploration: Experiment with using various distance metrics to gain an understanding of evolutionary distance; become familiar with phylogenetic trees. Part I can be used independently; Part II could be used independently by supplying an alignment file from the *Exploring Bioinformatics* website.
- Guided Programming Project: Implement the Jukes-Cantor model in a programming language of your choice.
- On Your Own Project: Develop a program to align two sequences and apply one of the distance metrics presented in this chapter to determine their evolutionary relatedness.

Nonprogramming courses:

- Web Exploration: Experiment with using various distance metrics to gain an understanding of evolutionary distance; become familiar with phylogenetic trees. Part I can be used independently; Part II could be used independently by supplying an alignment file from the *Exploring Bioinformatics* website.
- On Your Own Project: Download a completed program to align two sequences and choose a distance metric; then explore the effects of different distance metrics on alignments of interest.

Web Exploration: Evolution of Marine Mammals

Part I: Whales, Porpoises, and the Mammalian Phylogenetic Tree

Molecular evolution, or the application of bioinformatics to evolutionary problems, can often resolve controversies resulting from ambiguous morphological, physiological, or fossil evidence. An example is the evolution of marine mammals such as whales and porpoises introduced previously: They are so different from land mammals that it is difficult to decide where they might share a common ancestor with land mammals. In this section, we use distance metrics and a molecular clock to examine the genetic evidence and decide where the marine mammals belong.

This section of the project examines relatedness, or evolutionary distance, among whales, porpoises, and several terrestrial mammals. To evaluate distance, we need to align sequences of some conserved gene, our molecular clock. Many genes could be used as molecular clocks, and indeed different genes might be chosen for different purposes. The genes for the hemoglobin subunits, for example, are well conserved among vertebrate animals and have been used to examine evolutionary relationships in these groups. The gene encoding 16S rRNA (an RNA component of the ribosome) is found in every living creature from bacteria and fungi to animals and plants, so this gene can be used as a molecular clock across the whole tree of life. For this project, we use a gene encoding **casein**, the protein in milk; because all mammals nurse their young, we know this gene is conserved among the species we wish to investigate. Specifically, because casein is actually a family of proteins encoded by paralogous genes, we use the β-casein gene, which encodes the major milk protein.

The three major groups of mammals are the marsupials (kangaroos, opossums, and their relatives), the montremes (echidnas, the platypus, and many extinct relatives), and the placental mammals. Whales and porpoises clearly belong among the placental mammals, so let's start by downloading a β-casein sequence from a representative member of each group of placental mammals to get the big picture. Nucleotide sequences will allow us to use the nucleotide-based distance metrics discussed previously. GenBank accession numbers are given here for convenience, but it may be more instructive for you to search using key words to gain practice in using the genomic databases. It may help you to know that the standard gene name for the β-casein gene is *CSN2*. We use the rat (*Rattus norvegicus*; J00711) to represent rodents, the dromedary camel (*Camelus dromedaries*; AJ012630) to represent the broad group of herbivores, and the dog (*Canis lupus familiaris*; AB035080) to represent carnivores, three major groups of placental mammals. For each gene you retrieve, be sure you have obtained the complete *nucleotide* sequence of only the coding region (no introns or untranslated regions). Watch out for similar-sounding genes like the ones for casein kinase. Build a single text file with each of the sequences in FASTA format, each with its description line (starting with >).

Now, add sequences for whale (*Tasmacetus shepherdi*; JF701647) and porpoise (*Phocoenoides dalli*; JF701645) β-casein. Unfortunately, there are (at least as of this writing) no complete sequences for the casein gene for these mammals in GenBank; however, you will find sequences for the large exon #7, more than half of the total coding sequence. Do not be concerned about sequences of unequal length; we align the genes as our first step, and the regions not present in the whale and porpoise sequences will simply result in terminal gaps.

We use the site **phylogeny.fr** for this exercise. A huge number of phylogenetic programs are available for use on the Web (try a Google search if you'd like to get an idea of their diversity). The phylogeny.fr site is a great place to start because beginners can quickly generate a "one-click" phylogenetic tree, yet more advanced users can choose from a number of programs and modify all their parameters. Navigate to the site and choose the `One-click` option (under Phylogeny Analysis). Upload your file or paste your sequences (all of them, with their comment lines) into the input box. Click `Submit`. You will see the progress as the program aligns the sequences, makes adjustments, and produces a phylogenetic tree that should be very similar to the one shown in **Figure 6.4**. From this tree you can see immediately that whales and porpoises are more closely related to each other than to any of the other mammals we chose: They share a common ancestor that is much more recent than the ancestors they have in common with any other species. Furthermore, they are much more closely related to the camel than to the rodents or carnivores for the same reason. It seems clear that we should be looking at grouping the whales and porpoises among the herbivores.

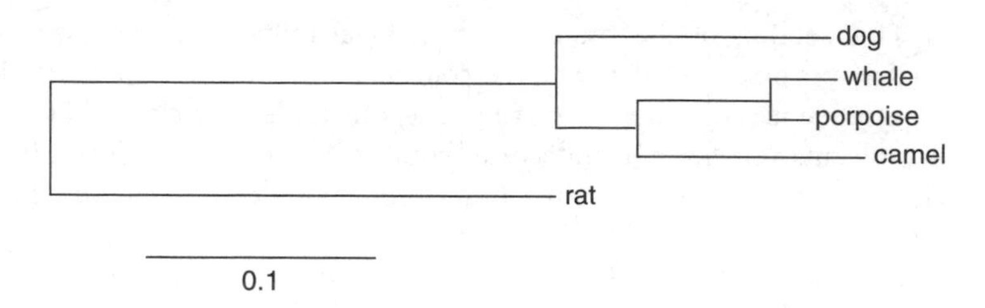

Figure 6.4 Phylogenetic tree showing the relationship of whales and porpoises to the major groups of mammals as generated by phylogeny.fr. Tree generated by phylogeny.fr; Dereeper A., et al., Phylogeny.fr: robust phylogenetic analysis for the non-specialist Nucleic Acids Research. 2008 Jul 1; 36 (Web Server Issue):W465–9. Epub 2008 Apr 19.

Take a look at the options below the tree. Notice that by default it is shown as a phylogram, meaning the branch lengths are drawn proportional to the calculated evolutionary distance. See how it looks as a cladogram or in an unrooted radial format. Try changing the outgroup used to root the tree by clicking `Reroot` and then clicking a species; you should quickly see why the root shown initially was chosen. Remember that clades can be rotated at their nodes; click `Swap` or `Flip` and then click a node or branch to see the effect; every tree that can be generated using swapping or flipping is equivalent. Finally (be sure you are showing a phylogram), click the radio button to display `Branch lengths` to see the evolutionary distances that were calculated by the program. By default, the distance metric used by phylogeny.fr is HKY85.

Web Exploration Questions

1. How many terminal nodes are there in your mammalian phylogenetic tree? How many internal nodes?
2. Which node represents the most recent common ancestor of whales and any terrestrial mammal?
3. Which sequence was used as the outgroup to root the tree as it is shown in phylogram view?
4. Notice that the whale and dolphin branches are not equally long (in phylogram view). What does this difference in length represent, biologically?
5. We have grouped the whales and dolphins with the herbivores but still have not answered the question posed at the start of the chapter about which terrestrial mammals are most closely related to them. Download sequences of the β-casein gene for more herbivores and use phylogeny.fr to explore this question further. Camels, hippos (*Hippopotamus amphibious*; U53901), giraffes (*Giraffa camelopardalis*; U53897), and horses (*Equus caballus*; NM_001081852) would make good representatives of some major herbivore groups. How should whales and dolphins be grouped? Which land mammal is their closest relative? Are they still more closely related to each other than to any of the land mammals you have examined?

Part II: Evolutionary Distance in the Mammalian Phylogenetic Tree

Now that you have gained some experience with a phylogenetic tree and how distance measurement leads to evolutionary groupings, let's examine how some different distance metrics behave. You have already seen the distances calculated by phylogeny.fr using the HKY85 model; now we compare some other models. To do this, we need aligned sequences (to know where there are identities and where there are substitutions); in the process of generating its phylogenetic tree, phylogeny.fr has done this for us, using a multiple alignment program (conceptually similar to ClustalW) called MUSCLE.

From your phylogeny.fr phylogenetic tree (you can use either the initial tree illustrated in Figure 6.4 or the tree you generated for exercise 4), click the tab labeled `Alignment`. The tabs allow you to look back and see what the program did to generate your tree; this tab shows you the multiple alignment done by MUSCLE. Notice the terminal gaps added to the sequences that include only exon 7; notice also the "extra" sequence in the middle of the dog gene not present in the others. How did phylogeny.fr deal with these sequence differences? To see, click the `Curation` tab; here, the program Gblocks identified the portions of the alignment suitable for distance calculation and tree-building (dark blue bars under the sequence); you can see that in essence it eliminated gapped positions, including the regions that were not present in all sequences (this is why it did not matter that we mixed complete sequences with exon 7 sequences). If there is an ambiguous alignment immediately adjacent to the gap, that position is also often excluded from the final alignment, on the grounds that the insertion or deletion event may be responsible for the observed ambiguity. You can tell by looking at the dark blue bars whether Gblocks has excluded a position next to a gap for your particular set of sequences.

On the Curation tab, click the link labeled `Cured alignment in FASTA Format` to get the equal-length aligned sequences resulting from curation by Gblocks in a nice, clean format. Now, use the **Evolutionary Distance Calculator** tool found on the *Exploring Bioinformatics* website to calculate the distances between pairs of sequences using the Jukes-Cantor, Kimura, and Tamura models. Make sure you omit the FASTA header when you paste your sequences into the input boxes.

Web Exploration Questions

6. Using the Jukes-Cantor model, which sequence is most similar to the whale sequence? Which is least similar? Do these results agree with the results you obtained from phylogeny.fr?
7. How do the distances change if you use the Kimura model? Based on what you observed for these particular sequences, how important is it to take the transition bias into account?
8. How do the distances change if you use the Tamura model? Do the distances remain proportional even if the absolute numbers change? Is G+C bias important for these sequences?
9. Choose one of the models and sketch a phylogenetic tree based on the distances your model calculated. Remember to make your branch lengths proportional to distance. Although we will see in Chapter 7 that tree-building programs use more sophisticated algorithms than this, assume in constructing your tree that the calculated distance is the distance from the tip of the branch belonging to species 1 to the tip of the branch belonging to species 2. How does your tree compare with the one generated by phylogeny.fr?

More to Explore: Solving Problems Using Phylogeny

Molecular phylogenetics can be used to solve many interesting evolutionary problems, and often a simple distance measurement is sufficient to resolve a question. If you are interested in this area, you may want to try more explorations on your own. For example, salamanders look somewhat like lizards but live more like frogs. To which are they more evolutionarily related? How would salamanders, frogs, turtles, lizards, and snakes be grouped? Are birds more closely related to these groups or to mammals? You should be able to find some ideas for potential molecular clocks for exploring reptiles and amphibians in the scientific literature.

Guided Programming Project: Using Distance Metrics to Measure Similarity

In this programming project, you will implement the Jukes-Cantor model to calculate the evolutionary distance between two sequences. You will need to count mismatches between two aligned sequences and then use the count in the Jukes-Cantor formula. The following pseudocode presents a solution for the Jukes-Cantor model assuming two aligned sequences are presented as input. Although we could choose to count gaps as mismatches, identified in the aligned sequences by a dash (-), the Jukes-Cantor model ignores gaps.

Algorithm

Evolutionary Distance Algorithm Using the Jukes-Cantor Model

Goal: To measure evolutionary distance between two nucleotide sequences

Input: Two aligned nucleotide sequences

Output: A numeric value representing the Jukes-Cantor distance

```
// Initialization

Input the two sequences: s1 and s2

difCtr = 0 // difference counter
lenCtr = 0 // length counter - gaps not counted in length

// STEP 1: Count differences between sequences ignoring gaps

for each i from 0 to length of sequence
    if (s1[i] not equal '-' and s2[i] not equal '-')
       add 1 to lenCtr
       if (s1[i] not equal s2[i])
          add 1 to difCtr

// STEP 2: Calculate and Output results

difCtr = difCtr / lenCtr // convert to fraction of substitutions
jukes = (-3/4 * log(1-(4/3*difCtr)))

output jukes
```

Putting Your Skills Into Practice

1. Write a program in the language used in your course to implement the given pseudocode. Your program should give the same results as the online calculator.
2. Suppose you wish to count gaps as mismatches. What change(s) would be needed to the algorithm? Why might you prefer this method over the traditional implementation of Jukes-Cantor?
3. As we saw in Chapter 3, a local alignment would ignore terminal gaps. Would it make sense to count internal gaps as a mismatch and ignore terminal gaps when calculating evolutionary distance, or should both types be treated similarly?
4. It would not make sense to try to count gaps as mismatches in the Kimura or Tamura model. Why not?

On Your Own Project: Alignment and Evolutionary Distance

Understanding the Problem

Measuring evolutionary distance is not a trivial task. It depends on a good alignment of the "molecular clock" sequences as well as on the choice of a distance metric appropriate to the data used. This project asks you to implement a more complete solution to the problem of calculating distance, carrying out sequence alignment and then offering a choice of distance metrics. Students in nonprogramming courses will have been able to develop all relevant skills except the actual programming by completing the Web Exploration exercises; however, if you would like to test the program resulting from this part of the project on your own sequences, a complete solution can be downloaded from the instructor section of the *Exploring Bioinformatics* website.

In the Guided Project, you implemented a commonly used distance metric for determining evolutionary relatedness. Because of the multitude of models that can be used, it would be beneficial to implement a program that allows a user to choose among multiple models. Additionally, it would be helpful if the user could use a single program to perform an alignment as well as calculate distance: In the Guided Project, the input sequences were limited to equal-length sequences, which is not very realistic. Because you already have good alignment programs, it would not be difficult to incorporate alignment of the sequences followed by a distance calculation.

Solving the Problem

Your solution should start with a program that aligns sequences using the Needleman-Wunsch algorithm (Chapter 3). The semiglobal alignment that does not penalize terminal gaps is probably the best choice, allowing for unequal input sequences like those we used earlier. After the sequences are aligned, calculate the distance according to whichever metric the user selects: Jukes-Cantor, Kimura two-parameter, or Tamura three-parameter model. These three models ignore gaps, so do not include any nucleotide that is aligned with a gap in computing distance. You will want to make this program user friendly and efficient. The user should be able to easily choose which model to use and should not have to rerun the program to use another model. Keep in mind that whereas you only needed to count the number of mismatches for the Jukes-Cantor model, the number of transitions and the number of transversions must be counted separately for the other two models.

Programming the Solution

Test your program initially with some short test sequences; hand-calculate (or use a spreadsheet) expected distances and see that it works properly. Then run your program using the sequences you used in the Web Exploration. Compare what happens if you use the curated sequences as input to what happens if the original sequences are used; if your program is behaving properly, the distances should be the same. The distances you get from your program should also match the distances you got with the distance calculator tool.

BioBackground: Measuring Evolution

Phylogenetic Trees

Figure 6.2C shows a phylogenetic tree for the some California salamander species. Two more generic phylogenetic trees are shown here: **Figure 6.5A** shows a **phylogram,** in which branch lengths are proportional to the calculated evolutionary distance between species, and **Figure 6.5B**

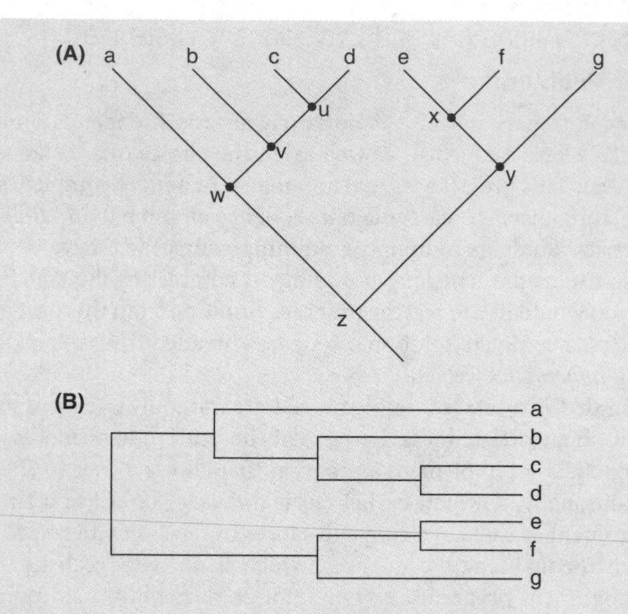

Figure 6.5 Phylogenetic tree for seven modern species (a–g) presented as (A) a phylogram and (B) a cladogram. Evolutionary time proceeds upward for (A) and rightward for (B); internal nodes (u–z) are denoted by dots in (A).

shows the same tree as a **cladogram**, in which all branches are brought out to the same point. Evolutionary time proceeds upward for the phylogram; the cladogram is rotated relative to the phylogram, so time proceeds left to right for the cladogram. The ends of the branches are **terminal nodes** (labeled a–g in the figure). Typically, terminal nodes represent modern species, but a tree could also deal with extinct species or could be intended to show the evolution of groups of organisms and not extend all the way to the species level. Branches are connected by **internal nodes** (labeled u–z); each node represents a common ancestor. Therefore, species a, b, c, and d all branched from a common ancestor w (we can say that w is the **most recent common ancestor** of a–d), but b, c, and d are more closely related, having branched more recently from common ancestor v. Species c and d are still more closely related, diverging recently from u. Branches can be rotated at nodes without changing the evolutionary relationships depicted in the tree.

The two trees shown are **rooted**: There is an "oldest" ancestor (z) common to all species in the tree. Usually, a tree can be rooted only by comparison with an **outgroup**, a reference group related to the species being studied but which all can agree is outside their group. For example, humans, chimpanzees, bonobos, and gorillas are all more closely related to each other than to orangutans, so orangutans can be used as an outgroup to root the tree for the rest. When the root cannot be determined, an unrooted tree can be drawn. An unrooted tree does not represent a unique evolutionary pathway, because multiple rooted trees can be derived from it.

A node and all its branches form a **clade**; the clade should be a **monophyletic group**, with all of the branches genuinely descended from the ancestor. **Polyphyletic** groupings arise from similarities that do not correspond to ancestry: For example, a polyphyletic group of warm-blooded animals might include both mammals and birds, even though birds have common ancestry with reptiles and share only a much more distant ancestor with mammals. Often, polyphyletic groups are accidental groupings based on incomplete information; genuine ancestry is frequently determined by molecular evidence. **Cladists** would argue that the goal is for every grouping in a phylogenetic tree to ultimately be monophyletic; in their view, evolutionary pathways are the standard by which relatedness should be measured. The primary concern of **pheneticists**, on the other hand, is to accurately quantitate the degree of similarity among species, so they are more relaxed about the possible existence of polyphyletic groups. They use statistical, quantitative, **distance-based** methods to measure relatedness, whereas cladists use **character-based** methods rooted in evolutionary theory.

Mutations and the Molecular Clock

In molecular phylogenetics, trees are drawn based on differences in DNA sequences, which arise by mutation. Phylogenetics, however, does not deal with the small changes that occur routinely in the DNA sequences of individuals: To look at the evolutionary history of a *species*, we need to consider only mutations that are passed on within the population until they become part of the species' genome or the normal genetic makeup of that species. The evolutionary biologist refers to these mutations as **substitutions**. This is a confusing term because it is also used by molecular geneticists to refer to any mutation that changes a nucleotide. In this chapter, we use this term specifically to mean a single-base change that over evolutionary time becomes part of the genome of a species.

The idea of a **molecular clock** is simple. If two species have the same gene (that is, each has an ortholog of the gene), that gene must have been present in their common ancestor. The two modern genes differ by substitutions that have occurred over evolutionary time that can be quantitated to give a measure of how much time has elapsed. **Figure 6.6** shows an example, using cytochrome *c* (a metabolic protein needed by nearly every organism) as a molecular clock. Unfortunately, simply counting the differences between two species may not give a true picture of how their DNA changed. Consider the three scenarios in **Figure 6.7**, each of which shows ACCTG changing to ACTTA over time. If these were orthologous sequences in two modern species, we would count two substitutions, but each scenario shows a very different path to that result. Many of these changes are hidden: They are not observed in the final sequence (for example, you have no way to know that the AC in the first scenario was at one point TT). Thus, simply counting the number of mismatches will not necessarily give a true measure of distance. The distance models discussed in this chapter were developed to improve the accuracy with which observed substitutions can be correlated with evolutionary distance.

Ideally, a molecular clock should "tick" at a constant rate: Substitutions should occur at a constant frequency regardless of the specific mutation or where it occurs. However, not every mutation is preserved over time and thus seen as a substitution. A particular mutation can either be advantageous to the organism, disadvantageous (deleterious), or have no significant effect

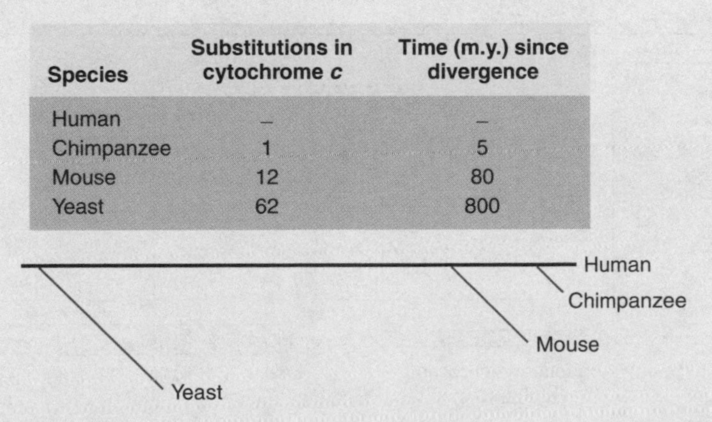

Species	Substitutions in cytochrome *c*	Time (m.y.) since divergence
Human	–	–
Chimpanzee	1	5
Mouse	12	80
Yeast	62	800

Figure 6.6 DNA sequence changes in the cytochrome *c* gene reflecting evolutionary distance (m.y. = millions of years).

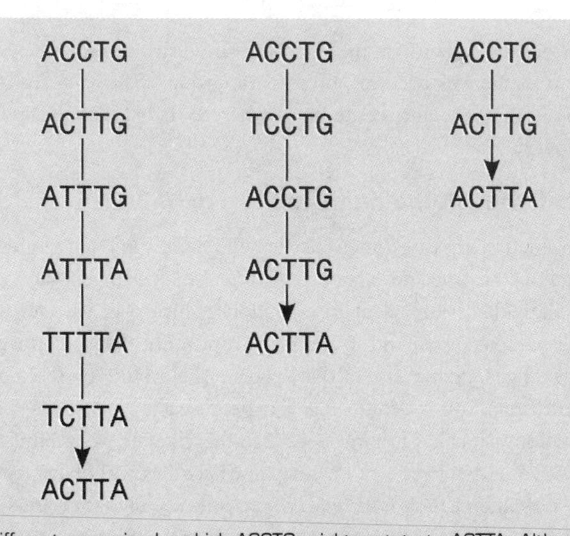

Figure 6.7 Three different scenarios by which ACCTG might mutate to ACTTA. Although the result is the same, the evolutionary pathways are very different, illustrating that a simple count of differences is insufficient to measure evolutionary distance.

(neutral). Deleterious mutations reduce the reproductive success of an individual; natural selection decreases their frequency in a population over time, and they do not become part of the genome. This means substitutions rarely occur in portions of genes that encode functionally important parts of a protein (**Figure 6.8**), because most mutations here are deleterious. These regions are said to be **functionally constrained**. However, silent mutations and those that lead to conservative amino-acid substitutions are more likely to be preserved. Less critical regions of proteins change faster, whereas introns and regions between genes can change even more rapidly.

There are also biochemical constraints on mutations. For example, **transition** mutations change one purine or pyrimidine to the other: G (purine) ↔ A (purine) or T (pyrimidine) ↔ C (pyrimidine). These are preserved as substitutions far more frequently than **transversion** mutations (G ↔ T, G ↔ C, C ↔ A, or A ↔ T). This is true because a chemical reaction or replication error that causes a mutation does not directly change both nucleotides in a base pair. The nucleotide on one strand changes, creating a mispair, and there is an opportunity for the error

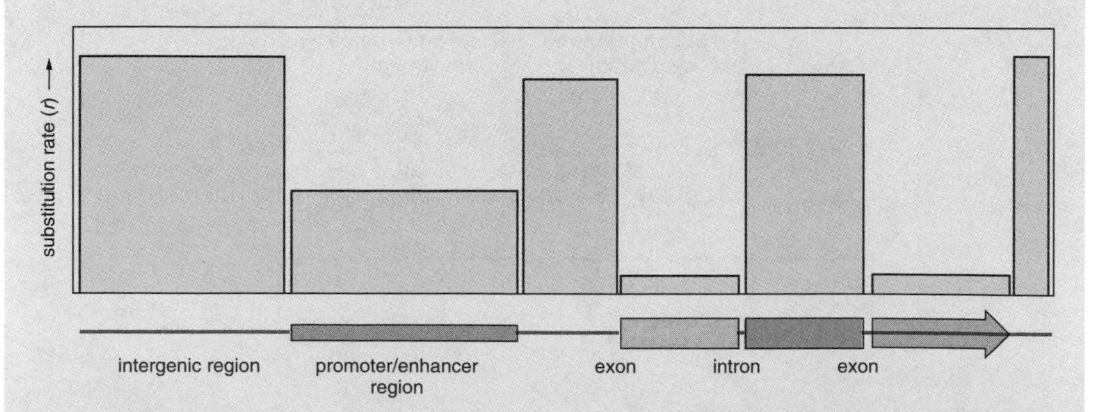

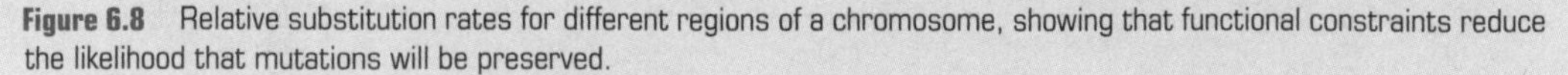

Figure 6.8 Relative substitution rates for different regions of a chromosome, showing that functional constraints reduce the likelihood that mutations will be preserved.

to be recognized by DNA repair enzymes before DNA replication puts the corresponding base on the opposite strand. A transversion requires that the mispair be a wide purine–purine pair or a narrow pyrimidine–pyrimidine pair, significantly altering the spacing of the DNA helix at this point. A transition, on the other hand, would result from a normal-width purine–pyrimidine mispair and is thus much more likely to escape detection.

References and Supplemental Reading

Introduction to Phylogenetics and Bioinformatics Methods

Baldauf, S. L. 2003. Phylogeny for the faint of heart: a tutorial. *Trends Genet.* **19**:345–351.

Distance Metrics

Jukes, T. H., and C. R. Cantor. 1969. Evolution of protein molecules. *In* H. N. Munro (ed.), *Mammalian Protein Metabolism*, pp. 21–123. Academic Press, New York.

Kimura, M. 1980. A simple method for estimating evolutionary rate of base substitution through comparative studies of nucleotide sequences. *J. Mol. Evol.* **16**:111–120.

Tamura, K. 1992. Estimation of the number of nucleotide substitutions when there are strong transition-transversion and G + C-content biases. *Mol. Biol. Evol.* **9**:678–687.

Phylogeny.fr

Dereeper, A., V. Guignon, G. Blanc, S. Audic, S. Buffet, F. Chevenet, J. F. Dufayard, S. Guindon, V. Lefort, M. Lescot, J. M. Claverie, and O. Gascuel. 2008. Phylogeny.fr: robust phylogenetic analysis for the non-specialist. *Nucleic Acids Res.* **36**:W465–W469.

Chapter 7

Tree-Building in Molecular Phylogenetics:

Three Domains of Life

Chapter Overview

Measuring evolutionary distance from a sequence alignment is only half the problem in phylogenetics. Given a complex dataset, a set of pairwise distance measurements can likely be compiled into any number of distinct trees. This chapter deals with the key problem of tree-building: how to use computational methods to obtain biologically relevant groupings of species in a phylogenetic tree. The value of a phylogenetic tree is in what we learn about evolution by observing groups (clades) with a common ancestor; we generate these groups computationally by means of what computer scientists refer to as clustering algorithms and/or by methods that search through possible trees to identify an optimal solution. The projects in this chapter will help students in both programming and nonprogramming courses understand how distance metrics we have already discussed are used by clustering algorithms to group related organisms. Through the use of Web-based tools, students will develop phylogenetic trees using both distance-based and character-based methods. Students in programming courses will develop their own solutions that implement two important distance-based algorithms.

Biological problem: Origins of genes in the bacteria, eukaryotes, and archaea

Bioinformatics skills: Agglomerative clustering, single linkage, UPGMA, neighbor joining, probabilistic methods in phylogenetics

Bioinformatics software: MUSCLE, Gblocks, BioNJ, PhyML, MrBayes (all at Phylogeny.fr), UPGMA

Programming skills: Hash table and nested hash table data structures

Understanding the Problem:

Rooting the Tree

In 1977, Carl Woese initiated a revolution in how biologists think about the living world. As phylogenetic thinking came to dominate systematics and taxonomy, evolutionary relationships among living organisms became the paramount criterion for classification. By the late 1960s, the "five-kingdom" system came into popular use (and sadly is still taught in many high-school curricula today): Linnaeus' plant and animal kingdoms, which obviously contained unrelated organisms, were divided into five kingdoms: plants, animals, fungi, protists, and bacteria. However, biologists also recognized the fundamental distinction in cell structure between the prokaryotes (bacteria) and eukaryotes (everything else). The waters were further muddied by the recognition that some prokaryotes living in extreme environments had rather different structures. With the advent of DNA sequencing and molecular phylogeny based on the universal 16S rRNA genes, Woese was able to recognize that these prokaryotes were as evolutionarily distant from the bacteria as the bacteria are from the eukaryotes. He proposed a higher level of classification, and we now recognize three **domains** *of living things (***Figure 7.1***): Eukaryotes, Bacteria, and Archaea, with the eukaryotes then being further broken down into kingdoms. However, there are still questions about the evolutionary relationships among the three domains and what these fundamental categories of living things might be able to tell us about the origins of life on earth. What can molecular phylogeny tell us about the three domains of life?*

The impetus for Woese's phylogenetic look at the prokaryotic world came from the growing recognition that certain prokaryotes found in hot springs, acidic pools, salt marshes, and other harsh environments were structurally very different from the more familiar bacteria that inhabit more temperate realms as well as our own bodies. They have cell walls like other prokaryotes but lack peptidoglycan, the carbohydrate universally present in all previously known bacteria. Their DNA is wrapped around histone proteins like the DNA of eukaryotic cells. Furthermore, they have some unique features of their own, like double-ended lipids that span their membranes. Many microbiologists believed these prokaryotes could represent the modern remnants of the ancestors of all living things; they were therefore termed "arachaebacteria" and later renamed the archaea to emphasize that they are unlike the bacteria. Once they were established as a distinct group, researchers soon began finding archaea everywhere, even in the human gut, and recognized their importance in the environment and in the evolution of life on earth.

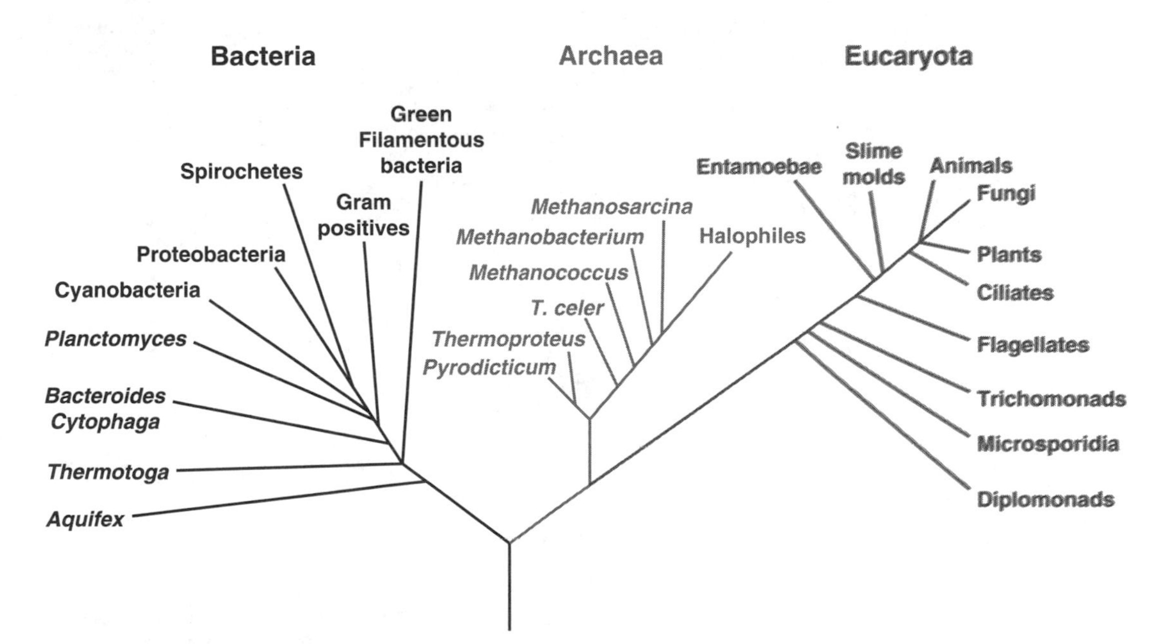

Figure 7.1 Phylogenetic tree for representatives of the three domains of life based on analysis of 16S rRNA sequences.

However, do the archaea really tell us what the original living cells were like? Some researchers still believe archaea represent the oldest evolutionary line, but that conclusion is far from clear. It is certain that the archaea have been around at least as long as the oldest bacteria, that their ability to survive in extreme environments suggests adaptations that would have been essential in the harsh conditions of the early earth, and that their genes clearly distinguish them from both bacteria and eukaryotes. Much remains to be learned about these organisms, and the application of bioinformatic methods to uncover their origins has led to some surprising results. We examine the relationships of the archaea to the other domains in this chapter's projects.

Bioinformatics Solutions:

Tree-Building

Looking at the relationships of the three domains of life using bioinformatic methods will require us to align orthologous genes and produce phylogenetic trees based on that information. In Chapter 6, we introduced the idea of a molecular clock and the value of molecular and bioinformatic methods in investigating evolutionary relationships. We considered how sequence diversity can be related to evolutionary distance and how various distance metrics can be applied to sequence alignments to model the evolutionary pathways that led to the observed substitutions. Ideally, any gene shared by two groups could be used to determine the evolutionary distance between them, but in practice different sequences have different functional constraints, and we sometimes find evidence of unexpected behavior over evolutionary time.

If you completed the Web Exploration exercises in Chapter 6, you even used these distance measures to draw a phylogenetic tree to show relatedness among mammals. Tree-building, however, is more complicated than using distance measures to draw a phylogenetic tree to show relatedness among mammals—or the rapid production of an attractive tree by the suite of programs at Phylogeny.fr—might lead you to believe. Given only four species, we can draw three different unrooted trees to show the relationships among them (**Figure 7.2A**). Distance data might help us choose one of these three, but in each case we

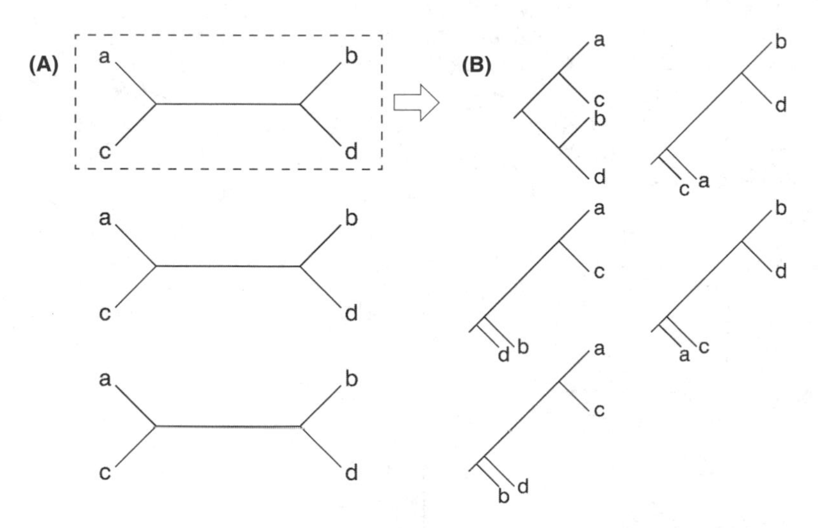

Figure 7.2 Possible phylogenetic trees for four species. (A) Three possible unrooted trees, showing relationships between species but not evolutionary pathways. (B) For the top unrooted tree, five possible rooted trees that preserve branch lengths and show evolutionary pathways.

can draw five rooted trees (**Figure 7.2B**), each maintaining the species relationships found in the unrooted tree but showing a unique evolutionary pathway. This means there are 15 different possible trees altogether for the four species. For 10 species, there are more than 2 million possible trees, and by the time we get to 50 species, there are a stunning 10^{74} possible trees. Thus, what phylogeneticists call "tree space" is intractably complex unless the dataset is extremely small; exhaustively drawing each possible tree and comparing it with the data is impossibly computationally intensive. Given that our goal is to construct a tree that represents biological reality by reconstructing to the extent possible the actual pathway of evolution, algorithms that are both computationally efficient and able to select an appropriate tree according to meaningful criteria are essential.

All tree-building methods depend on a multiple sequence alignment of the genes being considered. This is in itself a computationally difficult problem; Chapter 4 discussed heuristic methods by which ClustalW arrives at an alignment efficiently. It is then common for experienced researchers to examine the alignment by eye and make small adjustments, particularly to the positions of gaps. For example, the multiple alignment output might include a three-nucleotide gap in all the sequences, but that gap might be shifted left or right by a base or two in some sequences relative to others, when aligning the gaps would yield a better alignment overall. A multiple alignment editor such as **Jalview** (included in Phylogeny.fr's implementation of **MUSCLE** and the EBI implementation of **ClustalW**) or the desktop program **BioEdit** can be used for making these adjustments. Gapped positions can then be removed from the alignment using a program such as **Gblocks** (see Chapter 6). The result is a multiple alignment where every mismatched nucleotide or amino acid should represent (at least if our alignment algorithm is sufficiently good) the result of a substitution over evolutionary time.

There are two general ways in which bioinformatic programs can then attempt to select an optimal tree from the sequence data. **Distance-based** methods, as their name implies, apply a distance metric to the sequences and then use some form of **clustering algorithm** to decide how species should be grouped based on those distances. The UPGMA and neighbor-joining (NJ) algorithms are commonly used in distance-based methods; we explore those methods in detail in this chapter. **Character-based** methods are more probabilistic: They apply some model of evolution and then attempt to find the highest probability tree given that model and a particular dataset (alignment). For example, some models use parsimony: they apply the principle of Occam's razor ("the simplest explanation is the best one") and propose the evolutionary pathway that requires the fewest independent mutation events to generate the observed substitutions as the best one. Algorithms using Bayesian statistics to find an optimal tree are currently widely used in character-based methods. We do not specifically discuss character-based algorithms in this chapter (a comprehensive introduction to tree-building methods is beyond the scope of this text) but do use these methods in the Web Exploration.

BioConcept Questions

1. If all five rooted trees in Figure 7.2B are equivalent to the unrooted tree in Figure 7.2A, why is it so important to develop an algorithm for choosing among them? Describe in evolutionary terms in what important ways these trees are different.
2. Various distance metrics attempt to model what happens biologically as DNA mutates over evolutionary time. Yet, many researchers choose to use character-based tree-building methods that essentially ignore any calculation of distance. What limitations do you see in distance metrics that might keep us from accepting distance-based methods as the single best approach?

3. The distance metrics used in Chapter 6 apply specifically to nucleotide sequences. In this chapter's exercises, we use amino-acid sequence alignments as the basis for tree-building, and you may notice that we do not explicitly discuss distance metrics. In what way is a distance metric implicit in the alignment of protein sequences?
4. Suppose you are studying a group of organisms that are genuinely descended from a common ancestor and have many orthologous genes. Given a relatively constant rate of mutation and a relatively even distribution of mutations across the genome, we would expect that *any* of the orthologous genes could be used to construct a phylogenetic tree and that whatever gene we picked would give essentially the same results. It turns out, however, that not all genes are equal in terms of phylogenetic analysis. What factors can you think of that might account for differences between genes?

Understanding the Algorithm:
Clustering Algorithms

Learning Tools

Understanding clustering algorithms is one key idea in this chapter. To help with this, the *Exploring Bioinformatics* website has a link to a visual, interactive clustering simulation.

The goal of a phylogenetic tree is to reveal the evolutionary relationships among organisms, allowing us to classify (group) them according to genuine relatedness rather than superficial similarity. Thus, building a phylogenetic tree from a sequence alignment is in essence just grouping sequences according to their similarity as a means of inferring the evolutionary groupings of species. Whenever objects need to be grouped, computer scientists use clustering algorithms, which simply determine which objects are most similar and should be included in a group and which are less similar and should be excluded. **Hierarchical clustering** (**Figure 7.3**) is appropriate for a phylogenetic tree, because it places the most similar objects in groups and then relates those groups into larger clusters and then still larger ones—very much like the idea of common ancestors giving rise to broad groups of species that can then be subdivided into smaller groups with their own common ancestors. Specifically, we use a form of hierarchical clustering called **agglomerative clustering** that begins with individual objects (sequences representing species, in our case) and then merges the clusters until a single large group is formed.

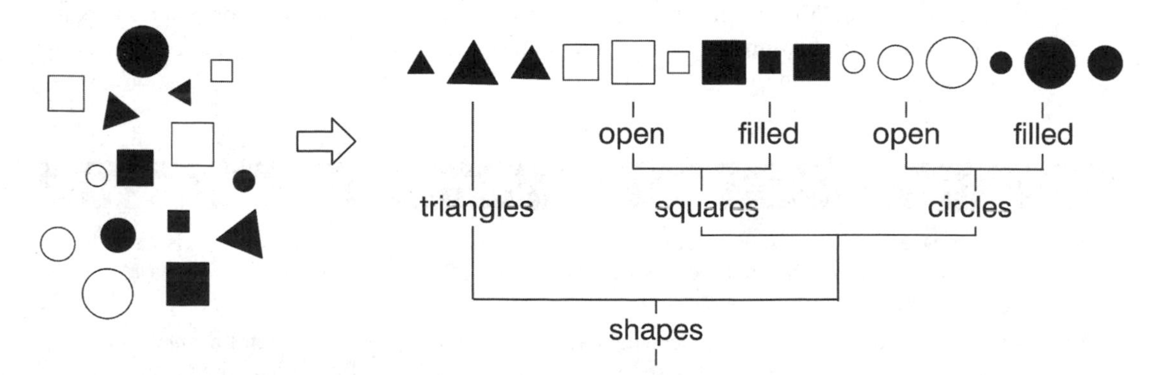

Figure 7.3 Example of hierarchical clustering. This is also agglomerative clustering if we start by grouping similar individual objects rather than by dividing the whole collection.

You know something about how to find the distances between individual sequences; clustering also requires a **linkage method**, which determines how the distance metric is applied when two *groups* are compared. After computing distances there is a **merge step**, in which those groups shown to be most closely related are brought together. The outcome of clustering is the information needed to draw the phylogenetic tree.

Let's use a small dataset as an example: Suppose we want to construct a tree for six species (A–F) that all diverged from a common ancestor. The most closely related species diverged from each other most recently and thus share a more recent common ancestor. After choosing an orthologous gene, aligning sequences, and applying a distance metric (remember that clustering is a distance-based method), we can construct the matrix shown in **Figure 7.4A** to show the distances between each pair of sequences. An agglomerative clustering algorithm works by sequentially merging the most closely related elements into **clusters** (or groups) until only one cluster remains. It starts with each individual element in its own cluster, and at each iteration the two closest clusters are determined and merged; for n elements, $n - 1$ iterations are required to complete the clustering. The key question we have not dealt with before is how to measure the distance between two clusters or between an individual element and a cluster. This is the linkage method, and we can choose from several linkage methods, depending on our assumptions about the data (**Figure 7.5**). **Single linkage** calculates the distances between each item in one cluster and each item in the other and chooses the smallest distance; it is suitable for elements that are not very tightly grouped. **Complete linkage** is the opposite: The largest individual distance value is chosen, which works best when the items are tightly grouped. **Centroid linkage** uses the distance between the centers of the clusters. The steps that follow show how the agglomerative clustering algorithm would produce a tree from the distances given in Figure 7.4, using the single linkage method.

(A)	A	B	C	D	E	F
A	0					
B	1	0				
C	3	2	0			
D	7	6	4	0		
E	17	16	14	10	0	
F	19	18	16	12	2	0

(B)	AB	C	D	E	F
AB	0				
C	2	0			
D	6	4	0		
E	16	14	10	0	
F	18	16	12	2	0

(C)	ABC	D	E	F
ABC	0			
D	4	0		
E	14	10	0	
F	16	12	2	

(D)	ABC	D	EF
ABC	0		
D	4	0	
EF	14	10	0

(E)	ABCD	EF
ABCD	0	
EF	10	0

Figure 7.4 Agglomerative clustering for six hypothetical species. (A) Distances between pairs of aligned sequences. (B–E) Successive iterations of the agglomerative clustering algorithm, merging the two closest clusters each time. Distances resulting from application of the single linkage method are shown in color.

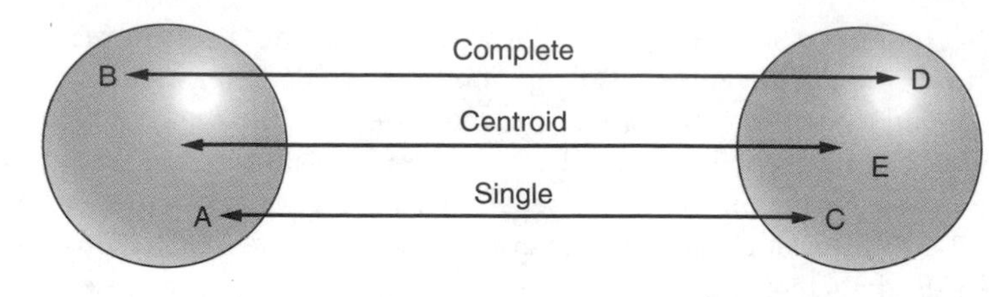

Figure 7.5 Three different linkage methods that could be used to compute the distance between two clusters.

Algorithm

Agglomerative Clustering Algorithm

1. Determine distances between sequences by alignment and a distance metric; for *n* sequences, create an *n* by *n* matrix of distance scores (Figure 7.4A). Each row and column of the matrix is a cluster, and each cluster currently contains just one element.
2. Ignoring the diagonal, find the cell that contains the smallest distance (representing the closest elements, in this case A and B) and group those elements to form one cluster. There are now *n* – 1 clusters. This is the merge step.
3. Redraw the distance matrix with the merged cluster (**Figure 7.4B**). Use the linkage method to determine the distance between the cluster and the other sequences. The distance from A to C is 3, and the distance from B to C is 2, so using the single linkage method, we choose the smallest and say that the distance from the cluster (AB) to C is 2. This calculation is repeated for the distance from (AB) to D, E and F. The distances resulting from the linkage calculation are shown in color in the figure.
4. Repeat steps 2 and 3 until only one cluster remains. In Figure 7.4B, we can see that both (AB) to C and E to F have a distance of 2, so we have to arbitrarily choose one to merge. If we choose to merge (AB) with C and again recalculate distance with the single linkage method, we get the matrix shown in **Figure 7.4C**. The next merge gives the matrix in **Figure 7.4D** and then the one in **Figure 7.4E**. The last step is to merge the two remaining clusters.

Now, how does this process relate to a phylogenetic tree? We can see the relationship better if we represent the clustering process in a computer-friendly conventional format known as **Newick format**. We first merged A and B, so we represent them with `(A,B);`. This cluster then merged with C and then eventually with D, which can be represented by `(((A,B),C),D);`. E and F merged with each other but not with any of the rest, so the final outcome is `((((A,B),C),D),(E,F));`. This very condensed representation of the data can be used to draw the cladogram in **Figure 7.6**. Each cluster has a common ancestor: A and B have the common ancestor shown by the internal node at *y*; *x* represents the common ancestor of A, B, and C; and so on. Notice that E and F have a common ancestor, *z*, but share no common ancestry with any of the other species except at the root of the tree, *v*.

The agglomerative clustering algorithm is used in many distance-based methods for calculating phylogenetic groupings. One of the first widely used tree-building methods applied agglomerative clustering with a linkage method called **UPGMA** (Unweighted Pair-Group Method with Arithmetic Mean), which calculates the distance between two clusters by averaging the distances (arithmetic mean) between each species in the cluster and every species in the other cluster. (UPGMA is in practice much like the centroid linkage illustrated in Figure 7.5 as far as clustering of sequences is concerned.) This method assumes a constant rate of evolution, so each species in a cluster contributes equally to

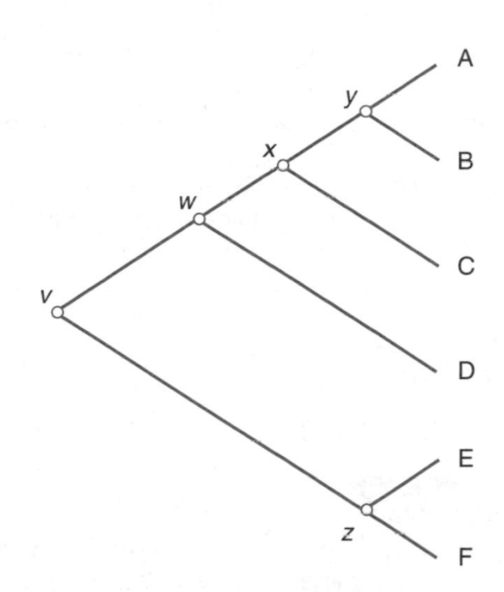

Figure 7.6 A phylogenetic tree showing the results of agglomerative clustering for six hypothetical species.

the new cluster value (unweighted). In the previous example, UPGMA would have given the distance from cluster (AB) to C as the average of the distances A–C (3) and B–C (2), or 2.5. More generally, if x and y are clusters containing n and m elements, respectively, and if x_i represents the ith element in cluster x and y_j represents the jth element in cluster y, the distance between the clusters is $d_{xy} = \dfrac{\sum_{i=1}^{n}\left(\sum_{j=1}^{m} d\left(x_i y_j\right)\right)}{n \times m}$.

In the Web Exploration and the Guided Programming Project, we look at the use of agglomerative clustering with UPGMA to build a distance-based tree. The same basic algorithm is also the basis for the NJ method discussed in the Web Exploration and the On Your Own Project. In the Web Exploration, we also look at some character-based methods that employ probabilistic models to find optimal trees.

Test Your Understanding

1. Given aligned sequences for four species with distances W–X = 1.8, W–Y = 0.8, W–Z = 2.4, X–Y = 1.8, X–Z = 2.4 and Y–Z = 2.4, cluster the sequences using single linkage and show the result in Newick format.
2. Apply the clustering algorithm to the distance data that you calculated for whales and their relatives in Chapter 6 (Chapter 6 Web Exploration exercise 6, 7, or 8). Do you get the same groupings as in the tree you drew from those data (Chapter 6 Web Exploration exercise 9)?
3. Try the UPGMA linkage method instead of the single linkage method for our sample dataset presented previously in Understanding the Algorithm. Do you get the same groupings? The same distances?
4. In the sample dataset used in Understanding the Algorithm, at the second merge we had a choice of either merging cluster (AB) with C (which we chose to do) or merging clusters E and F (which we ignored); both choices had a distance value of 2. Use the clustering algorithm to determine how the tree would have come out if we had chosen E and F instead. Would it have been different? Would this always be the case? In other words, does the arbitrary choice of one grouping when there are two possibilities have the potential to affect our view of the evolutionary relationships?

5. The tree in Figure 7.6 is drawn as a cladogram, not a phylogram: that is, the branch lengths are not strictly proportional, although the evolutionary pathways are shown correctly. Try putting branch lengths onto the tree, using the data in Figure 7.4A. What problem do you encounter? How would you explain this difficulty, biologically? (*Hint: what assumption are we implicitly making when we calculate distances between clusters?*) In the On Your Own Project, you will see how the NJ algorithm deals with this important complication by changing the way the distances between clusters are calculated.

CHAPTER PROJECT:

Placing the Archaea in the Tree of Life

Learning Objectives

- Understand how groups of organisms are clustered to develop a phylogenetic tree
- Recognize the difficulty of choosing a "best" phylogenetic tree and various approaches to that problem, including distance- and character-based methods
- Gain experience using Web-based software to develop trees using different algorithms
- Understand how molecular phylogenetics can help unravel relationships among the three domains of living things
- Identify some potential pitfalls of molecular phylogeny

Suggestions for Using the Project

This project is designed to be used either in courses that require programming skills or in nonprogramming courses. Following are suggestions for modules of the project that instructors might choose to use in these two types of courses. Instructors should also feel free to ask questions of their own that use these same skills.

Programming courses:

- Web Exploration: Gain experience with Web-based tools to build phylogenetic trees from sequence data, compare various tree-building methods, and develop a set of sequences for use with the programming projects.
- Guided Programming Project: Implement a clustering algorithm and extend the solution to give a workable program to determine phylogenetic relationships using the UPGMA method.
- On Your Own Project: Implement the NJ method to deal with unequal rates of evolution, and compare the results with the UPGMA method.

Nonprogramming courses:

- Web Exploration: Gain experience with Web-based tools to build phylogenetic trees from sequence data, and compare various tree-building methods.
- On Your Own Project: Identify modifications to the clustering algorithm that would allow for unequal rates of evolution; compare trees built by UPGMA and by NJ.

Web Exploration: Molecular Clocks and the Archaea

As described previously, due to molecular phylogenetics we realized that the diverse species of archaea in fact represented a coherent clade and that the archaea as a group are as different from the bacteria as they are from the eukaryotes. Many questions remain unanswered, however, including what the archaea might tell us about the origins of life on earth. Their adaptation to extreme environments (like the harsh conditions of 4 billion years ago) and the finding that their structures are similar in some ways to bacteria but in others to modern eukaryotes has suggested to some researchers that the archaea might be the closest living relatives of the first living things. However, interpretation of the molecular data is not always straightforward. In this project, we develop a phylogenetic tree using representatives of the three domains, examine the effect of different tree-building methods, and then look at what happens when different "clock" genes are used. The Phylogeny.fr site will be our primary tool for this exercise, because it provides a convenient and consistent framework for using several different phylogenetic tools.

Developing the Dataset

We need a sequence alignment to serve as the basis for our phylogenetic tree, and that means we need a molecular clock—in this case, a gene conserved across all three domains. It might surprise you to learn that humans and bacteria have recognizably similar proteins, but indeed they do. A good example is an accessory factor involved in the translation process that helps bring amino acid–carrying tRNA into the ribosome. This protein is called EF-1α in eukaryotes and EF-Tu in prokaryotes but is structurally and functionally similar in both: a good example of a protein conserved all the way from bacteria to humans and thus a suitable molecular clock for comparing species across all three domains of life. Because we are looking at such long time spans and because DNA sequences change faster than protein sequences, we use the EF-1α/EF-Tu *protein* rather than the DNA sequence of its gene.

Start with a file of representative sequences. For the eukaryotes, let's use human and yeast (*Saccharomyces cerevisiae*) EF-1α. Search the NCBI **Protein** and/or **Gene** databases for these proteins or download them from the *Exploring Bioinformatics* website. Save the sequences in FASTA format in a single text file, separated by the comment lines for each sequences (no blank lines). Change the comment line to something readable, like "Human_EF-1a," but remember it must be a single line and some programs do not like spaces. For bacteria, two rather different well-studied species would be *Escherichia coli* strain K-12 and *Bacillus subtilis*. Remember the protein is EF-Tu in prokaryotes. For the archaea, *Methanosarcina acetivorans* and *Haloarcula marismortui* represent two distinct groups.

A Distance-Based Tree Using UPGMA

Let's start by building a tree using UPGMA as an example of a straightforward distance-based linkage method. UPGMA is still commonly used by multiple sequence alignment programs but has become less common in tree-building programs. This method is not an option in the Phylogeny.fr suite of phylogenetic software, but we can use Phylogeny.fr to align sequences, calculate UPGMA distances with EMBOSS, and then return to Phylogeny.fr to benefit from the flexible tree rendering of TreeDyn.

Navigate to **Phylogeny.fr**, but this time choose `A la Carte` under `Phylogeny Analysis`. This option will give you more control over the steps of the analysis. Choose the programs you will use: MUSCLE for alignment, Gblocks for curation, ProtDist/FastDist + BioNJ (a distance-based method) for tree construction, and TreeDyn for tree visualization. Choose to run the workflow step by step and click `Create workflow`. You should now see an input box for your sequences; paste them there (or upload your file) and click `Submit` to run MUSCLE and produce a multiple sequence alignment.

The final tree will be based on the multiple alignment, so it is valuable to verify its quality at this stage. As you scroll through the sequence, can you find any specific evidence to suggest the sequences are aligned appropriately? For example, what does the alignment suggest about the similarity of the two representatives of each domain to each other versus their similarity to the other domains? After examining the alignment, click `Next step` to go on to curation. At this point, you can choose whether to hand-adjust the alignment; to do so, click `Edit stage input data` to see the multiple sequence alignment in an editable form. Press `F2` to get a black editing cursor, and then press `delete` to remove a gap or `space` to add a gap where you believe you can improve on the alignment. Most likely these spots will be in areas where gaps have been added, especially if they have not been added in the same place across all the sequences.

When you have finished, click `Submit` to allow Gblocks to curate the sequences and then proceed to the phylogeny step. Notice the distance metrics (substitution matrices) available to you; some should sound familiar. Continue to the phylogeny results page. Below the tree (which we ignore for now), you should see several output options, one of which is a distance matrix in Phylip format. Click this link to see the distances between all possible pairs of sequences: We can use this to create a tree by the UPGMA method in an external program. Save this matrix to a text file. Keep your Phylogeny.fr window open; carry out the next step in a new browser window or tab.

An agglomerative clustering program to build trees using UPGMA can be found at **emboss.bioinformatics.nl**. Find `fneighbor` in the list at the left, under `Phylogeny distance matrix`. This program accepts a distance matrix in Phylip format as input; upload your distance matrix file. Change the tree to `UPGMA`; the other parameters can be left at their defaults. Run the program. On the output page, you should see the data for a tree in Newick format (notice that specific branch lengths can be incorporated within this format, as well). The TreeDyn program at Phylogeny.fr can use this as input, giving us a nicer, more configurable tree. Copy the Newick formatted data to the clipboard.

Back at Phylogeny.fr, click `Next step` to get to the Tree Rendering tab. Click `Edit stage input data` to feed TreeDyn the UPGMA tree data. Paste the UPGMA tree data into the input box and run the program to see your tree. As you examine the tree, consider it both qualitatively and quantitatively. Qualitatively, the hypothesized pathway of evolution is shown by the patterns of branching and grouping. You would expect the two members of each domain to cluster together (share a more recent common ancestor); do they? Which group branches off first? What does this tell you about the hypothesized relationship of the domains? Quantitatively, examine the branch lengths. Remember that this is a phylogram, so branch lengths are meaningful. What do they tell you about the evolutionary time between the branch points? Which branchings are more ancient and which more recent? What do the branch lengths tell you about the assumptions of the program? Notice that this tree has a root, but where the tree *should* be rooted is unclear—we do not really know what "the" ancestral organism was like, and we do not have an agreed-upon outgroup. Therefore, you may get a more realistic tree if it is unrooted; click one of the radio buttons labeled `Radial` to look at it this way. What would you conclude about these groups of organisms, based on this (admittedly very limited) analysis? Save or print the tree for later comparison.

Neighbor-Joining Algorithm

The On Your Own project discusses in some detail a variation of agglomerative clustering called the **neighbor-joining (NJ) algorithm**. NJ is still a distance-based method, but it models evolution differently. A strength of Phylogeny.fr is that it is easy to rerun a phylogenetic scenario with a different algorithm. Click the `Phylogeny` tab and choose either `BioNJ` or `Neighbor` (two implementations of the NJ algorithm). The same curated multiple alignment and even the same distance calculations will be used, but the NJ algorithm will be applied to build the tree. Again, examine the resulting tree both qualitatively and quantitatively and look for differences as compared with the UPGMA tree. Can you see the important difference in the program's assumptions?

Character-Based Algorithms

Character-based algorithms consider individual characters—nucleotides or amino acids—in building a tree. For example, if at a particular position in the alignment four of six sequences have `A`, it is probable that `A` represents the **ancestral state**, or the hypothesized sequence of the common ancestor of all the modern sequences. The default tree-building algorithm at Phylogeny.fr is PhyML, a character-based algorithm that uses **maximum likelihood**. Maximum likelihood applies some model of evolution (which might take into account transitions and transversions or other known biases in the data) and then identifies trees with the highest likelihood given the model. For example, in a coin flip, if your model is that the coin is normal, 50% heads would be a high-likelihood result and 100% heads would be an extremely low-likelihood result; if the model is a two-headed coin, the reverse would be true.

The likelihood model can be further extended to use **Bayesian statistics**. Bayes' theorem involves an initial prior probability leading to the computation (based on an evolutionary model) of a posterior distribution of trees with high likelihood given the dataset. There is often minimal *a priori* information, so the prior distribution may be merely the distribution of all trees; the algorithm can then iterate repetitively using the outcome of one computation as the prior distribution for the next. (See References and Supplemental Reading if you are interested in knowing more about these statistical methods.)

Using the same curated alignment as before, use the PhyML method to draw a tree at Phylogeny.fr. Again compare your tree qualitatively and quantitatively to the other trees you have drawn. Then, try MrBayes, an algorithm based on Bayesian statistics. Here, you need to set some limits or the computation can take a very long time. Limit the number of generations (iterations) to 1,000 and sampling to every 100 generations. Even with those limits, expect this analysis to take some time; you may wish to submit the job and request an email when it is done.

Web Exploration Questions

1. In what important way is a tree computed using the UPGMA algorithm different from a tree computed by the NJ algorithm? Which do you believe better models evolution, and why?
2. Summarize concisely what you learned about the relationships among the three domains from your trees. Were the trees you developed by different methods consistent in terms of branching orders and evolutionary pathways? How consistent were they in terms of branch length?
3. It would make sense that if one highly conserved protein works as a "molecular clock," then any other similarly conserved protein would give the same results. To test that assumption, generate a phylogeny with a different highly conserved protein, the heat-shock protein Hsp70 (also known as DnaK in bacteria). Download the amino-acid sequence of the Hsp70 protein for the same six organisms (NP_002145, AET14830, DNAK_ECOLI, DNAK_BACSU, YP_306886, DNAK_HALMA), align the sequences, examine and curate the alignment, and produce trees using NJ and maximum likelihood methods. Summarize the results of this analysis and discuss anomalies between the two molecular clocks. What did you learn about the reliability of evolutionary hypotheses based on molecular data from this exercise?

More to Explore: Generating Datasets

Thus far, you have looked at molecular phylogeny using small datasets built by looking up individual genes. Larger datasets increase reliability: In a small dataset, one or two sequences that contain sequencing errors or are for some reason far from typical, misidentified, or incomplete could readily lead to spurious conclusions. However, text searching is not the easiest way to assemble a larger dataset. Instead, BLAST could be used to search by similarity for sequences similar to one known sequence of interest that can then be used to build the dataset for phylogenetic analysis. Additional tools have been developed specifically to accomplish this kind of task, including BLAST Explorer, which is included in the Phylogeny.fr workspace. BLAST Explorer makes it easy to identify proteins similar to a query sequence and choose from among them the sequences to include in a phylogenetic analysis; this method allows the use of sequences that may not have been annotated as orthologs of your query. You could explore further (or an instructor could assign further exploration) by using BLAST Explorer to collect additional EF-1a or Hsp70 sequences.

Guided Programming Project: Phylogenetic Trees Using Agglomerative Clustering

The programming projects in this chapter implement distance-based algorithms. In the Guided Programming Project, you will develop a program to perform agglomerative clustering using the single linkage method. The skills exercises will ask you to expand your program by producing the final tree in Newick format, allowing a user to choose between single and UPGMA linkage, printing branch lengths, and allowing the program to handle sequence input data. The On Your Own project will lead you to modify the solution further by implementing the NJ method.

As you saw in Understanding the Algorithm, hierarchical clustering is a matter of determining distances between clusters using a linkage method, merging the two closest clusters, and iterating until all clusters have been merged. Initially, each sequence (species) is an individual cluster, with the distances between clusters calculated by alignment and the application of some distance metric. For this project, we assume that the input for our program is a set of calculated distances between sequences. You will read these data in from a Phylip-formatted input file. The discussion and pseudocode that follow use single linkage, paralleling the example given earlier, but this is easily modified to use UPGMA (see Putting Your Skills Into Practice).

Let's take a moment to consider the data structures we might need. In Understanding the Algorithm, a distance matrix was used to represent cluster distances. We could use a two-dimensional array to hold this matrix, but it might be more efficient to use a nested hash structure. What happens, however, when we want to merge two clusters? Assume we merge clusters A and B. We could remove these two elements from the hash table and replace them with a merged element whose key is AB. But we need the original distances between A, B, and the other clusters when we apply our linkage method. Therefore, we might want to hold the original distances in one nested hash structure and use another nested hash structure to represent the working cluster distances, which would change as we merge. At the start of the algorithm, the original distances could be stored in a nested hash structure similar to the following (only a partial set is shown; keys C, D, E, and F are not included):

Hash of Hash Table of Original Distances

```
key = A, value = {key = A, value = 0}
                 {key = B, value = 1}
                 {key = C, value = 3}
                 {key = D, value = 7}
                 {key = E, value = 17}
                 {key = F, value = 19}
```

```
key = B, value = {key = A, value = 1}
                 {key = B, value = 0}
                 {key = C, value = 2}
                 {key = D, value = 6}
                 {key = E, value = 16}
                 {key = F, value = 18}
```

This structure would not change during the program, so we can always reference original distances. A copy should be made of this structure and used to represent merging clusters, similar to those in Figure 7.4. After the first iteration, the structure representing merging clusters would look as follows, assuming we use single linkage and merge clusters A and B (only a partial set is shown; keys D, E, and F are not included):

Hash of Hash Table of Merging Clusters

```
key = AB, value = {key = AB, value = 0}
                  {key = C, value = 2}
                  {key = D, value = 6}
                  {key = E, value = 16}
                  {key = F, value = 18}

key = C, value = {key = AB, value = 2}
                 {key = C, value = 0}
                 {key = D, value = 4}
                 {key = E, value = 14}
                 {key = F, value = 16}
```

We would continue to work with this nested hash structure, reducing the size by one with each iteration. At the end, we would be left with two keys in our nested hash structure, which would represent the final two clusters to merge. The following pseudocode presents a solution to cluster a set of items using the approach just described. This implementation assumes the data file is a Phylip-formatted file and with each iteration the merging clusters are printed.

Algorithm

Agglomerative Clustering Algorithm to Determine Evolutionary Relatedness

Goal: To cluster a set of data items

Input: A set of sequence distances in a Phylip formatted file

Output: Clusters merged at each step

```
// Initialization - Read in data and build nested hash structures
Open input file containing sequence distances: infile
numSeq = read first line of infile
clusterNames = array of size numSeq
distances = array of size numSeq
i = 0
for each line of data in infile
```

```
    clusterNames[i] = first value in line
    distances[i] = remaining data in line split using space as
    delimeter

// Build nested hash structure of original and cluster distances
originalDist = nested hash structure
clusterDist = nested hash structure
for each i from 0 to numSeq-1
    for each j from 0 to numSeq-1
        originalDist[clusterNames[i]][clusterNames[j]] =
            distances[i][j]
        clusterDist[clusterNames[i]][clusterNames[j]] =
            distances[i][j]

// STEP 1: Cluster

while numClusters > 2
    shortestD = shortest distance in clusterDist
    shortestI = outer key of shortest distance in clusterDist
    shortestJ = inner key of shortest distance in clusterDist

    // merge clusters I and J
    newClusterName = shortestI + shortestJ
    remove shortestI from clusterNames
    remove shortestJ from clusterNames
    remove shortestI keys and nested keys from clusterDist
    remove shortestJ keys and nested keys from clusterDist

    singleLinkage(clusterDist, newClusterName, originalDist,
        clusterNames)
    append newClusterName to clusterNames

    output "merging clusters" shortestI and shortestJ

    numClusters—

output remaining two clusters

// function to calculate distances between new cluster and all
// other clusters using single linkage
function singleLinkage(clusterdist, newClusterName, originalDist,
clusterNames)
    for each cluster in clusterNames
        smallestD = maximum integer
        for each c1 in cluster
            for each c2 in newClusterName
                if originalDist[c1][c2] < smallestD
                    smallestD = originalDist[c1][c2]
        clusterDist[newClusterName][cluster] = smallestD
        clusterDist[cluster][newClusterName] = smallestD
```

Putting Your Skills Into Practice

1. Write a program in the language used in your course to implement the given pseudocode. Test your program using the sample data values for the six species (A–F) used as an example in Understanding the Algorithm. You can create your own distance matrix data file or **download Phylip-formatted data** (see Web Exploration) from the *Understanding Bioinformatics* website. Be sure your program correctly deals with the format of the data file. Ensure that the program merges the clusters as expected.
2. Although an implementation of this pseudocode shows which clusters are merged at each iteration, a representation of the final evolutionary tree in Newick format would be much more useful. Modify your program to output a tree in Newick format; as discussed earlier, for our sample data, the output should be `((((A,B), C), D),(E,F));`.
3. Modify your program so it allows the user to choose between the single linkage and the UPGMA linkage method.
4. As you may have observed when you obtained a UPGMA-based tree for input to TreeDyn in the Web Exploration, Newick format also allows for branch lengths to be explicitly specified. Adding branch lengths would not only convey additional information to the user but would also allow your program to output data that could be used directly by TreeDyn or another tree-rendering program. Modify your program to calculate branch lengths and include them in the Newick format output. Remember that these agglomerative clustering methods assume constant rates of evolution, so at each node (for example, where A diverges from B), the distances (from A to the node and from B to the node) should be the same.
5. Currently, your program takes a distance matrix as input. A more flexible program would allow you to input sequence data, calculate distances, and then output the clustered data. To do this realistically would require a multiple sequence alignment algorithm, which is beyond the scope of this project. However, you already have programs that can do global alignment (Chapters 3 and 5) and apply distance metrics to pairwise alignments (Chapter 6); you could incorporate distances calculated by these methods into your phylogenetic tree. This requires two modifications to the program. (1) Read in nucleotide sequences from a text file and store them as the hash *value* of each species. Use alphabetic characters to represent the *key* for each species or cluster. (2) Align sequences and calculate distances, using either a nucleotide alignment with a choice of distance metrics (start with the code from the On Your Own Project in Chapter 6) or a protein alignment with a substitution matrix.

■ On Your Own Project: The Neighbor-Joining Method

Understanding the Problem: Determining Branch Lengths

The agglomerative clustering algorithm discussed in Understanding the Algorithm, particularly when coupled with the UPGMA linkage method, was at one time widely used in constructing phylogenetic trees and is still used in many multiple sequence alignment algorithms. However, simple agglomerative clustering is rarely used in tree-building today because of its limitations, notably the fact that it is **ultrametric**: It assumes a constant rate of evolution or a molecular clock that "ticks" at a constant rate. In the phylogenetic tree shown in Figure 7.6, for example, note that the distance from A to the node at y is the same as from B to y. You should have observed similar results for the branch lengths when you constructed a tree using the UPGMA method in the Web Exploration. There is a biological basis for this assumption: Because the two modern species A and B have been evolving for the same amount of time since they diverged from their common ancestor (y), the distance (i.e., number of substitutions) should be the same along each branch.

Unfortunately, in reality, distances between sequences may not be ultrametric. As we saw in Understanding the Algorithm, our simple example tree fails when we attempt to label branch lengths. Our example resulted in the grouping ((A,B),C);, for instance, given the distances A–B = 1, A–C = 3, and B–C = 2. Assuming a constant rate of evolution, the distance from A to *y* and from B to *y* should be equal, 0.5 each. Then, the distances from A to C and from B to C should also be equal—but they are not! Therefore, although UPGMA is a convenient and easy-to-implement linkage method, it is not suitable for building phylogenetic trees under all conditions.

Solving the Problem

The **NJ method** is an alternative that does not require the assumption of a constant rate of evolution across all species. The NJ method is a variation of the agglomerative clustering technique and can be applied to a set of sequences for which distances have been calculated using any desired metric. As before, there is a merge step in which the two closest clusters are merged. The difference is in the linkage method: NJ calculates a **transformed distance** value when calculating the distances between the remaining clusters at each iteration. This allows the branch lengths to correspond to the observed distance between species, even when those branch lengths are not ultrametric, accounting for differences in the rate of evolution.

Using NJ, each iteration of the clustering algorithm thus begins by calculating an *r* value for each cluster, representing the corrected net distance between it and all other clusters. This is essentially the average distance between a given cluster, *x*, and each other cluster (*i*); if there are *n* total clusters, we can use the following formula:

$$r_x = \frac{\sum_{i=1\ldots x-1,x+1\ldots n} d_{ix}}{n-2}$$

The value d_{ix} is the distance between cluster *x* and cluster *i* as determined by the previous iteration (or the initial distance matrix, for the first iteration). This distance is determined for every cluster *i* other than *x* itself and summed.

These *r* values are then used to compute transition distances (*td*) to be used in determining which cluster to merge at the merge step. The following formula shows how this is done given clusters *x* and *y*, where *x* != y:

$$td_{xy} = d_{xy} - r_x - r_y$$

The cluster pair that has the smallest transition distance is merged. Once the clusters are merged, new distances are calculated between the newly formed cluster (*K*) and all other clusters (the distances between unmerged clusters do not change). After a merge of clusters *i* and *j*, the distance from the new cluster to any cluster *x* is given by

$$d_{Kx} = \frac{d_{ix} + d_{jx} - d_{ij}}{2}$$

As before, this process repeats for additional clusters; we stop when only two clusters remain and join the last two based on calculated distance between them (see the next section). Finally, the branch lengths within the tree must be calculated; because the distances from an

Table 7.1 Initial distances for the neighbor-joining example.

	A	B	C	D	E
A	0				
B	5	0			
C	11	10	0		
D	12	11	7	0	
E	11	10	6	3	0

ancestor to its descendants need not be the same, the distance from cluster i to j must be calculated as two branch lengths, from each of the clusters to their shared ancestor K:

$$d_{iK} = \frac{d_{ij} + r_i - r_j}{2}, \; d_{jK} = \frac{d_{ij} + r_j - r_i}{2}$$

Now that we have the formulas, let's see how they work with a simple example. Suppose we have sequences from five species with initial distances as shown in **Table 7.1**. The first step is to calculate transformed r values. For our first iteration, these are A = 13, B = 12, C = 11.34, D = 11, and E = 10. Using these values, we can compute transition distance values for our first iteration, resulting in the transition matrix in **Table 7.2A**. The transition matrix is used to determine which clusters to merge. Because the lowest value in the transition matrix is in the cell represented by clusters A and B, these two clusters are merged. The new distance matrix is then populated with the initial matrix distances, except for the distances between the newly created cluster, represented by AB, and the other clusters, which must be calculated. These distances are shown in **Table 7.2B**.

Before moving on to the next iteration, let's look at the partial tree represented by the merge of clusters A and B. This merge implies these two species have a common ancestor (AB), and to obtain the branch length from the common ancestor to each species, we apply the branch length formula just given:

$$d_{A(AB)} = \frac{d_{AB} + r_A - r_B}{2} = 3, \; d_{B(AB)} = \frac{d_{AB} + r_B - r_A}{2} = 2$$

This partial tree can now be drawn as shown in **Figure 7.7A**. Notice that the two branch lengths are unequal, something that would not have been possible using the UPGMA method.

Table 7.2 First transition matrix (A) and recalculated distance matrix (B) for the neighbor-joining example.

A

	A	B	C	D
B	−20			
C	−13.34	−13.34		
D	−12	−12	−15.34	
E	−12	−12	−15.34	−18

B

	AB	C	D	E
AB	0			
C	8	0		
D	9	7	0	
E	8	6	3	0

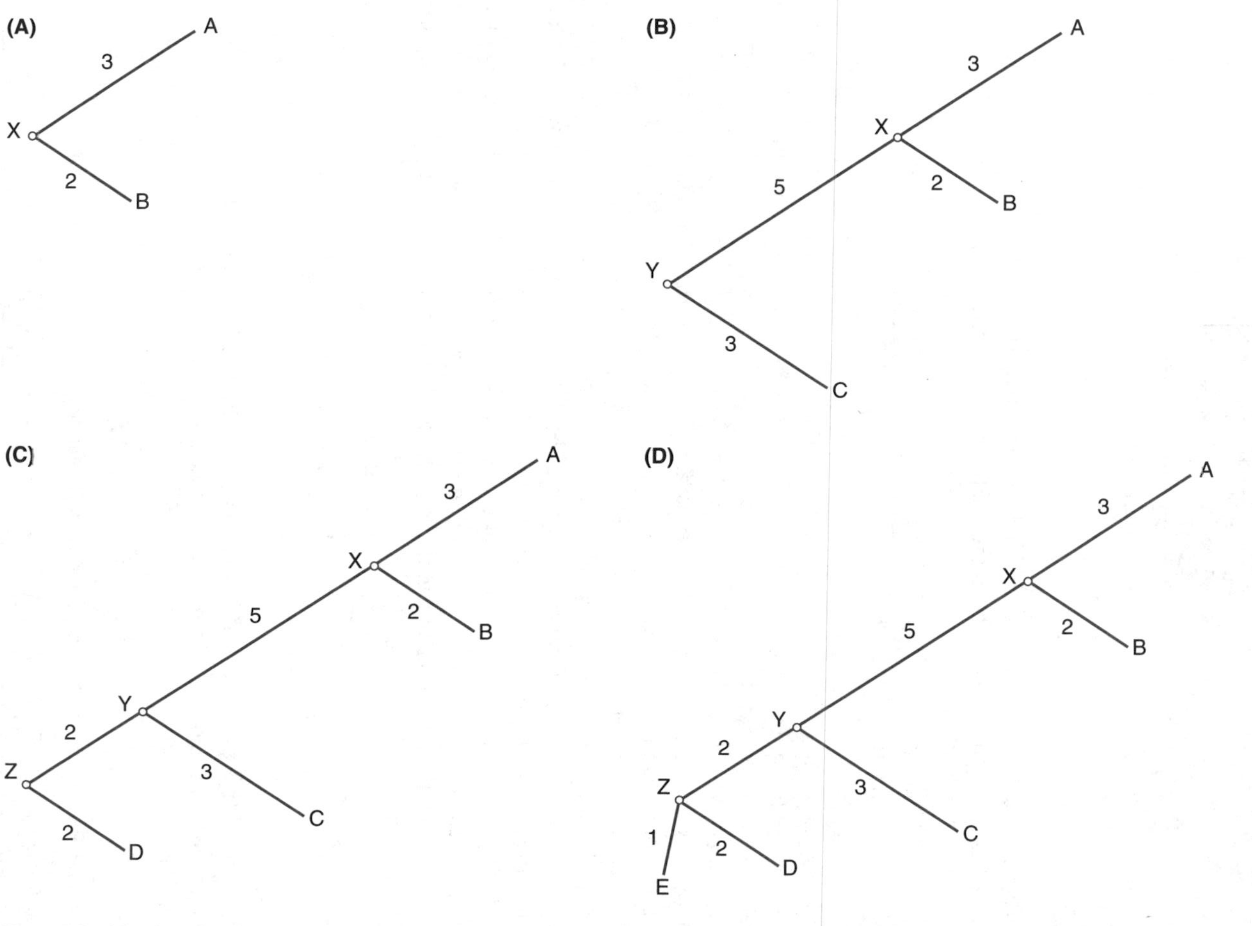

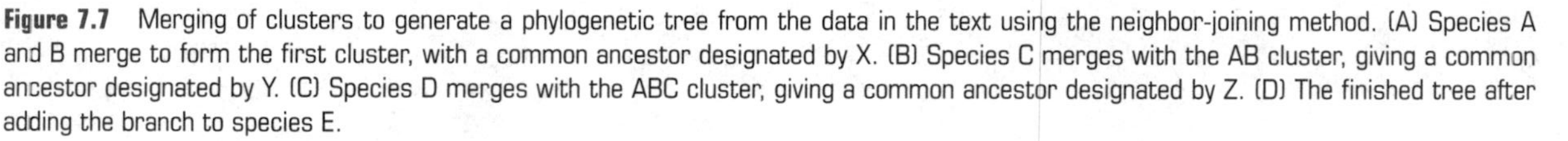
Figure 7.7 Merging of clusters to generate a phylogenetic tree from the data in the text using the neighbor-joining method. (A) Species A and B merge to form the first cluster, with a common ancestor designated by X. (B) Species C merges with the AB cluster, giving a common ancestor designated by Y. (C) Species D merges with the ABC cluster, giving a common ancestor designated by Z. (D) The finished tree after adding the branch to species E.

Table 7.3 Transition matrix (A) and recalculated distance matrix (B) after the second merge in the neighbor-joining example.

A

	AB	C	D
C	−15		
D	−13	−13	
E	−13	−13	−15

B

	ABC	D	E
ABC	0		
D	4	0	
E	3	3	0

Our next iteration begins by recalculating transformed r values: AB = 12.5, C = 10.5, D = 9.5, and E = 8.5. **Table 7.3** shows the new transition matrix (A) and new distance matrix after the second merge (B). In this iteration, two cells contain the lowest value in the transition matrix. We can choose to merge cluster AB with C or cluster D with E; here, we arbitrarily choose to merge AB with C. The new distance matrix is populated with the previous iteration's distances, except for the distances between the newly created cluster, represented by ABC, and the other clusters. Calculating branch lengths and adding the results of this merge to our partial tree results in the tree shown in **Figure 7.7B**.

With the next iteration, we obtain the transition matrix in **Table 7.4A**. According to this transition matrix, we could now merge any of the remaining clusters, because they have the same value. We choose to merge ABC and D, and again we recalculate distances (**Table 7.4B**) and branch lengths and then add our newly merged clusters to our partial tree (**Figure 7.7C**). Because we are now left with only two clusters, we can simply attach these two clusters using our distance information. In our example, notice that the final distance matrix (Table 7.4B) conveniently gives us the distance between species E and the cluster ABCD (or common ancestor of species A–D), and we get the final tree shown in **Figure 7.7D**. If we were merging two clusters at this point, the last distance we need for our tree would be the distance between two internal nodes (ancestral species), and we could calculate this by going back to the original distance matrix, finding the distance between a species in one cluster and a species in the other, and then subtracting the already calculated branch lengths to get the distance between the remaining internal nodes.

Notice that the NJ method has produced an unrooted tree, whereas UPGMA produced rooted trees. The NJ branch length formula allows for the calculation of unequal branch lengths. If you compare the distances in the final phylogenetic tree (Figure 7.7D) with our original set of distances (Table 7.1), you will see that the tree matches the original distances, demonstrating the additivity property of the NJ method.

Given the equations and example presented here, you should now be able to use the NJ algorithm to construct a phylogenetic tree with calculated branch lengths for the six sample species whose distance matrix is given previously in Understanding the Algorithm. How does the tree thus generated differ from the tree shown in Figure 7.6?

Table 7.4 Transition matrix (A) and recalculated distance matrix (B) after the last merge in the neighbor-joining example.

A

	ABC	D
D	−10	
E	−10	−10

B

	ABCD	E
ABCD	0	
F	1	0

If your course involves programming, your instructor may ask you to implement the NJ algorithm as described next. If it does not, a completed program implementing NJ can be downloaded from the instructor section of the *Exploring Bioinformatics* website and used to complete the exercises at the end of the Programming the Solution section without programming.

Programming the Solution

Using your solutions to the Guided Programming Project exercises as a starting point, implement the NJ method in the programming language of your choice. Depending on the exercises your instructor chose previously, you may have a program to carry out agglomerative clustering given a distance matrix or a more comprehensive program to generate a Newick format tree from nucleotide or amino-acid sequence data. Any of these solutions can be readily modified to implement NJ or offer NJ as a choice of method for the user.

The initial steps (reading sequence or distance files, aligning sequences, calculating initial distances, etc.) will not change, but you will need to make changes to the decision process in the merge step and the calculation of intercluster distances thereafter, as well as a calculation of final branch lengths. Use the formulas given in the chapter to make these calculations. You will also notice some differences in the data that need to be stored. In the guided project, a nested hash table was used to hold cluster information. This was important, because we needed to keep track of each cluster element's value to determine distance. However, the NJ method recalculates distances at each iteration from the previous cluster distances. For troubleshooting purposes, you may wish to print out the clusters merged and the branch lengths as each merge occurs, but the final program should output results in Newick format, including branch lengths.

Run your program on the following test data set using a simple nucleotide count as your distance metric and NJ as the linkage method:

```
(A) TCAT, (B) TCCT, (C) TCCC, (D) GCGT, (E) GCTT
```

You should end up with the following tree: `((C:1,B:0):0.5,(A:0.5,(D:0.5,E:0.5):1));`, after merging D with E, DE with A, and C with B.

Then, try your program with the data from Understanding the Algorithm. You should get the same results as when you worked out the tree by hand. Compare your outcome with the results using UPGMA as a linkage method. Can you explain why there are differences? Finally, test your program on the eIF-1α and Hsp70 sequences from the Web Exploration. Which algorithm do you believe gives you the best picture of the actual evolutionary pathways?

Connections: What Is a Species?

Chapter 5 included the example of two salamander populations that became separated by California's Central Valley and had evolved into subspecies. Assuming continued separation, these subspecies may eventually become two distinct species. But just how do we define a species? One long-used biological definition is that two organisms are members of the same species if they are able to mate and have fertile offspring. Perhaps, however, you can already see problems with this definition. All domestic dogs, for example, are considered to be members of a single species—indeed, a single subspecies, *Canis lupus familiaris*—but it is obvious that successful mating between a St. Bernard and a chihuahua is unlikely.

Where we find similar but distinct kinds of birds, such as the readily distinguishable Eastern Bluebird and Mountain Bluebird, do we have one species or two? What do we do about the many kinds of organisms that have no sexual reproduction? What about plants, where in some cases two quite different plants can mate and yield a new type of plant with twice as many chromosomes? (This happened naturally at least twice in the history of our modern red wheat.) And perhaps most puzzling of all, what about bacteria and archaea, where we find enormous biochemical and metabolic diversity despite very limited visually distinguishable features and a complete lack of genuine sexual reproduction?

Bioinformatics and molecular evolution are central to research aimed at untangling difficulties in the concept of a species and in classifying organisms throughout the living world. Where morphology, ecology, physiology, and even biochemistry cannot resolve the question, bioinformatics can quantify differences in DNA and protein sequences and establish standards for how different two organisms need to be in order to be considered two species. Evolutionary journals are currently full of articles in which bioinformatic tools are used to investigate questions such as these, frequently resulting in splitting what was thought to be one species into two, or the reverse—sometimes producing heated debates. As more and more DNA sequences and complete genomes become available, we can anticipate ongoing progress in this area.

References and Supplemental Reading

Carl Woese's Original Paper on the Evolutionary Distinctiveness of Archaebacteria

Woese, C. R., and G. E. Fox. 1977. Phylogenetic structure of the prokaryotic domain: the primary kingdoms. *Proc. Natl. Acad. Sci. U.S.A.* **74**:5088–5090.

Proposal for a Three-Domain Classification System

Woese, C. R., O. Kandler, and M. L. Wheelis. 1990. Towards a natural system of organisms: proposal for the domains Archaea, Bacteria, and Eucarya. *Proc. Natl. Acad. Sci. U.S.A.* **87**:4576–4579.

UPGMA and NJ Methods

Gronau, I., and S. Moran. 2007. Optimal implementations of UPGMA and other common clustering algorithms. *Inform. Process. Lett.* **104**:205–210.

Saitou, N., and M. Nei. 1987. The neighbor-joining method: a new method for reconstructing phylogenetic trees. *Mol. Biol. Evol.* **4**:406–425.

Probabilistic Methods for Tree-Finding

Archibald, J. K., M. E. Mort, and D. J. Crawford. 2003. Bayesian inference of phylogeny: a non-technical primer. *Taxon* **52**:187–191.

Felsenstein, J. 1981. Evolutionary trees from DNA sequences: a maximum likelihood approach. *J. Mol. Evol.* **17**:368–376.

MUSCLE Multiple Sequence Alignment

Edgar, R. C. 2004. MUSCLE: multiple sequence alignment with high accuracy and high throughput. *Nucleic Acids Res.* **32**:1792–1797.

BLAST Explorer

Dereeper A., S. Audic, J. M. Claverie, and G. Blanc. 2010. BLAST-EXPLORER helps you building datasets for phylogenetic analysis. *BMC Evol. Biol.* **10**:8–13.

Chapter 8

DNA Sequencing:
Identification of Novel Viral Pathogens

Chapter Overview

In addition to the value of DNA sequencing for identifying genes and examining whole genomes, new technologies now permit "deep sequencing" of transcriptomes, metagenomes, and environmental samples. Bioinformatics is essential for assembly of short sequences into complete gene or genome sequences and for applications that use the short sequences themselves. By completing the projects in this chapter, students will understand how sequence data are read, some uses of sequences produced by high-throughput next-generation sequencing methods, the problem of sequence assembly, algorithmic approaches to constructing a full-length sequence from an array of short sequences, and the use of coverage as a measure of assembly quality. Additionally, students in programming courses will write programs to create test sequence data with a desired level of coverage and write a miniassembler program.

Biological problem: Identification of unknown causes of viral disease

Bioinformatics skills: Manipulating and mapping short sequence reads, assembling sequences into contigs, measures of quality

Bioinformatics software: Galaxy, Megablast, SRA and Trace databases, CAP assembler

Programming skills: Generating random string fragments, Overlapping strings, Traveling Salesperson Problem

Understanding the Problem:
Deep Sequencing of Clinical Samples

It might surprise you to know that diarrhea is the second most common cause of death in children under age 5, killing an estimated 2 million children worldwide each year. Although many people in countries with access to clean drinking water and reliable sanitation may consider this disease a mere annoyance, globally, billions of people lack these basic services. Indeed, diarrhea is third among causes of death for both children and adults in low-income countries, accounting for nearly 7% of fatalities. Most deaths from diarrheal disease result from dehydration, and the chronic or recurrent diarrhea common in many parts of the world is also an important cause of malnutrition. In recent years, several new causes of diarrhea have

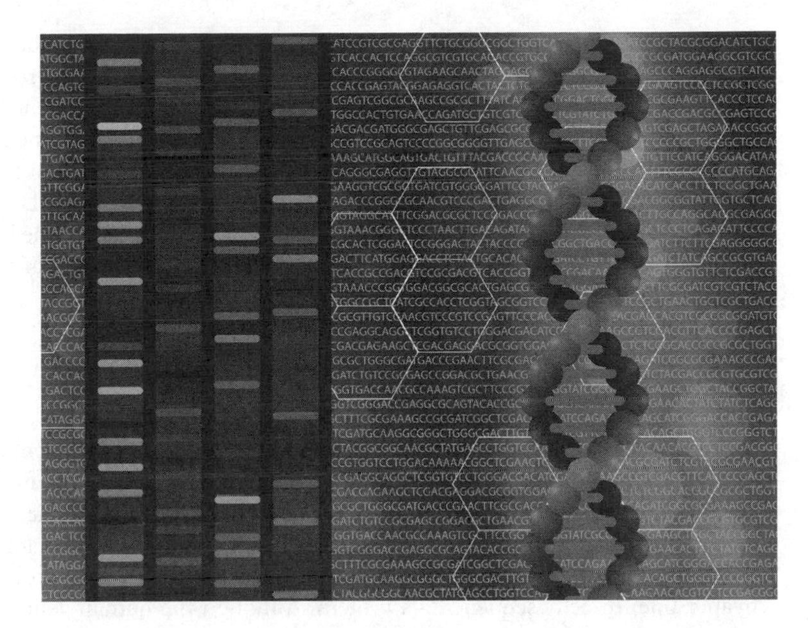

Figure 8.1 Automated sequencing of shotgun sequences and high-throughput next-generation techniques have enabled advances in genome and metagenome sequencing. A computer-generated image of automated sequencing output is shown here. © The Biochemist Artist/ShutterStock, Inc.

been identified, including cosavirus, klassevirus, and an entirely new genus of parvoviruses. Importantly, these new viruses have been identified not by traditional culture methods but by metagenomics (see References and Supplemental Reading). New "deep sequencing" methods (**Figure 8.1**) *applied to any and all DNA found in a human clinical sample not only tell us about what bacteria and viruses are present but have led to the identification of previously uncharacterized species, including novel pathogens.*

Identification of the specific microbe responsible for a given disease has been a difficult problem ever since Robert Koch and Louis Pasteur pioneered the germ theory of disease in the late 1800s. Indeed, given an uncomplicated case of diarrheal disease, it is more efficient for a physician to simply treat dehydration and determine whether antibiotic intervention is warranted than to pursue time-consuming and expensive procedures to identify a specific causative organism. The same is true for many other common diseases—upper respiratory syndromes, fevers, skin problems, and so on. Thus, the full spectrum of pathogens that can cause these diseases remains undetermined, and this is particularly true for viral pathogens because of the difficulty of isolating and culturing unknown viruses. Unexpectedly, DNA sequencing has become an unexpected resource for solving problems of this kind and for examining genomes, measuring gene expression, characterizing ecosystems, and more.

Initially, sequencing was limited by technology and cost to individual genes of interest cloned into plasmid vectors but quickly progressed to sequencing of entire genomes, potentially allowing researchers to define all the functions of a cell and even an entire organism in terms of its genes and their interactions with the environment. The publicly funded International Human Genome Project (IHGP) began in 1990 with a plan to obtain the complete sequence of the human genome—3,000,000,000 nucleotides of information—by mapping and sequencing an ordered set of genome segments. Eight years later, a competitor, Celera Genomics, a private company headed by Dr. Craig Venter, entered what became an acrimonious race. Despite the IHGP's sizeable head start, both groups announced draft genome sequences in 2000. The key to Celera's success was to eliminate the time required to develop orderly arrays and simply sequence random genome fragments, relying on bioinformatic techniques and computational power to assemble these short "shotgun" sequences into complete chromosome sequences (see References and Supplemental Reading). Further advances in sequencing technology have taken this approach to the extreme: so-called next-generation sequencing techniques generate huge numbers of sequences in parallel, but they are as short as tens of bases each. Sophisticated assembly software can join these bits of sequence into full-length DNA sequences with a high degree of accuracy. With these technologies constantly pushing the boundaries of faster, cheaper sequencing, in what new ways might we use DNA sequencing?

Bioinformatics Solutions:
Assembly and Mapping of Short Sequence Reads

DNA sequencing is the process of determining the order of the nucleotides that make up a piece of DNA. This is the laboratory technique that generates not only all the DNA sequences you've been working with throughout this text (for more detail on sequencing techniques, see the BioBackground section at the end of the chapter) but most of the amino-acid sequences as well, because computational "translation" of a nucleotide sequence is much faster and cheaper than directly sequencing a protein. Although the human genome was not the first to be sequenced (among cellular organisms, that honor belongs to the yeast *Saccharomyces cerevisiae*), it has generated the most interest: Its far-reaching potential has been compared with the invention of the printing press. We remain a long way from knowing the function of every gene in the human genome, but we have all the raw data: the nucleotide sequences of all 23 distinct human chromosomes and all the 20,000+ genes they carry.

Although dideoxy sequencing was used in both cases, Celera genomics was able to complete the sequencing of the human genome in a fraction of the time required by the IHGP by pioneering a faster **shotgun sequencing** technique (**Figure 8.2**). The Celera approach was fast because many DNA fragments could be sequenced at once, but it created a major computational problem because it produced many short DNA sequences whose relationship to each other was unknown. With algorithms capable of accurately assembling these sequences into the sequences of complete chromosomes, Celera opened the door to rapid genome sequencing. As this technique gained momentum, dozens of other genomes were completed, including bacteria, vertebrate and invertebrate animals, plants, fungi, and viruses.

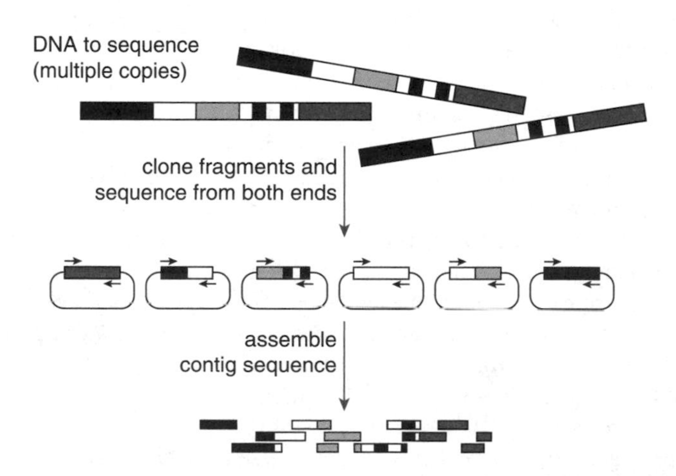

Figure 8.2 Schematic representation of shotgun sequencing. The DNA to be sequenced is fragmented, random fragments are cloned into plasmids, and the fragments are then sequenced from both ends. Computational assembly of many fragments allows the complete sequence of the original DNA to be reconstructed.

Huge benefits have already been reaped from genome sequencing, including better understanding of biological processes, identification of genes responsible for disease, development of improved therapies, and industrial and agricultural applications.

However, translating shotgun sequence data into quality genome sequence requires high **coverage**: Each segment of the genome must be sequenced many times over to generate enough overlapping fragments to assemble the complete genome. The advent of next-generation sequencing techniques (see BioBackground) drastically increased the rate at which sequence could be obtained. In 454 sequencing, for example, a million individual sequence reads can be done in a single run, and Illumina and SOLiD technology can multiply that by 1,000 times. However, in maximizing data throughput, these techniques sacrifice **read length**, or the lengths of the DNA sequences they identify. Read lengths in dideoxy sequencing can be 800 nucleotides long or longer, but that number drops to 500 nucleotides for 454 sequencing and less than 100 nucleotides for Illumina and SOLiD. These sequencing techniques are therefore only as good as the bioinformatics software that allows us to analyze and interpret them.

The short sequences generated by next-generation sequencing are used in two general ways: assembled into genomes or used directly to identify RNAs, organisms, or functional segments. Genome assembly has progressed to the point that we can begin to contemplate applications such as the rapid and inexpensive determination of each *individual* human's complete genome sequence. Meanwhile, short sequence reads from cellular, environmental, or clinical samples are used to determine the complete set of mRNAs produced in a given tissue or under a given condition (**RNA-seq**) or to identify all the organisms present in a particular environment without the need to isolate or culture them (**metagenomics**). These latter applications are often referred to as **deep sequencing** techniques. Deep sequencing of DNA present in a stool sample, throat swab, or skin wash can be used to identify the microbes normally present in the human body (the **microbiome**) as well as any pathogenic organisms that may be present. In the future, doctors may be able to use deep sequencing to take a microbial "census" of patient tissues. Applications like these depend on bioinformatics to provide algorithms for reliable assembly of massive amounts of fragmentary data into meaningful sequences and for mapping sequence reads relative to genomes.

In this chapter's projects, you will examine and use some of these bioinformatics applications to work with sequencing data. You will see how DNA sequencing data are presented, identify viruses from short sequence reads, experiment with assembly programs, and (if your course includes programming) write your own miniassembly program.

BioConcept Questions

1. In Sanger sequencing, why does a newly synthesized strand of DNA terminate when DNA polymerase inserts a dideoxy nucleotide? How are these terminated DNA strands used to "read" the nucleotide sequence of the original DNA molecule?
2. Why is shotgun sequencing so much faster than the directed approach originally taken by the IHGP? Why is it more dependent on computer power and bioinformatics?
3. If the entire human genome were cleaved into a single set of small, non-overlapping fragments, we could not determine the genome sequence by sequencing the fragments. Explain why this is the case.
4. How do next-generation sequencing techniques extend and improve on the shotgun sequencing technique? What are their disadvantages?
5. Complex genomes often contain many repeated sequences. For example, there are many STR (short tandem repeat) sites in the human genome, where a short sequence such as GATA might be repeated anywhere from a few to dozens of times. Why would an STR region potentially pose a problem for sequencing? Are next-generation techniques more or less susceptible to errors resulting from repeated sequences than older technologies?

Understanding the Algorithm: Determining Overlap in Sequence Assembly

Learning Tools

To better understand the problem of sequence assembly and the importance of the depth of coverage, you can download **Assembly exercise.pdf** from the *Exploring Bioinformatics* website. This file contains three copies of a short sequence representing threefold sequencing coverage that can be printed, cut into pieces, and reassembled.

In this chapter's Web Exploration, you will gain experience using both sequence assembly tools and alignment-based tools for metagenomic analysis of short sequence reads from a clinical sample. In the Guided Programming Project and On Your Own Project, we focus on programs for assembling sequences, so this section explores assembly algorithms.

The problem of **sequence assembly** is similar to the problem you would encounter if you ran this page through a paper shredder. Each fragment of the page might contain just a few words or perhaps just a handful of letters, and reassembling the complete text would be a daunting task. In shotgun sequencing, however, each part of the genome is represented more than once, as if you made several copies of this page and then shredded them together. Therefore, you might find pieces with `ach fr` and `ent o` and not see how to put them together, but then you might discover a piece with `Each fragm` and another with `nt of th` that came from different copies of the page. Even if you never found the missing piece with `agm` that fits between the first two, you could conclude that a segment of the original text read `Each fragment of th`. In this way, overlapping short DNA sequences can allow us to build up the original sequence of a long piece of DNA (**Figure 8.3**).

A **sequence read** is a single piece of data from a DNA sequencing reaction; whether it is an 800-nt fragment from Sanger sequencing or a 40-nt fragment from a next-generation platform, it can be represented as a string of nucleotides. Thus, given a large number of strings representing nucleotide sequences, a sequence assembly program looks for overlaps

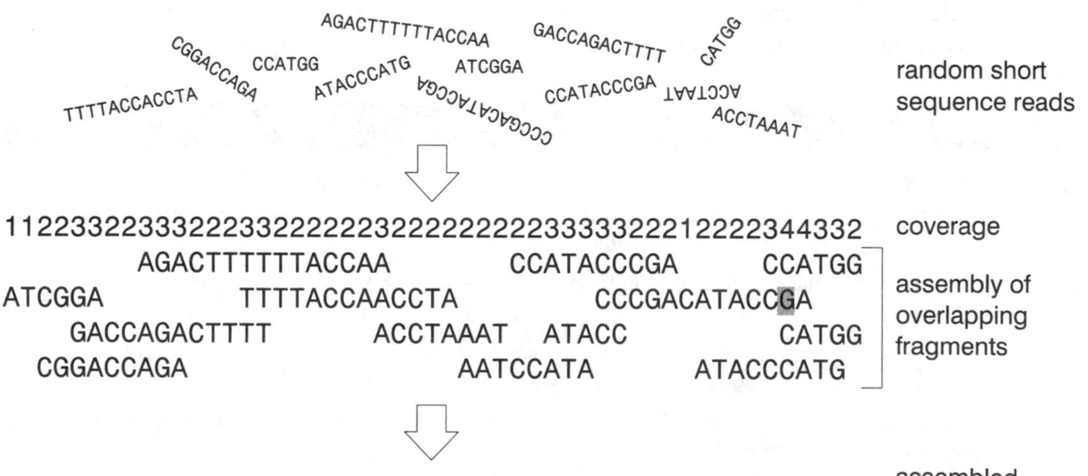

Figure 8.3 Assembly of short sequence reads into a longer contiguous sequence (contig). Overlaps are used to order the fragments, and coverage shows how often each nucleotide in the contig has been sequenced, a measure of the quality of the assembly. The highlighted G nucleotide appears to be a sequencing error.

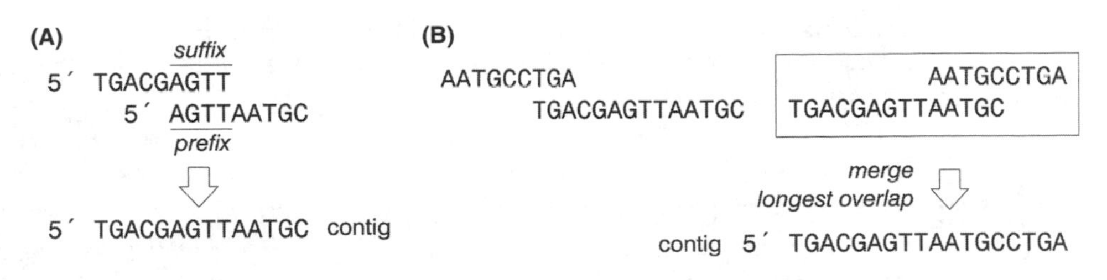

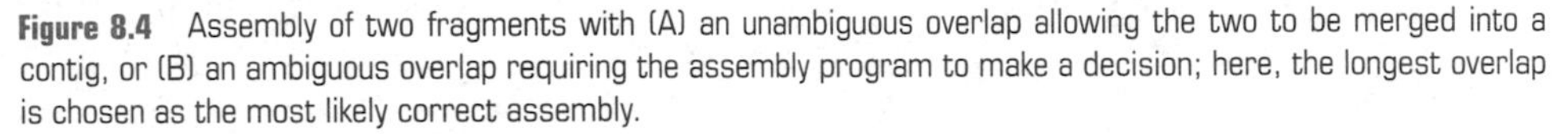

Figure 8.4 Assembly of two fragments with (A) an unambiguous overlap allowing the two to be merged into a contig, or (B) an ambiguous overlap requiring the assembly program to make a decision; here, the longest overlap is chosen as the most likely correct assembly.

to decide which strings should be joined together. Joining sequence fragments builds longer sequences called **contigs**, and when enough overlaps have been found, the contig represents an entire mRNA, plasmid, or chromosome. This process is far from trivial, however: Overlaps may be small, and the repeated sequences common to complex genomes (more than 50% of the human genome consists of repeated sequences of various kinds) introduce the possibility of misassembly. Furthermore, a fragment could come from either strand of the DNA molecule, so the assembler has to try assembling a given fragment and also its reverse complement. Furthermore, inaccuracy in the sequencing reactions themselves or in the base-calling software (see BioBackground) means that it may be unclear whether an overlap is genuine.

Sequence assembly is therefore one of the most complex bioinformatics problems. In fact, it is still considered an open problem—one not completely solved—because no existing algorithm can reassemble fragments with complete accuracy in all situations. We are thus forced to rely on technical solutions to increase accuracy, such as continuing to sequence random fragments until every section of the target DNA has been sequenced multiple times. Then, errors become noticeable as bases fail to align properly (Figure 8.3). The assembly program keeps track of **coverage**—the number of times the nucleotide at each position has been sequenced—as a measure of the reliability of the assembly at each point.

To see how an assembly algorithm would work, let's consider the problem of assembling just two fragments. If two fragments overlap, the "suffix" (right or 3′ end) of one fragment must overlap the "prefix" (left or 5′ end) of the other fragment so that the base positions in the overlapping region match (**Figure 8.4A**). Then, the two fragments can be merged. However, what if there is more than one way they could overlap? Consider the sequences `AATGCCTGA` and `TGACGAGTTAATGC`: These could overlap in two different ways, as shown in **Figure 8.4B**. Which is the correct one? A common initial criterion for an assembly program is simply to choose the largest overlap as the one that most likely represents a correct assembly. Assuming the sequences are not identical and neither is a substring of the other, the longest possible overlap is one less than the length of the shorter sequence. We can therefore start with this maximum length and see if we can find an overlap this long. If not, we can look for an overlap one base shorter and so on, stopping the search as soon as a matching overlap is found. Then, we know we have identified the longest possible overlap and can merge the sequences (Figure 8.4B). Algorithmic steps to accomplish this are as follows.

Algorithm

Determining Largest Overlap Algorithm

1. Start with two sequences: s1 and s2.
2. Set n = size of the smallest sequence – 1 (n will represent the largest overlap).
3. Compare `n` suffix characters from s1 with `n` prefix characters from s2. Also compare `n` suffix characters from s2 with `n` prefix characters from the s1.

4. Count matching bases in the prospective overlap region. If the number of matches in either set equals n, the largest overlap has been found: merge sequences to yield the contig sequence.
5. If the number of matches is less than n, subtract 1 from n. If n is 0, there is no overlap; otherwise, go to step 3.

Given the sample sequences provided, this algorithm would first look at eight-base overlaps (the short sequence is nine nucleotides) and then seven and six. At $n = 5$, a match would be found with AATGC in the prefix of the short sequence matching AATGC in the suffix of the long sequence (Figure 8.4B), and the two would merge to form the contig TGACGAGTTAATGCCTGA.

Of course, real sequence assembly is much more complicated: We have not considered the opposite strand or allowed for possible imperfect matches due to sequencing errors, and we have considered only two fragments, instead of the millions or even billions that can result from next-generation sequencing. The exhaustive matching of pairs of fragments will quickly become so computationally intensive as to be impractical, so heuristics must be used. One heuristic solution is a "greedy" algorithm: Given the choice of overlapping fragment A with fragment B, B with C, or C with A, the program makes the "educated guess" that the largest overlap is the best and proceeds without trying every possibility. The On Your Own Project provides a more detailed explanation of using this heuristic for sequence assembly.

Test Your Understanding

1. Suppose two sequence reads give GGGGCAGGCC and GCCCCGG. What would be the sequence of the contig produced using the algorithm just given?
2. Now suppose you would like your algorithm to account for the possibility that the sequences could come from either strand of the DNA. How would you modify the algorithm to accomplish this? Would the contig resulting from the two sequences in question 1 change as a result?
3. The algorithm presented assumes that the strings cannot be identical and that one cannot be contained completely within the other (one cannot be a substring of the other). But this is a somewhat arbitrary constraint, particularly when comparing a short sequence with a longer contig that has been built. How would you change the algorithm to allow for substrings and identical sequences?
4. Real sequencing data are "noisy:" They can contain incorrect characters due to sequencing errors (for example, the accuracy of most next-generation methods decreases as the fragment length increases) or to ambiguities leading to incorrect base-calling. How would you modify the algorithm so that a perfectly matching overlap is not required but merely one that exceeds some threshold value? How would incorporating this change affect the number of comparisons that must be made between two sequences?

CHAPTER PROJECT:

Identifying Viruses Through Metagenomic Analysis of Clinical Samples

To sequence, for example, the human genome, one might imagine extracting DNA from a sample of human cells free from contamination by bacteria or other sources of nonhuman DNA. However, what if we were to extract and sequence DNA of *any* kind that might happen to be in a soil sample, water sample, or stool sample? The resulting sequence would give us information about the genomes of all the different organisms present in that environment: We call this mixture a **metagenome**. This information could be used in a number of ways: We might use specific primers to sequence only diagnostic DNA segments, such as the genes for ribosomal RNA that are present in every organism and commonly used in phylogenetic analysis (see Chapters 6 and 7). Or, a biotech company might try to get a broad sample of protein coding genes and look for novel enzymes that might have practical applications. Or, we might use the metagenome to find evidence of microbes that live in association with humans, potentially proceeding from there to build a complete genome of a previously unknown organism. This is how several new viruses that cause diarrheal disease were actually identified, and we use some of these same techniques in this chapter's projects.

Learning Objectives

- Understand how short, random DNA sequences can be assembled to generate sequences of genes and genomes
- Appreciate the difficulty of accurate assembly and the dependence of sequencing on strong, efficient bioinformatics algorithms
- Gain experience with metagenomic uses of next-generation sequencing
- Know the various sources of inaccuracy, biological and computational, in sequence assembly and how quality data and coverage can increase accuracy
- Understand how to produce test data that simulate sequence reads and the value of these simulated data

Suggestions for Using the Project

The Web Exploration Project for this chapter allows students to deal with DNA sequence data in three distinct ways; the three parts of this project can be used independently depending on the focus desired by the instructor. The Guided Programming Project leads students to write code to generate simulated sequence data that are then used with the miniassembler in the On Your Own Project; instructors can provide either or both of these solutions in finished form for use in nonprogramming courses.

Programming courses:

- Web Exploration: See the output of Sanger sequencing data and understand base-calling, assemble a small sequence read dataset, and map metagenomic sequence data to known organisms. Parts I, II, and III can be used independently.
- Guided Programming Project: Develop a simulator to produce test data resembling the output of various sequencing platforms.
- On Your Own Project: Understand greedy algorithms for heuristic assembly of sequence data; develop a miniassembler to assemble sequencing data.

Nonprogramming courses:

- Web Exploration: See the output of Sanger sequencing data and understand base-calling, assemble a small sequence read dataset, and map metagenomic sequence data to known organisms. Parts I, II, and III can be used independently.
- Guided Programming Project: Download executable code for a sequence data simulator and use it to further experiment with the Web-based assembler.
- On Your Own Project: Understand greedy algorithms for heuristic assembly of sequence data; download executable code for a miniassembler and test with data from the Guided Programming Project and/or Web Exploration.

■ Web Exploration: Analysis of Virus Sequences in the Human Metagenome

The Web Exploration for this chapter is divided into three independent parts. In the first section, we look at sequence traces for dideoxy sequencing of a virus genome to better understand the nature of automated DNA sequence, how base-calling works, and some potential sources of error in sequence data. We then use a small sample of actual next-generation sequencing data taken from a metagenomic experiment to identify the organisms present in a stool sample based on short, random DNA sequence reads. Finally, we use an assembly program to see how sequence reads can be built into a contiguous virus genome sequence.

Part I: DNA Sequence Traces and Base-Calling

Automated dideoxy sequencing (see BioBackground) was one of the major innovations that made genome sequencing possible. However, it changed the nature of raw sequence data from bands on a gel to a computerized record of light wavelengths and intensities. These data can be output as an **electropherogram**, more commonly called a **DNA trace** (**Figure 8.5**), in which the fluorescence emitted by each dideoxynucleotide is represented by a color and the intensity represented by peak height. A researcher can examine the trace by hand to determine the sequence (e.g., a T for each red peak). However, this is extremely tedious even for a short sequence and certainly impossible for an entire genome. Thus, sequencing software also includes a **base-calling** program (**Phred** is a popular example) that interprets the color and intensity data and outputs an actual sequence of nucleotides.

Today, dideoxy sequencing is done inexpensively by many companies and universities. A researcher submitting DNA to be sequenced usually receives not only FASTA-formatted sequence files but also the sequence trace itself. Although the base-calling programs have good accuracy, there are always ambiguities: Is a broad peak one base or two? Is a weak peak an actual base or an artifact? In a small sequencing project, the reliability of the sequence can be improved by checking the accuracy of the base-calling using a trace viewer. **Chromas** is a commonly used desktop trace viewer that comes in a free "light" version. For our purposes, however, we can look at some sequence traces stored in the **NCBI Trace Archive**, a database of dideoxy sequencing projects.

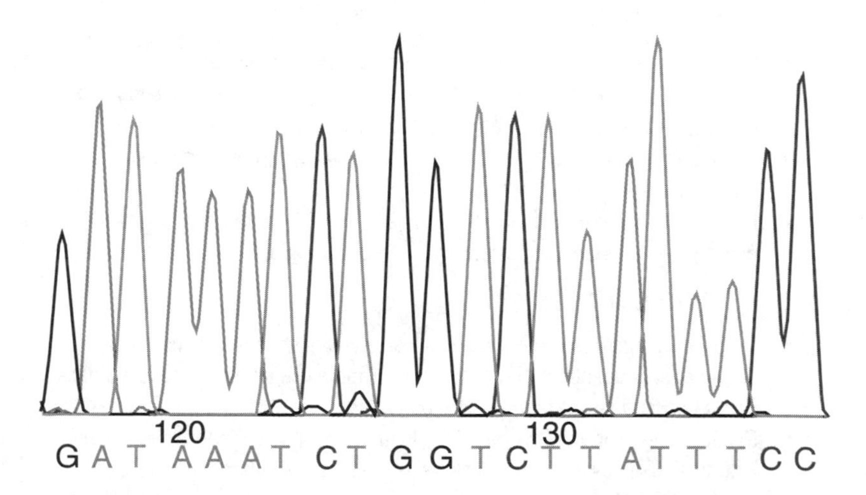

Figure 8.5 A sample electropherogram or "DNA trace" that would be generated by automated Sanger sequencing. The different color shades represent the four distinct fluorescent nucleotides detected, while the peak heights represent the intensity of detection of that particular fluorescence. At the bottom of the figure is the DNA sequence as determined by an automated base-calling program.

Navigate to the Trace Archive database. A difficulty in using this database is that it does not use the standard NCBI Entrez search interface. To locate some sequences to examine, click the tab labeled `Obtaining Data` and then the option `Registered Species` to see a list of species for which there is sequence in the database. You should find an entry for `Human Gut Metagenome`; clicking this entry creates a query in the search field above; click `Submit` to see the results. These sequences come from a metagenomic project in which DNA taken from the human gut (via a fecal sample) was sequenced to identify the microbial species present. You can see that the data consist of a great number of comparatively long reads. Although they have already been edited to remove the least reliable data from each sequence, you may be able to see some spots where bases could not be accurately determined, indicated by N.

Change the display to show the sequencing traces rather than the FASTA file. Click and drag the trace itself or click in the bar just above it to move through the sequence. It should be clear how the quality of the sequence changes along the read, from tall but indistinct initial peaks to a region where the sequence is very easy to read, to much lower peaks farther on. Notice that the base-calling software can determine bases far past where we can distinguish peaks (though a more sophisticated trace reader allows changing the scale to increase the viewable size of the peaks). Examine any Ns that occur in the sequence; can you manually call the base that the software could not call? Look for some runs of bases, such as three or more Gs or As in a row; can you see why these can be hard to call? Do you agree with the base-calling program? What other areas of the sequence appear to be difficult to determine precisely? You can also change the display to show quality, an estimate of reliability for each nucleotide, or to show information about the sequence run.

Web Exploration Questions

1. Looking at the DNA sequence traces, what conditions appear to cause the base-calling program to output N rather than designating a specific base?
2. How many nucleotides of sequence was the base-calling program able to read for the traces you examined?
3. Why does the lowest quality sequence occur at the beginning and the end of the sequence run?
4. Although each dideoxy sequencing run produces a sequence trace, in a large metagenomic or genome sequencing project, it would not be practical to examine each trace and manually assign difficult bases. How can the sequences returned by an automated base-caller be used reliably in such a project?

Part II: Metagenomic Analysis of the Human Virome by Next-Generation Sequencing

To sequence a genome, many-fold coverage of every nucleotide is necessary for high accuracy. However, there are many uses of sequencing in which *individual* reads provide valuable information. Notably, for a metagenomic project intended to sample all the organisms in an environment, individual reads can be compared with sequence databases to identify known organisms or distinguish novel ones. Data from many such projects can be found in public databases such as NCBI's **Sequence Read Archive** (SRA) database. One example is a project led by Gary D. Wu at the University of Pennsylvania in which DNA from fecal samples was sequenced with the intent of examining the microbial population (**microbiome**) of the human gut under various dietary conditions. Although the original intent of the study was to relate the microbiome to Crohn's disease, these same data were also mined to examine the **virome**, or viral population, of the gut. Here, we examine a small sample of data from this research project to see how metagenomic data can be analyzed.

Although there are many freely available programs, most software for analysis of next-generation sequencing data must be downloaded and run on a desktop computer, because of the complexity of working with millions or billions of short sequences (see More to Explore, for some programs you might be interested in using). Indeed, our reason for examining only a portion of the available metagenomic data is to keep processing time reasonable for a course project. One notable Web-based tool that can be used for metagenomic analysis (as well as many other kinds of sequence data analysis) is **Galaxy** (see References and Supplemental Reading), a flexible interface that can be used to run many different kinds of bioinformatics programs.

On the Galaxy main page (try `Galaxy sequence analysis` if you are using a search engine to find it), you will see three panes: a tool list on the left, a pane with parameters for the current tool in the center (initially, some available tutorials are displayed here), and a history pane on the right showing pending and completed analyses. Let's start by downloading data from the virome study. In Galaxy, this can be done by accessing EBI's interface to the SRA database: In the tool pane, choose `Get Data` and then `EBI SRA` to open a search interface in the second pane. The accession number for the data from the virome project is `SRS072363`; enter this in the search box and submit the search. You should see the SRA database entry for a sample from one specific subject in the study; at the bottom of the page, you should see listings for two specific files of Illumina sequencing data in FASTQ format. The far-right column in the file listing table is headed `Galaxy`; click on file 1 in this column to import these data into Galaxy. You will see a message in the center pane and then a task added to the history pane; the task will turn yellow when the server starts on it and green when it is complete. Once complete, you can click the task to see a "preview" of the data it contains within the history pane or the eye icon to see the data file itself in the center pane. Notice in the description of this file the large number of sequences it contains.

Next-generation sequencing techniques automate sequencing and base-calling even more fully than in automated dideoxy sequencing. Although raw sequence data can be viewed (454 sequencing, for example, generates a flowgram similar in principle to an electropherogram), the enormous number of reads and the automation drastically decrease the value of any manual examination of the data. Instead, it is common to summarize both the called bases and data on the quality of the read within a single file in FASTQ format (**Figure 8.6**). Like a FASTA file, a FASTQ sequence file starts with a comment line, in this case beginning with @, to identify the sequence. The next line of the file is the sequence itself. There is then a line starting with + where an additional comment may optionally be added. The last line gives a quality score for each nucleotide, encoded as an ASCII character. The quality score range depends on the sequencing software; older Illumina software used a quality score from –5 to 62, whereas Sanger format uses a quality score based on the Phred algorithm, from 0 to 93.

Converting the imported data to Sanger FASTQ format is needed for many of the Galaxy tools; to accomplish this, find `FASTQ Groomer` under `NGS` (next-generation sequencing)

```
@EAS100R:3:90:836:2213#0
TCGATGATTTGGGGTTCAAATCCATTTGTTCAA
+
%%%%)!''*((((***+))%)**55CCF>>>>>
```

Figure 8.6 Example of next-generation sequencing data in FASTQ format. The first line is a comment marked by @ and identifies the sequence (including the instrument, run, specific cell, etc.). The second line is the sequence itself. The third line is an additional comment line marked by +, and the fourth line is the quality score for each base encoded by calculating a Phred quality score, adding 33 and using the ASCII character corresponding to that number (so, % = ASCII 37 = Phred 4;C = ASCII 67 = Phred 34).

QC and manipulation in the tools pane. Notice that this tool will work on an item from your history, in this case the imported sequence data. Be sure the input data type is Sanger (your data are from Illumina 1.8, which uses the same FASTQ quality score system as FASTQ files for Sanger sequencing) and execute the task. You can expect this task to take a fairly long time to process (maybe hours if the load on the server is high). However, you can put additional tasks into the queue while you are waiting and they will be completed in order once this step is done.

If you look at the actual sequences in the imported or groomed data file, you will notice that many are runs that consist only of Ns, indicating that no useful sequence data were obtained. Others may be very short, and others may have very low-quality scores. Let's limit our analysis for this project to runs that yielded a reasonable amount of good-quality sequence. To do so, look in the same category of tools for Filter FASTQ reads by quality score and length. Use the FASTQ Groomer output and set minimum length to 50 nucleotides and the quality cut-off value to 20, which represents a 99% probability that the base has been called correctly. Run the analysis and note how many sequences were discarded.

Now that we have used the quality data to develop a subset of sequences we want to pursue further, we can convert the FASTQ data to a simple FASTA file of sequences with an identifying line. The complex identifiers in the FASTQ file are not really needed; let's give each of our sequences a simple identifier like GutVirome-1, GutVirome-2, and so on. Galaxy has tools for manipulating complex genome files that perform these actions easily. First, convert the FASTQ data to a table, using the FASTQ to Tabular converter tool. The output is in columns: sequence identifier, the sequence itself, and the quality data. Now add a column to the table, using the Add column tool found under Text manipulation. In the Add this value field, type GutVirome and then change Iterate to Yes; this adds a column of data containing sequentially numbered labels as suggested earlier. Finally, generate the FASTA file using Tabular-to-FASTA (under FASTA manipulation), with the new fourth column (c4) as the title column and the second column (c2) as the sequence column. The resulting FASTA data should look very familiar to you.

Using the FASTA file, we can now do the actual metagenomic analysis. We want to compare each remaining sequence read with the entire database of known sequences and identify the source of the sequence: human DNA, known or novel bacterial species, known or novel virus, and so on. Galaxy includes MegaBLAST as a tool that can perform this search; essentially, it will carry out a BLAST search for every sequence in your FASTA file, using parameters optimized to allow for small differences due to sequencing errors. Choose Megablast under NGS: Mapping, set the FASTA data as input, nt as the target database, a word size of 16, and a minimum percent identity threshold of 80%. Note at the bottom of this pane how the MegaBLAST output will look and then execute the database search. This process may also take some time; when it is complete, you should see that the number of lines has grown drastically, because any of the sequence reads can match multiple database sequences.

How can one deal with such a large set of results—to say nothing of the enormous amount of data we would have obtained had we started with all the sequence data from all the study subjects? One way to summarize the results is by retrieving from the database the taxonomic information (species, genus, family, order, etc.) for each matched sequence. Results can then be grouped on this basis to reveal whether the sequenced DNAs belong to viruses, bacteria, human cells, or other organisms—even those that do not match a known species can be classified into larger groups. Under Metagenomic analyses, choose Fetch taxonomic representation. Set the name column to c1 (the identifier you gave the sequences) and the GIs column (GenBank gene identifier) to c2; these accession numbers will be used to retrieve the taxonomic information. Run the analysis.

There are now a number of possible ways to examine the data further. To look at all the virus sequences in the dataset, for example, filter the data to show only the lines in which the Superkingdom column contains viruses. Similarly, well-chosen filters can allow you to look at bacteria or fungi or other organisms (you may need to look at some sample data to decide on filter terms). Another way to look at the data is to generate a phylogenetic tree of the organisms identified by the alignments: first run `Find lowest diagnostic rank` on the taxonomic data and then `Draw phylogeny` to get a PDF file showing the tree.

Web Exploration Questions

5. How does the number of viral sequences found in the sequence runs you analyzed compare with the number of bacterial sequences? Are there fungal sequences? Protists? Do these relative numbers make sense in terms of the human gut environment and the roles of these organisms?
6. Some of the species represented among the gut sequences might seem surprising. What seemingly unlikely species were identified, and what are some possible reasons for these results?
7. What are the most commonly found viral sequences? Why is this the case?
8. How could viruses that are normal residents of the gut community be distinguished from those that might be pathogens?
9. How could novel viruses be distinguished from related viruses that have already been characterized?

Part III: Assembling the Sequence of a Novel Virus

Whereas metagenomic analysis can be conducted using individual short sequence reads resulting from next-generation sequencing of clinical samples, determining the genome sequence of any organism requires assembly of sequence reads into contigs with sufficient depth of coverage to detect and correct errors. The depth of coverage required is lower for sequencing methods producing long reads and much higher for techniques producing very short reads. Once a genome of interest has been identified from a metagenomic sample, it may be possible to identify enough reads from that genome to begin assembling its sequence. With the identification of portions of the genome, specific primers can be designed based on the now known sequence and used on the same metagenomic DNA samples to fill in the gaps in the genome. This process has been used to identify a number of novel pathogenic human viruses in recent years. For example, klassevirus, a new human virus in the picornavirus family, was identified in this manner from stool samples taken from children with diarrhea who tested negative for known diarrheal viruses (see References and Supplemental Reading). Viruses such as these may turn out to be important causes of human disease that have escaped detection until now.

As with metagenomic analysis, most assembly programs that can handle genomic sequence data, especially next-generation sequencing data, are intended to run on powerful desktop machines (see More to Explore for some desktop programs you could use for assembly). For this project, we use **EGassembler**, a Web implementation of the CAP3 (*c*ontig *a*ssembly *p*rogram) assembler (see References and Supplemental Reading). From the *Understanding Bioinformatics* website, you can download **reads.txt**, a file that contains 2,500 simulated 454 sequencing reads in FASTA format, representing the genome of an unknown virus identified in metagenomic samples. These sequences range in length from 100 to 500 bases and contain between 1 and 10 random substitutions or single-nucleotide deletions each, representing the errors inherent in sequencing data.

Navigate to the EGassembler page and either upload the sequence file or copy and paste the sequences into the input field. Notice that in addition to the CAP3 assembler itself, EGassembler includes software to scan for low-quality sequence (e.g., sequences containing many Ns) and remove sequences matching databases of organelle and cloning vector DNA as well as highly repetitive sequences. For our purposes, turn off the options other than sequence cleaning and the assembly step itself and then run the program. You should immediately see the results of sequence cleaning; you can view a .cln file to identify reads that were discarded and then examine these reads in the original sequence file.

In a few minutes, the link to the results should become functional. From the results page, you can view (1) the contig or contigs that resulted from the assembly of your sequence reads; (2) any "singletons," which are reads that could not be assembled into the contigs or that were not used in creating the contig; and (3) an alignment of the individual sequence reads showing how they led to the generation of the consensus contig sequence.

Web Exploration Questions

10. How many sequence reads were rejected in the sequence cleaning process? Can you determine why they were rejected?
11. Use BLAST to compare your contig sequence with known sequences in GenBank. The assembled sequence should match one known sequence with a high degree of similarity. What have we sequenced? How long is its genome?
12. Because next-generation sequencing produces random short reads, there is no guarantee that even 2,500 reads would be sufficient to completely sequence a particular genome. Did the sequence reads you assembled cover the entire genome or do gaps remain? To fill any gaps, would it make sense to simply run more sequencing reactions, or are there other approaches that should be considered?
13. Looking at the contig alignment file in the EGassembler results, you should be able to see hundreds if not thousands of small sequencing errors among the sequence reads. Was the assembler able to generate a correct contig sequence (as compared with the known sequence in the database) despite these errors? Explain how the sequence errors were accurately corrected. Were all errors caught, or did some remain in the final contig sequence?
14. You used the default parameters for the CAP3 assembler in your EGassembler run. In a real sequencing project, however, you might want to change variables such as the overlap percent identity cut-off (the minimum percentage of nucleotides that must be identical in the overlapping region of two fragments). By default, CAP3 is quite tolerant of sequencing errors (and in fact automatically compensates for some of the common problems of high-throughput sequencing, such as low-quality sequence at the beginning and end of fragments). To see how these parameters affect the assembly, try setting the overlap percent identity cut-off to 100%. What happens to your contig? Does the quality of your alignment change? (You can choose Step-by-Step Assembly at the top of the page to access more parameters.)

More to Explore: Sequencing Tools

DNA sequencing has become such an important part of molecular biology and bioinformatics that a large number of software tools for analyzing sequencing information are available, both proprietary and otherwise. As mentioned previously, the sizes of data files containing millions or hundreds of millions of sequencing reads and the processing power required to analyze them reduce the desirability of Web-based interfaces, so many of the freely available programs must be downloaded and installed on one's own computer. **Table 8.1** lists a number of sequence analysis programs that you might be interested in working with in the future.

Guided Programming Project: Sequencing and Assembly

The goal of this guided project is to better understand sequencing data and how they are handled computationally in two ways: by developing a program to generate fragments of a known

Table 8.1 Some sequence assembly and analysis software.

Program	Description
Sequence assembly	
Velvet	Assembler optimized for very short sequence reads
Oases	Extension of Velvet for transcriptome assembly
IDBA-UD	Assembler optimized for uneven coverage
SSAKE	Short-read assembler based on a greedy algorithm
CABOG	Celera software for small and large genome assembly
SOAPdenovo	Assembler capable of human genome-size assembly
Mapping of sequence reads to reference genomes (metagenomics)	
Bowtie	Fast alignment of sequence reads to human genome
BWA	Aligns sequence data with a reference sequence
MAQ	Maps sequence reads and identifies variants
SOAPaligner	Maps short oligonucleotides onto reference sequences

sequence that effectively simulate actual sequencing data and by using a simple assembly algorithm to assemble pairs of error-free sequence reads. In the On Your Own Project, you will carry this further, developing a miniassembly program capable of a more complex assembly.

Simulating Sequencing Data

The accuracy of any sequence assembly or metagenomic read-mapping program must be tested, and it is often convenient to have a set of test data that closely matches real sequencing data but has a known solution. Instead of using a contig assembled from actual sequencing data (which could be subject to assembly errors), a sequence simulation algorithm is commonly used to generate test fragments of a known DNA sequence that are designed to mimic the results of a particular sequencing platform (see References and Supplemental Reading). These simulated sequence reads should be random segments of the known sequence (representing random "shotgun" sequence data) whose size is appropriate for the sequencing technology being simulated (**Table 8.2**); we can ask the user to supply a desired minimum and maximum fragment length. Simulated sequencing errors and variable sequence quality can be introduced to increase the realism of the simulation.

In a real sequencing project, fragments of the DNA to be sequenced are produced by random processes. Thus, our program should randomly choose a substring of the input DNA string that falls within the specified size range. However, we need to make sure we

Table 8.2 Read lengths for major sequencing technologies.

Sequencing Platform	Typical Read Length	No. of Reads per Run
Sanger	500–900 bp	1–96
454	200–300 bp	400,000
Solexa	36 bp	3.4 million
Illumina	100 bp	3 billion
SOLiD	35 bp	1.7 billion

generate enough overlapping fragments to cover the whole genome. The original shotgun sequencing genome projects tried to achieve about eightfold coverage of the entire original sequence: that is, each base position in the original sequence should appear in at least eight fragments. Next-generation sequencing methods, with their shorter reads, typically work with 30-fold coverage, while an application such as identifying rare mutations with a high degree of confidence may require 1,000-fold coverage. We should allow the user to input a desired coverage value, simulating the ability of a user to "tune" the coverage in a sequencing project. The pseudocode that follows describes an algorithm to simulate sequence reads by generating fragments of an input sequence to achieve a desired coverage level.

Algorithm

Sequence Read Simulator: Generating Fragments for Sequence Assembly

Goal: To generate random fragments from an input sequence.

Input: A single nucleotide sequence, user-defined minimum fragment size, maximum fragment size and coverage fold

Output: A set of fragments

Note: substring is assumed exclusive, thus substring(1,4) includes positions 1, 2, 3 only

```
// Initialization
Input the sequence: s1
Input the minimum and maximum fragment size: fMin, fMax
Input the coverage fold expected: fold
for each i from 0 to length of s1 - 1
    coverage[i] = 0    // holds coverage count of nucleotides

// STEP 1: Generate a set of fragments for the input sequence

numFrags = 0
do
    randLength = random number between fMin and fMax, inclusive
    randStart = random number between 0 and (length of s1-randLength)
    frags[numFrags] = s1.substring(randStart, randStart+randLength)
    numFrags++
    // update coverage
    for each i from randStart to (randStart+randLength-1)
         coverage[i]++
while (!coverageMet(coverage, fold))
output frags

// function to determine if coverage met
function coverageMet(coverage, fold)
    i = 0
    met = true // assume coverage met
    while (i < coverage length and met==true)
         if coverage[i] < fold
            met = false
         i++
    return met
```

Assembling Pairs of Sequence Reads

The goal of an assembly program is to produce one contig from a set of sequence fragments; here, we implement one small but important step in the assembly process: assembling each *pair* of fragments in a set of sequence reads into a contig using the overlap algorithm (see Understanding the Algorithm). The program will need to use a nested loop to iterate through the set of sequence reads, attempting to overlap each fragment with every other fragment and looking for the largest overlap. The output for each pair of fragments should include the original fragments, the resulting contig, and the number of characters in the overlapping region. The following pseudocode shows a solution for finding the largest overlap.

Algorithm

Fragment Overlap Generator: Finding Overlaps Between Pairs of Fragments

Goal: To determine the largest overlap between pairs of fragments

Input: Set of fragments

Output: The fragments, the resulting contig, and length of the overlapping region for each pair of fragments in the input file

```
// Initialization
Input the fragments and store in an array: frags
numFrags = number of fragments read

// STEP 1: Determine overlap for each pair of fragments

for each i from 0 to numFrags-1
    for each j from i+1 to numFrags
         f1Len = length of frags[i]
         f2Len = length of frags[j]
         minLen = minimum of f1Len and f2Len
         overlap = 0
         frag1 = frags[i]
         frag2 = frags[j]
         k = minLen - 1
         while k >= 1 and overlap == 0
             // compare suffix of frag1 to prefix of frag2
             if frag1.substring(f1Len-k, f1Len) == frag2.substring
                 (0, k)
                // create contig
                contig = frag1.substring(0, f1Len-k) + frag2
                overlap = k
                output frag1, frag2, contig, overlap
             else if frag2.substring(f2Len-k, f2Len)
               == frag1.substring(0, k)
                contig = frag2.substring(0, f2Len-k) + frag1
                overlap = k
                output frag1, frag2, contig, overlap
             k--
```

Putting Your Skills Into Practice

1. Implement the Sequence Read Simulator and Fragment Overlap Generator programs described in the pseudocode, using whatever language is used in your course. Test your programs with a short sequence to validate them, but note that it will be difficult to obtain adequate coverage if your sequence is too short.
2. Experiment with different short sequence lengths, coverage values, and minimum/maximum fragment sizes. How many "sequence reads" did it take to get the level of coverage you specified? How does that change if you change the fragment size? Was your overlapping assembly program able to match the correct fragments to generate a set of contigs found in the original sequence? Then, try running your program on a larger sequence, such as the klassevirus genome sequence you can download from the *Exploring Bioinformatics* website.
3. Modify your sequence read simulator to output the coverage values for your sequence. Where in the sequence do the highest coverage values occur? Can you explain this pattern? Does the pattern change if you change the fragment length or coverage parameters or the size of the input sequence? Does this pattern accurately simulate what would happen in a real sequencing experiment, or is it merely a computational artifact? If you wanted more even coverage, how could you modify your program?
4. You may have noticed that it is possible for your sequence read simulator to generate a fragment that is a substring of (entirely contained within) another fragment. Fragments that are substrings of other fragments are considered "singletons" and are often eliminated from the assembly process, because they do not add any additional information and can even decrease the efficiency of the assembly process. Modify either program to remove all fragments that are entirely substrings of other fragments so they are not used when finding overlaps.
5. The real sequencing process is prone to misreads; these occur with high frequency at the beginning and end of a sequence read, where sequencing is difficult for technical reasons, but can be found randomly throughout the sequences when the data sent to the base-calling software is ambiguous. Sometimes the sequencing reactions fail and a particular fragment is unreadable (usually represented by all Ns). Make your sequence read simulator more realistic by modifying its code to introduce random changes (inserted, deleted, or changed base) or Ns at a low rate. (For a more challenging exercise, make the likelihood of such changes higher at the ends of the sequence.) Then, modify your overlap generator so it looks for matches that exceed some configurable threshold but does not require an exact match (for example, if the matching threshold is 75%, then at least 75% of the characters in the overlapping region must match). Can you still get accurate assembly?
6. Our simulator program only considered a single input string, representing one DNA strand. Real sequencing data could come from either strand, and in fact pairs of sequences from opposite ends of a DNA fragment, one from each strand, are often generated. The assembler cannot know in advance which strand a fragment came from, so it would have to try each fragment *and* its inverse complement to determine which assembled best. Modify your simulator so it chooses whether to output a selected fragment or its reverse complement (use your code from Chapter 2) and your fragment generator so it will try both strands.
7. Repeated sequences pose a major problem for sequence assembly programs (indeed, some repeat-intensive regions of the human genome, such as the areas around centromeres, have yet to be sequenced). Test your overlap generator program with the following sequences (assume their positions relative to the original sequence are as shown). Considering the results, discuss the difficulty the repeat problem presents in determining a best overlap. Keep in mind that the length of a repeated sequence can often be much longer than the possible size of a fragment read.

```
Original sequence:    GGATAGATATATATATATATCGACTTC
Test fragment set 1:  GGATAGATATAT
                                      ATATGCACTTC
Test fragment set 2:  GGATAGATATATATAT
                                  ATATATATAT
```

■ **On Your Own Project**: A Mini-Assembly Program

The Guided Programming Project introduced you to the problem of finding overlaps between pairs of fragments, and the Putting Your Skills Into Practice exercises should have helped you recognize the additional complexity introduced by sequencing errors and repeated sequences. In this project, we develop a miniassembly program capable of assembling multiple overlapping fragments into a contig using a "greedy" algorithm based on the traveling salesperson problem. We also look at the role of coverage in correcting errors and in determining which overlap is best when multiple options exist. Instructors of nonprogramming courses can download a **completed miniassembly program** from the *Exploring Bioinformatics* website that students can use in completing the exercises under Programming the Solution later in the chapter.

Understanding the Problem

Assembling a contig requires identifying overlaps among sequence reads and then determining how best to piece together the overlapping fragments. However, a single fragment may overlap with many other fragments, making it difficult to choose which pair to merge. **Table 8.3** shows a simple example: For each fragment in a simple hypothetical sequencing project, the fragments that can overlap its suffix are shown along with (in parentheses) the length of each overlap. This output could be produced by a simple modification of the program you wrote for the Guided Programming Project. The suffix of fragment 1, for example, overlaps the prefixes of four other fragments: fragment 2 by three characters (TTG) and fragments 3, 4, and 7 by one character each (G in each case).

Once the overlaps are identified, how do we merge the fragments? One approach is to simply start with the first fragment, merge with a matching one (fragment 2 in Table 8.3 would work), choose another fragment that matches the growing contig (fragment 3 in this case), and so on until all fragments are chosen. In the example, fragments 7, 5, 6, 1, 2, 3, and 4, merged in that order, would form a contig. But how did we know where to start? Would other choices have given a different path or led us to a dead end?

To develop an algorithmic solution to this problem, let's look at the data in the form of a **graph**, which in computer science is a data structure showing relationships among elements: **Figure 8.7** shows a graph of the data in Table 8.3. Each fragment is represented by a numbered node, with directional arrows representing overlaps between fragments. The suffix of the fragment at the tail of the arrow overlaps the prefix of the fragment at its head, and the arrow is labeled with the length of the overlap. The contig is then generated by finding a path in the graph that passes through each node once.

Table 8.3 Overlaps for a hypothetical set of sequence reads.

Fragments	Overlaps (Length)
1. TACCTTG	2 (3), 3 (1), 4 (1), 7 (1)
2. TTGAT	1 (1), 3 (3)
3. GATATGG	4 (2), 7 (1)
4. GGAG	3 (1), 7 (1)
5. CTCTA	1 (2), 6 (3)
6. CTAGT	1 (1), 2 (1)
7. GCTCT	1 (1), 2 (1), 5 (4), 6 (2)

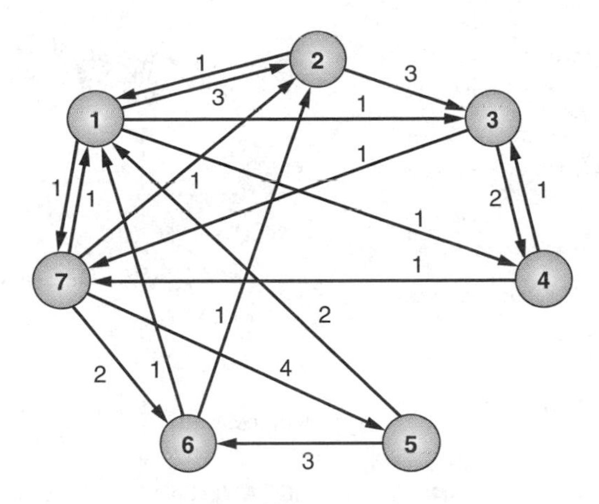

Figure 8.7 Graph representing overlaps between fragments as paths between nodes.

The assembly problem is closely related to a very famous problem called the **Traveling Salesperson Problem** (TSP), usually described as finding the shortest flight path between a set of cities so that each city is visited only once and the path begins and ends in the same city. The possible flight paths and distances are fixed, so the problem can be represented as a graph where cities are nodes and arrows are flight paths, much like Figure 8.7. The good news is that there is a solution to the TSP, but the bad news is that it can take an enormous amount of computational time to find it: if n is the number of cities (or sequences), the number of possible paths is $n!$. This is a truly huge number if we consider the 3 billion–base pair human genome covered 30-fold by 100-base pair sequencing reads! Worse, we do not know which fragment comes first and (if it is a new sequence we are assembling) we have no way to verify the correctness of the solution, unlike the traveling salesperson, who at least knows the starting and ending city and that the goal is the shortest path. Fortunately, as you saw previously in Understanding the Algorithm, using heuristics will help.

A **greedy** algorithm is one way to solve the TSP in a reasonable time. This is a heuristic that when faced with a decision "greedily" chooses the option that appears to best serve its goals. Because the goal of the TSP is the shortest path, a greedy algorithm would always choose the arrow with the shortest distance at any decision point. Unfortunately, this approach does not guarantee a solution: It is possible to arrive at a node with no arrows leading away from it.

For the assembly problem, because we are unable to determine in advance which overlap is the correct one (i.e., the one that leads to assembling the original sequence) at any node, we could greedily choose the arrow representing the longest overlap. In this problem, we do not have a predetermined starting node, but we can be greedy here as well and start with the largest overlap among all the pairs of fragments. But does this make sense biologically? As the length of the overlap increases, the probability that it is genuine and not a chance match increases: A fragment that ends in A will overlap *any* fragment that starts with A (one of four just based on chance), whereas a fragment ending in ACTG will find a chance match just one time in 256, and the probability of a chance eight-base overlap is only one in 65,536. Therefore, by always greedily choosing the largest overlap, we can reasonably expect to end up with the shortest common **superstring**—a string that includes all the fragments in the smallest total number of characters, which would be our contig.

Remember, however, that the greedy algorithm does not guarantee finding the original sequence; sequencing errors and repeated sequences are problematic because they make it more difficult to correctly determine the overlaps in the first place. Because we do not

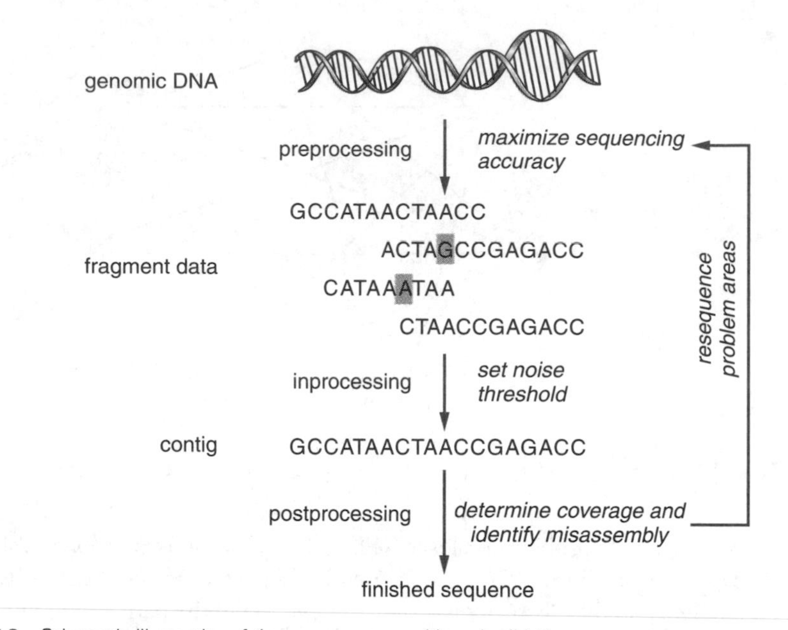

Figure 8.8 Schematic illustration of the sequence assembly and validation process.

know the original sequence, we cannot be sure our program has found the correct solution (though successful testing with good simulated data will increase our confidence), and error correction becomes very important. There are three general ways to correct errors (**Figure 8.8**). **Preprocessing** error correction means fixing problems in the data *before* processing that data; this might be done by improving sequencing techniques, increasing read lengths to reduce the impact of repeats, hand-calling bases in questionable areas (in a small enough project), or analyzing the output data and eliminating reads or regions with poor quality scores. **Inprocessing** modifies an algorithm to better handle errors in the data, such as setting a threshold match value in the assembly algorithm to deal with misreads, which we can easily implement (see Putting Your Skills Into Practice, exercise 5). An inprocessing solution for repeated regions is to use **mate-pair reads**, pairs of sequence reads from the two ends of a DNA fragment. If these reads aren't found on opposite strands within a short distance of each other in the final assembly, the assembly is incorrect. **Postprocessing** validates the output *after* the algorithm has run; taking this approach, we will use coverage statistics to identify possible areas of misassembly resulting from repeated regions.

Given the sequence fragments and resulting contig sequence in **Figure 8.9A**, even with a misread (GTCTA, rather than GTCTC), a consensus sequence can be successfully built. However, even though all the fragments overlapped, the contig does not match the original sequence: fragment 3 was misassembled due to the repeated sequence TCGTAG. How could we recognize this without knowing the original sequence? Examining coverage is one method, because coverage should be relatively constant across the sequence. Repeated sequences can match more fragments than they should, producing a high-coverage peak (**Figure 8.9B**) that could be flagged as a possible location of misassembly. Determining coverage for each base position is simply a matter of counting the number of fragments that overlap that position; in Figure 8.9A, coverage values would be (1,1,2,3,3,2,3,3,2,2,1,2,1,1,2, 2,1,1) for an average of 1.8. The repeat region shows noticeably higher coverage even in this small sample. Regions with low coverage values, on the other hand, should be considered unreliable simply because they may not have been sequenced enough to correct misreads or

(A) original sequence GTCTCGTAGGAGTCGTTCGTAG

fragments GTCTA CTCGT TAGGA AGTCG CGTT TCGTAG

assembly GTCTA TAGGA
CTCGT AGTCG
TCGTAG CGTT

contig GTCTCGTAGGAGTCGTT

(B)

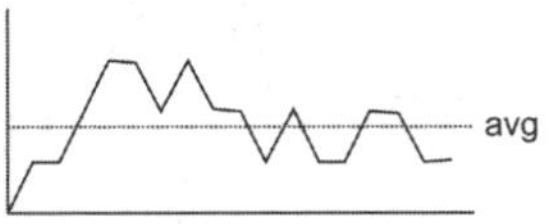

avg

coverage 11233232212112211

Figure 8.9 Sample sequencing project: (A) Fragments are generated, sequenced, and assembled, but a repeated sequence results in misassembly; (B) Peak of coverage shows possible location of misassembly.

other problems. Statistical calculations can be done to establish minimum and maximum coverage values for high reliability.

Solving the Problem

At this point, you should be able to see how an algorithm would be built to tackle the difficult problem of sequence assembly using our TSP-based "greedy" approach. First, use what you learned from the Guided Programming Project to determine the overlaps, implementing a threshold percentage for matching the overlaps (see Putting Your Skills Into Practice, exercise 4) and think carefully about how to organize and store the overlap information so that it is easy to retrieve as you begin merging sequences.

Figure 8.10 steps through the process of merging the test sequences in Table 8.3 using their overlap data and a greedy algorithm. The arrow linking nodes 7 and 5 has the largest overlap (**Figure 8.10A**, blue arrow), so using the greedy algorithm, our first merge is the suffix of 7 with the prefix of 5. Remember that our final path must visit each node only once; node 7 now leads to node 5 and therefore cannot lead to any other node, so we can eliminate any other arrows leading away from node 7 (Figure 8.10A, dashed arrows). In sequence terms, we have overlapped GCTCT with CTCTA to give the contig GCTCTA, so we cannot overlap the 3′ end of GCTCT with any other fragment. Similarly, node 5 has now been visited, so we can eliminate any other arrows leading into node 5 (there are none in this case).

Now, we choose the next-longest available overlap. The overlap is 3 nt for 1→2, 2→3, and 5→6, so we could choose any of these; by simply taking the first one, we would choose 1→2 (**Figure 8.10B**, blue arrow). Again, other paths leading away from 1 or into 2 are eliminated (dashed arrows). Proceeding in this fashion, we would choose the paths from 2→3 (**Figure 8.10C**) and then 5→6 (**Figure 8.10D**), eliminating potential choices as we proceed. Now the longest overlap remaining is between 3 and 4 (**Figure 8.10E**); once this is chosen, the only remaining paths are 4→7 and 6→1. Again, arbitrarily choosing the first one gives the result in **Figure 8.10F**: a complete path through all seven nodes in the order 1, 2, 3, 4, 7, 5, 6. Once you have the final path, you can easily obtain the final contig by overlapping the fragments in order of the path. Therefore, in our example, you would overlap fragment 1 with fragment 2. The resulting contig would then overlap with fragment 3. That contig would overlap with fragment 4 and so on. This corresponds to the assembly

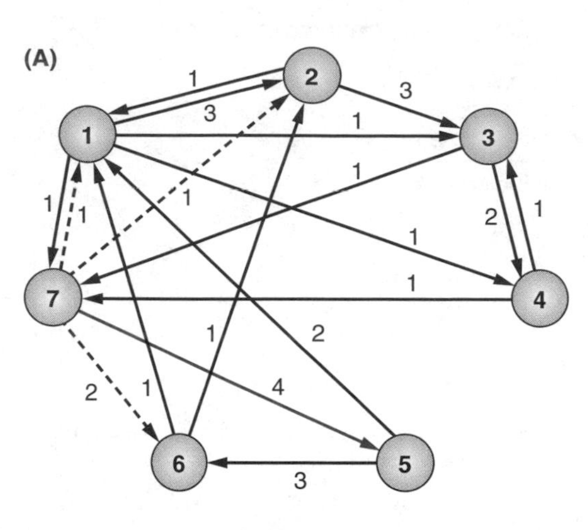

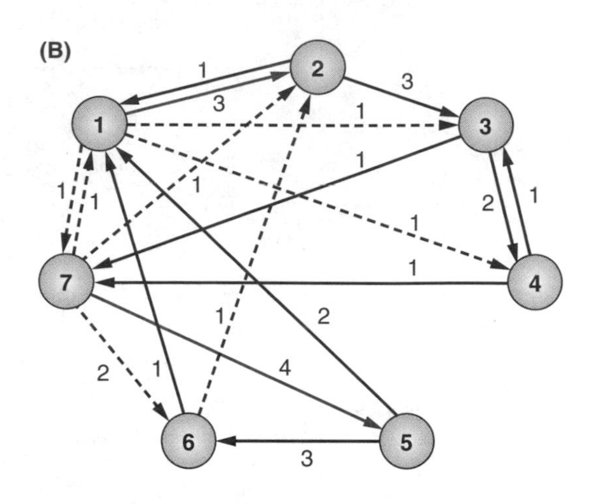

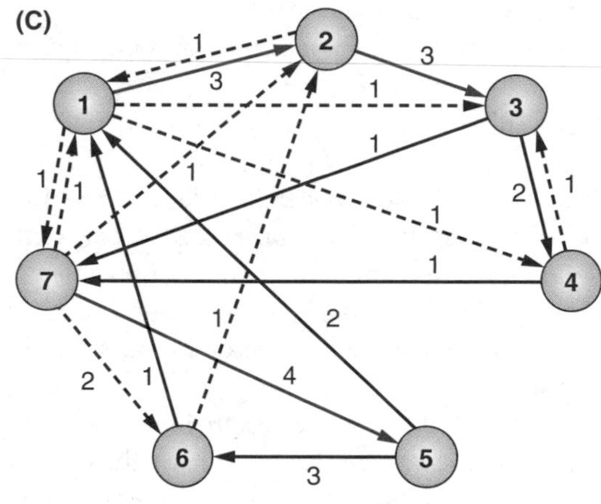

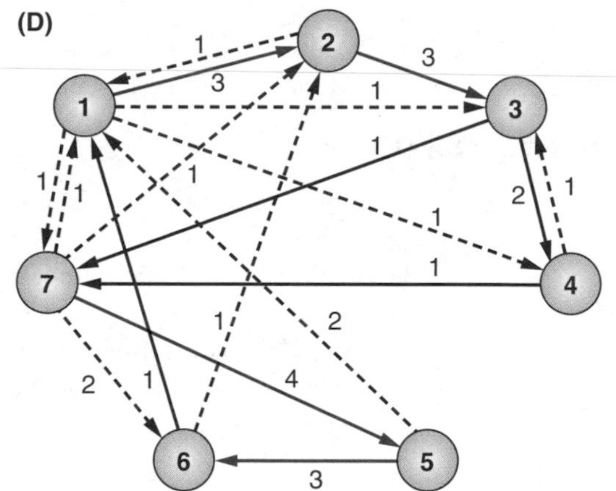

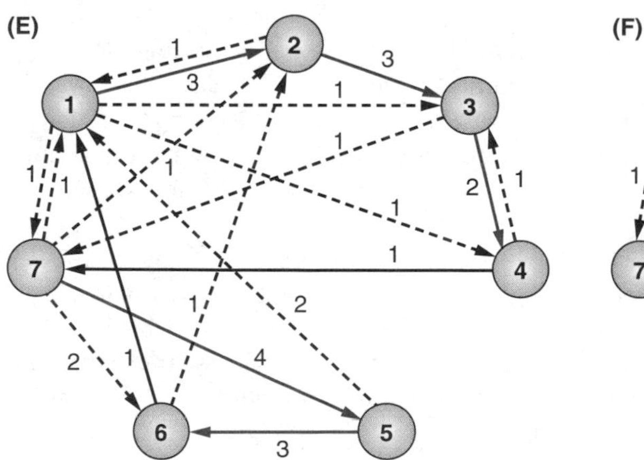

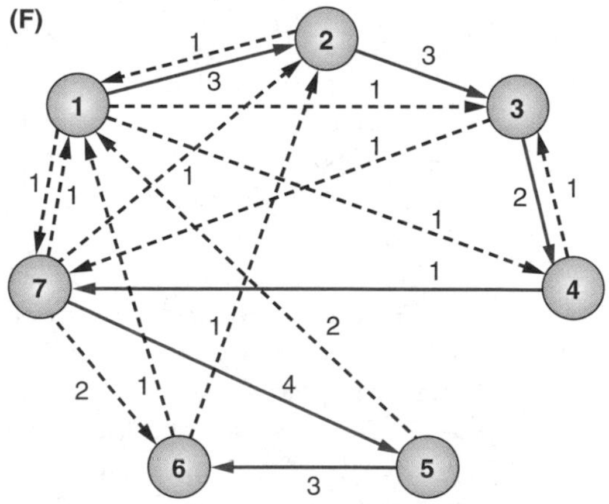

Figure 8.10 Steps in finding a path to a sequence alignment. From left to right and top to bottom, each graph shows a link (heavy arrow) between two sequence fragments that would be chosen using a "greedy" algorithm. Dashed arrows show paths that can be discarded once a choice is made.

of all the fragments into the contig TACCTTGATATGGAGCTCTAGT. Note, however, that the algorithm gives the path but does not specify *how* the developing contig overlaps with the next fragment, which will not necessarily be by the same number of nucleotides as the original fragments.

Programming the Solution

Now you should have enough information to extend your overlap-finding program to become a full-fledged mini-assembler, using the "greedy" algorithm as described here. As each fragment is chosen, keep track of where it fits in the growing contig so you can calculate coverage; use this information to flag any unreliable sequences or likely repeats once you have built your contig. Check manually to see if your algorithm can correctly assemble a set of fragments with good overlaps based on a short test sequence before you start implementing your solution in your language of choice.

Test your program on the short sequences in Table 8.3; do you get the contig described earlier? Then use your **sequencing simulator program** (instructors can download this from the *Exploring Bioinformatics* website for nonprogramming courses) to generate fragments for some longer sequences (try 200 nucleotides or so at a time from the **klassevirus sequence**, for example) with more coverage. Does your program correctly assemble the fragments? How much coverage is necessary for it to do so reliably? How does the average fragment length affect its accuracy? Does the program ever fail to find a solution (and did you think to have it let the user know of this failure)?

To see how your program handles repeats, introduce some into your test sequences—for example, put 10 consecutive repeats of GCATC in the middle of a 100- or 200-nucleotide sequence, generate fragments that are 10 or 20 nucleotides long, and then see how your program handles the assembly and whether your coverage values correctly identify problem areas. Then try a more realistic sequence, such as the complete klassevirus genome. Does your program work equally well here, or does it encounter problems? What do your coverage values tell you about the reliability of various regions of your contig sequence?

Connections: The Future of Genome Sequencing

The rate at which genes, genomes, and metagenomes can be sequenced continues to expand rapidly, whereas the cost continues to decline. As a result, sequencing is being used in ways we never previously imagined. Not only will the sequencing of individual human genomes soon become practical (the so-called $1,000 genome is nearly within the reach of several companies as of this writing), but we are sequencing the genomes of the entire human microbiome and applying sequencing technology to the identification of targets for transcription factors, mutations resulting in complex genetic disorders, and genetic diversity of endangered animals. As sequencing moves from the research lab to the hospital lab, we will see it used for genetic screening, cancer diagnosis, and preimplantation diagnosis of embryos. Individualized medicine will likely become a reality, with drugs tailored to the individual genetic makeup of a particular patient. Ecologists, evolutionary biologists, pathologists, forensic scientists, and many others will also benefit.

Of course, just obtaining the sequence is not the end of the story. To make the sequence useful, improved bioinformatics techniques to identify genes (Chapters 9 and 10) are needed, especially as small RNAs and other unexpected findings challenge our definition of genes. Sequence alignment will also continue to be a major player as genomics moves increasingly into the interpretation phase. And along with the rapid pace of scientific change will come a need to consider the wise and ethical use of these vast volumes of data: Should we diagnose genetic diseases we cannot yet treat? Should insurance companies have access to risk data based on sequence analysis? What would constitute fair and equitable access to new medical technologies that may be highly effective but at least initially extremely expensive? Continued advances in sequencing technology will no doubt provide both new answers and new questions in the near future.

BioBackground: Sequencing DNA

It is helpful for both developers and users of sequence analysis software to understand how DNA sequencing is done. The strengths and limitations of a particular sequencing technology affect the nature and quality of the DNA sequences obtained, which in turn impact how those sequences should be treated by assembly or mapping software. This section does not attempt to be a complete manual on DNA sequencing, but we discuss three commonly used sequencing platforms to aid in understanding how the fragments analyzed by sequencing software are obtained. Good sources of further information are listed in References and Supplemental Reading.

Automated Sanger Sequencing

The sequencing method developed by Fred Sanger in 1975 was not the first, but it was far better suited to the rapid sequencing of long DNAs than the laborious chemical cleavage methods that preceded it. The technique became widely used and by the time the human genome project began had been improved by the use of fluorescent nucleotides and automated.

Sanger sequencing (or **dideoxy sequencing**) harnesses DNA polymerase, the enzyme that normally replicates DNA in the cell. The DNA molecule to be sequenced serves as the template for DNA polymerase, and a short single-stranded **primer** binds to the template and serves as the starting point. DNA polymerase can then synthesize multiple copies of a single strand of DNA complementary to the template (**Figure 8.11**). However, there is a twist: In addition to providing ordinary nucleotides (**dNTPs**) to be joined into the new DNA strand, fluorescent **dideoxy nucleotides** are added. Dideoxy nucleotides lack the 3′ –OH group to which the next nucleotide in the chain would be joined; when a dideoxy nucleotide is added to a growing DNA strand, synthesis stops. Thus, if low concentrations of dideoxy A, C, G, and T nucleotides (**ddNTPs** for short), each fluorescing a different color, are added to a reaction containing polymerase, primer, template, and dNTPs, a set of DNA fragments will be generated, each of which ends in a dideoxy nucleotide that can be identified by its fluorescence.

In automated Sanger sequencing, the fragments are placed on a gel-like matrix in a tiny capillary tube and an electric current is applied. The DNA fragments, being strongly negatively

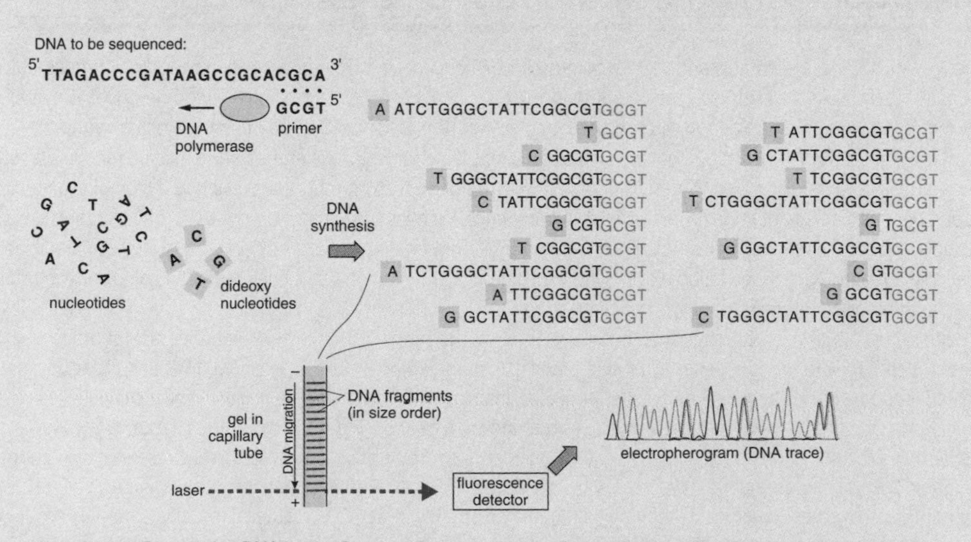

Figure 8.11 Sequencing DNA by the Sanger (dideoxy) method. Dideoxy nucleotides that terminate fragments are shown in boxes.

charged, move through the gel toward the positive pole, with smaller fragments moving faster. A laser excites each fluorescent nucleotide as the fragments move past it, and a computer-connected reader determines which base the fragment ends with by the color of the fluorescence (Figure 8.11). Each succeeding fragment is one nucleotide longer than the one before it, and the pattern of fluorescence color and intensity allows the DNA "trace" (Figure 8.5) to be constructed. Sanger sequencing cannot read bases extremely close to the primer, as a fragment of some reasonable length is needed to resolve properly in its passage through the gel. High quality can typically be maintained for some 500–800 nucleotides from a single capillary tube, and 384 such tubes can be run simultaneously on a single instrument.

Shotgun Sequencing

In **directed sequencing**, a primer is used to obtain sequence from a particular template, and then a new primer can be synthesized to match the just-read bases from the end of that sequence and the process repeated. Thus, there is no ambiguity regarding what part of a long template has been sequenced, but the process is slow even if multiple templates are sequenced at the same time. **Shotgun sequencing** and computerized assembly revolutionized this process: A long DNA is fragmented by mechanical shearing or enzymatic digestion into many short pieces, each of which is joined to a cloning vector (plasmid). Because the vector sequence is known, the sequence from each end of each fragment can be obtained using primers that match vector sequences (**Figure 8.12**). When many random fragments have been sequenced, there should be overlapping sequences, allowing for computerized assembly. Using this technique, it is not necessary to wait for new primers to be synthesized: Fragmenting and cloning can go on at the same time as sequencing of already cloned fragments. The concept of shotgun sequencing is also used in all next-generation sequencing methods, but the need for the cloning step has been eliminated—for example, by direct analysis of uncloned fragments or PCR amplification of random DNA regions.

454 Sequencing

The first widely used next-generation sequencing method was developed by 454 Life Sciences (now owned by Roche) in 2004; this **pyrosequencing** method is popularly referred to as 454 sequencing. As with any shotgun sequencing method, the DNA to be sequenced must be fragmented, either chemically, mechanically, or by enzymatic digestion; fragments of 300–800 bp are suitable for 454 sequencing and must be made blunt (no single-stranded overhangs) on each end. Short oligonucleotide adapters of known sequence are then joined on to each end of the fragments; one adapter has a biotin molecule that can be reacted with a bead coated with

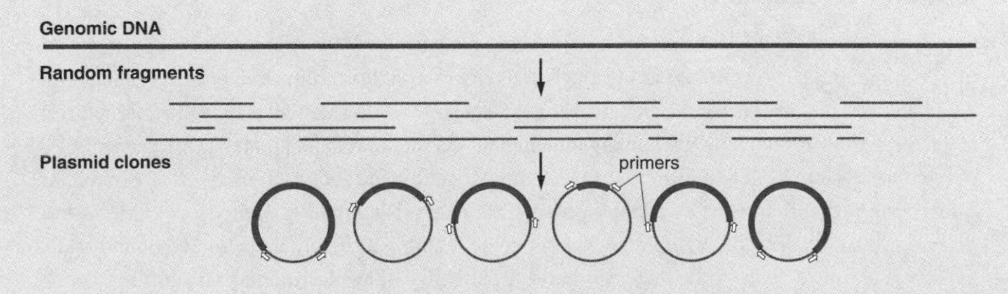

Figure 8.12 Shotgun sequencing: A large genomic DNA is broken into random fragments, which are cloned into plasmid vectors. Primers complementary to the vector allow sequence to be obtained from both ends of the cloned fragments.

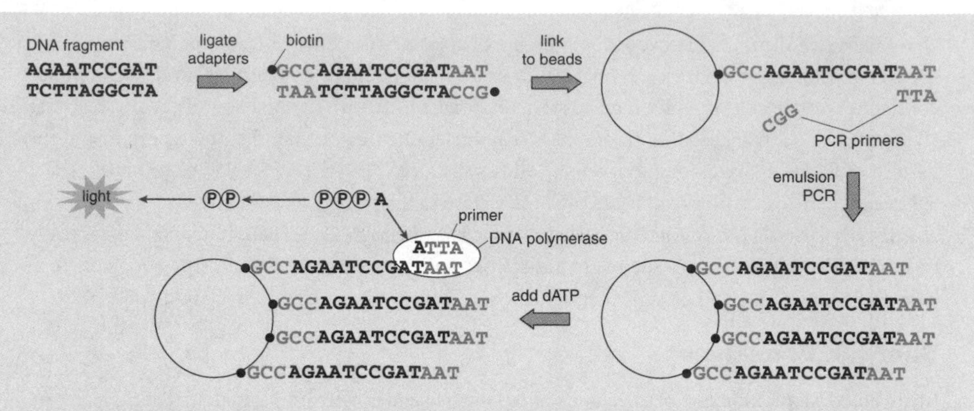

Figure 8.13 Sequencing DNA by the 454 (pyrosequencing) method: adaptors are ligated to DNA fragments, immobilized on beads, and amplified by PCR. Solutions of single nucleotides are added and light resulting from an enzymatic reaction involving the pyrophosphate cleaved from the nucleotide when it is added to the DNA chain is detected as evidence that a particular nucleotide was incorporated. Reactions can be done on 1.6 million beads in parallel.

streptavidin. Single-stranded DNA fragments with adapters thus become immobilized on the beads (**Figure 8.13**).

Next, the immobilized fragments are amplified. Beads are captured in individual oil droplets containing PCR reagents, and primers matching the adapters are used to generate some 10 million copies of the original fragments, all attached to a single bead. The beads are then transferred to individual wells, each holding only 75 pl of volume, of a PicoTiter plate capable of holding 1.6 million individual beads. Primers are then bound to the adapter sequences, and DNA polymerase can then add nucleotides complementary to the single-stranded template much as in Sanger sequencing (Figure 8.13). A solution containing a single nucleotide is "flowed" over the plate, and reagents bound to the beads react with the diphosphate (pyrophosphate) released from the polymerization reaction to produce a tiny emission of light. A camera monitors each well and detects the light, indicating that a particular nucleotide was successfully added to the growing chain (and was thus complementary to the template strand) in a particular well. The process then repeats with each of the other three nucleotides in turn and then the whole cycle of four nucleotides repeats. Recording which nucleotides are added in which order to the DNA in each well generates a sequence read, and the reads can then be assembled by computer to produce the complete sequence of the original DNA.

Illumina Sequencing

Solexa announced its high-throughput sequencing platform in 2006; this company was acquired by Illumina, and the technology is variously referred to as Solexa, Illumina, or SBS sequencing. As in 454 sequencing, adaptors are added to the ends of DNA fragments; they are then bound to primers that in this case are already attached to a slide, and PCR creates local clusters of a particular DNA molecule. Fluorescent nucleotides are then added, each nucleotide capable of fluorescing a distinct color, and DNA polymerase can incorporate a single nucleotide into a growing complementary DNA strand. A laser removes a blocking group from each nucleotide, allowing its fluorescence to be visualized and the identity of the last-added nucleotide in each cluster thus determined. A new batch of nucleotides is then added. As before, the sequence of a fragment is generated by monitoring the order in which the different colors (wavelengths) of fluorescence appear in each cluster.

SOLiD Sequencing

Both the 454 and Illumina methods (and, in fact, the Sanger method) involve "sequencing by synthesis," with a polymerase enzyme adding detectable nucleotides sequentially to a new strand. SOLiD sequencing, developed by Applied Biosystems and available since 2008, relies instead on the ability of two-nucleotide fluorescent probes to hybridize with (bind) the DNA template and be ligated to a growing chain by DNA ligase. Fragments of the DNA to be sequenced are linked to adapters, joined to beads, and amplified by emulsion PCR much as in 454 sequencing. Each two-base pair emits fluorescence at a distinct wavelength. The primer determines the position at which probes can hybridize, and after several cycles of hybridization, ligation, and cleavage, a new probe is used that is one nucleotide shorter, requiring a different set of probes to bind the same sequence to increase accuracy. Sequence reads produced by the SOLiD platform are very short, only 50 nt long, increasing the dependence of this technology on accurate and efficient assembly algorithms and powerful computers.

References and Supplemental Reading

Diarrheal Disease as a Worldwide Health Problem

The United Nations Children's Fund and The World Health Organization. 2009. *Diarrhoea: Why Children Are Still Dying and What Can Be Done.* WHO Press, Geneva.

Metagenomics and Metagenomic Discovery of New Viruses

Mokili, J. L., F. Rohwer, and B. E. Dutilh. 2012. Metagenomics and future perspectives in virus discovery. *Curr. Opin. Virol.* **2**:63–77.

Phan, T. G., N. P. Vo, I. J. Bonkoungou, A. Kapoor, N. Barro, M. O'Ryan, B. Kapusinszky, C. Wang, and E. Delwart. 2012. Acute diarrhea in West African children: diverse enteric viruses and a novel parvovirus genus. *J. Virol.* **86**:11024–11030.

Thomas, T., J. Gilbert, and F. Meyer. 2012. Metagenomics—a guide from sampling to data analysis. *Microb. Inform. Exp.* **2**:3.

Sequencing of the Human Genome

Venter, J. C., M. D. Adams, G. G. Sutton, A. R. Kerlavage, H. O. Smith, and M. Hunkapiller. 1998. Shotgun sequencing of the human genome. *Science* **280**:1540–1542.

Galaxy

Blankenberg, D., G. Von Kuster, N. Coraor, G. Ananda, R. Lazarus, M. Mangan, A. Nekrutenko, and J. Taylor. 2010. Galaxy: a web-based genome analysis tool for experimentalists. *Curr. Protoc. Mol. Biol.* **89**:19.10.1–19.1021. Wiley-Blackwell, Hoboken.

Giardine, B., C. Riemer, R. C. Hardison, R. Burhans, L. Elnitski, P. Shah, Y. Zhang, D. Blankenberg, I. Albert, J. Taylor, W. Miller, W. J. Kent, and A. Nekrutenko. 2005. Galaxy: a platform for interactive large-scale genome analysis. *Genome Res.* **15**:1451–1455.

Goecks, J., A. Nekrutenko, J. Taylor, and The Galaxy Team. 2010. Galaxy: a comprehensive approach for supporting accessible, reproducible, and transparent computational research in the life sciences. *Genome Biol.* **11**:R86.

Next-Generation Sequencing

Metzker, M. L. 2010. Sequencing technologies—the next generation. *Nat. Rev. Genet.* **11**:31–46.

Rothberg, J. M., and J. H. Leamon. 2008. The development and impact of 454 sequencing. *Nat. Biotechnol.* **26**:1117–1124.

Shendure, J., and J. Hanlee. 2008. Next-generation DNA sequencing. *Nat. Biotechnol.* **26**:1135–1145.

Trapnell, C., and S. L. Salzberg. 2009. How to map billions of short reads onto genomes. *Nat. Biotechnol.* **27**:453–457.

Sequence Assembly

Flicek, P., and E. Birney. 2009. Sense from sequence reads: methods for alignment and assembly. *Nat. Methods* **6**:S6–S12.

Huang, X., and A. Madan. 1999. CAP3: A DNA sequence assembly program. *Genome Res.* **9**:868–877.

Masoudi-Nejad, A., K. Tonomura, S. Kawashima, Y. Moriya, M. Suzuki, M. Itoh, M. Kanehisa, T. Endo, and S. Goto. 2006. EGassembler: online bioinformatics service for large-scale processing, clustering and assembling ESTs and genomic DNA fragments. *Nucleic Acids Res.* **34**:W459–W462.

Miller, J. R., S. Koren, and G. Sutton. 2010. Assembly algorithms for next-generation sequencing data. *Genomics* **95**:315–327.

Klassevirus

Greninger, A. L., C. Runckel, C. Y. Chiu, T. Haggerty, J. Parsonnet, D. Ganem, and J. L. DeRisi. 2009. The complete genome of klassevirus—a novel picornavirus in pediatric stool. *Virol. J.* **6**:82.

Generation of Simulated Sequence Reads

Balzer, S., K. Malde, A. Lanzén, A. Sharma, and I. Jonassen. 2010. Characteristics of 454 pyrosequencing data—enabling realistic simulation with flowsim. *Bioinformatics* **26**:i420–i425.

Chapter 9

Sequence-Based Gene Prediction:

Annotation of a Resistance Plasmid

Chapter Overview

Assembling a genome sequence (Chapter 8) does not by itself reveal key information such as where the genes are within that sequence. This chapter and the next one focus on gene prediction: how to identify possible genes within a genome sequence. In this chapter, sequence-based methods suitable for gene prediction in prokaryotes are explored and their value and limitations in eukaryotic gene discovery examined; the next chapter will take up the more complex gene prediction methods needed for eukaryotic genome annotation. Students in both programming and nonprogramming courses will be introduced to algorithms for gene prediction. Using a variety of Web-based tools, students will be able to use sequence-based methods for gene prediction in prokaryotes. Students in programming courses will implement sequence-based algorithms for gene prediction in prokaryotes. The On Your Own Project will then examine the extent to which these algorithms can be applied to eukaryotes.

Biological problem: Prediction and annotation of genes in a resistance plasmid sequence

Bioinformatics skills: Sequence-based ORF finding and promoter prediction

Bioinformatics software: NCBI ORF Finder, NEBcutter, EasyGene

Programming skills: Pattern-matching algorithms, modularization, functions

Understanding the Problem:

Gene Discovery

We have come a long way since the preantibiotic days when the risk of infection made surgery often more dangerous than the condition it was intended to cure. However, despite our many medical advances and modern methods of controlling infectious agents, in the United States approximately 1.7 million individuals per year acquire infections while hospitalized. Of these hospital-acquired, or **nosocomial**, *infections, some 99,000 cause or contribute to the death of the patient. Control of nosocomial infections is difficult because of the high concentration of infectious agents in the hospital environment, the already compromised or immunodeficient state of the patients, and the close contact of medical personnel with many patients per day. Furthermore, the use of invasive measures such as surgical procedures, catheters, and intravenous tubes may grant pathogens access to areas of the body that are normally well protected. Among*

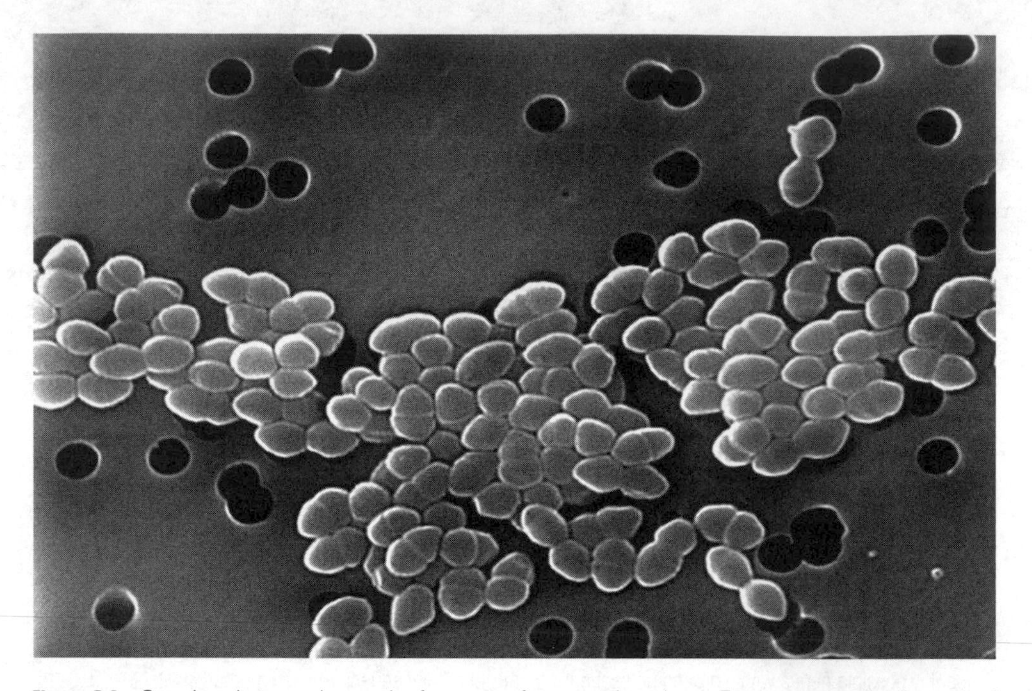

Figure 9.1 Scanning electron micrograph of a group of vancomycin-resistant Enterococcus cells. Courtesy of Janice Haney Carr/CDC.

the most common agents of nosocomial infection are Enterococcus *species* (**Figure 9.1**), *normally harmless residents of the human colon that can seize an opportunity to enter other parts of the body where they can be highly pathogenic. To make matters worse, many* Enterococcus *isolates are highly antibiotic resistant—even to "last resort" drugs such as vancomycin—and capable of transferring multiple resistance genes horizontally on large plasmids. Sequencing of plasmid DNA from these resistant strains is one way to learn more about the nature of the resistant organisms and their potential to spread resistance.*

Once a genome, chromosome, plasmid, or other large piece of DNA has been sequenced, the processes of **gene discovery** (also called **gene prediction**) and **genome annotation** begin. By itself, a DNA sequence is just a bunch of As, Cs, Gs, and Ts with no obvious meaning; to use that sequence to cure a genetic disease or understand how a specialized cell type develops, we have to find the genes within that sequence and understand their functions. Many people are surprised that we still cannot say exactly how many genes there are in the human genome, let alone identify all their functions. The presence of introns, the existence of surprisingly short or long genes, and the difficulty of definitively identifying promoters and translational start sites are among the complexities involved. Furthermore, although we tend to focus on protein coding genes, genomes also include protein binding sites, genes for noncoding RNAs, regions important to chromatin structure, methylation sites, and more.

Gene discovery is one of the major applications of bioinformatics to genomics. Although we often think of gene discovery as it applies to the analysis of major genome sequencing projects, it is also important on a smaller scale. Consider, for example, the major medical problems created by the horizontal transfer of antibiotic resistance. Often, resistance is due to large multi-drug resistance plasmids that by horizontal transfer can make another cell simultaneously resistant to many antibiotics—in some cases, even to *all* the classes of antibiotics in current use, including such "last resort" drugs as vancomycin.

In this chapter, we apply gene prediction methods to a large plasmid isolated from a highly resistant *Enterococcus faecium*, a bacterium that is naturally resistant to some antibiotics, including the penicillin family, and can readily acquire additional resistance. The plasmid we examine was isolated from a patient with a life-threatening postsurgical abdominal infection. Using gene prediction methods, we can identify potential resistance genes within this plasmid sequence and annotate them by looking for conserved sequences, thus determining what resistances the bacteria have and potentially how best to treat infection.

Bioinformatics Solutions:
Gene Prediction

Back in Chapter 2, we considered a gene to be a coding sequence within an mRNA (or cDNA) sequence; an AUG start codon and a UAG, UGA, or UAA stop codon identified this sequence, and the genetic code table allowed us to find the amino acid encoded by each three-nucleotide codon in between. This coding sequence is called an **open reading frame** (**ORF**). However, finding a gene is not as simple as finding an ORF. An ORF-like sequence could occur accidentally in noncoding DNA. Therefore, long ORFs are usually considered more likely to be real genes—but we also do not want to miss short but genuine genes that encode short proteins (sarcolipin, the shortest known protein in mice, is only 31 amino acids long). Additionally, genes for untranslated functional RNAs (tRNAs, rRNAs, snRNAs, and others) have no coding sequence. Predicting which sequences serve as promoters can help us recognize actual genes, but this is in itself complex, especially in eukaryotes. **Introns** introduce a huge amount of difficulty in eukaryotic genomes: An average protein coding sequence in the human genome is only about 1,500 base pairs long, but an average complete gene (typically including four to five introns) is nearly 10 times that long.

No method exists yet that can comprehensively and unambiguously identify all the genes in a DNA sequence; indeed, the problem is usually approached from multiple directions by applying a variety of methods. Commonly used computational approaches to this problem fall into several categories: algorithms based on alignment, sequence, content, or probability.

Alignment-based algorithms. If a region of a newly sequenced genome is orthologous to a previously identified gene in a well-studied organism such as mice, zebrafish, fruit flies, nematodes, or even bacteria or yeast, that would be good evidence that it is a gene. Indeed, even if no specific orthologous gene has yet been identified, strong conservation of a genome region over evolutionary time is strongly suggestive of its functional importance. **Alignment-based** algorithms look for genes based on conserved sequences; the alignment tracks in the UCSC genome browser (Chapter 1) gave you some idea of the value of this kind of comparison.

Sequence-based algorithms. Searching for ORFs is an example of a **sequence-based method** of gene prediction: A simple ORF-finder program would look for the sequence AUG (the start codon) followed by some amino-acid codons and a UAG, UGA, or UAA stop codon. More complex variations would take into account additional sequence clues such as promoter sequences and intron–exon boundaries. These functional regions of DNA would be identified based on the development of **consensus** sequences (see BioBackground at the end of this chapter) that can then be computationally identified in a genome. Sequence-based methods do not require similarity to other organisms, but they can only find genes that include sequences matching known patterns, and they have difficulty with sequence patterns that are relatively loose, like the sequences at the boundaries of exons and introns. Sequence-based methods are the focus of this chapter.

Content-based algorithms. **Content-based methods** do not look for specific sequences but rather for patterns such as nucleotide or codon frequency that are characteristic of coding sequences in a particular organism. These methods can identify novel genes and find coding regions that would be missed by sequence-based methods. One tool used in the Web Exploration in this chapter includes a content-based method (codon frequency); content-based methods will be discussed in more detail in the next chapter.

Probabilistic algorithms. More sophisticated gene discovery methods may combine elements of both sequence-based and content-based gene prediction in algorithms that model the probability that a given sequence is part of a gene. Hidden Markov models and neural network algorithms are two major examples of probabilistic solutions; these will be discussed in the next chapter

In this chapter's Web Exploration and Guided Programming Project, we see how sequence-based methods work and use them to identify genes involved in antibiotic resistance and virulence within the sequence of a large bacterial plasmid. In the On Your Own Project, we apply similar methods to eukaryotes and explore their limitations. A good understanding of gene structure is essential to the development and use of computational methods for gene discovery. The BioBackground section in Chapter 2 introduced the structure of genes, and that introduction is extended in this chapter's BioBackground section, along with an introduction to how the sequences of promoters and other functional sites are identified.

BioConcept Questions

1. Why are long ORFs sometimes considered to be the same as genes? In what ways is this definition insufficient?
2. How does RNA polymerase find the transcriptional start site of a gene in prokaryotes? How can we use this information in a gene prediction algorithm?
3. How does RNA polymerase find the transcriptional start site of a gene in eukaryotes? Why is it more difficult to develop an algorithm to find a eukaryotic promoter than a prokaryotic promoter?
5. How does a prokaryotic ribosome find the correct start codon within an mRNA? How can we use this information in distinguishing which ORFs are genes?
6. Why can't we use a similar strategy to distinguish which ORFs are genes in eukaryotes?
7. A simple ORF-finding program would do a very poor job of predicting the amino-acid sequences of the proteins encoded in the human genome. Discuss why this is the case.
8. How might you identify a gene encoding a functional RNA (that does not encode a protein)? How does the discovery of key functions for very small RNA molecules complicate the issue?

Understanding the Algorithm:
Pattern Matching in Sequence-Based Gene Prediction

Learning Tools

If you want to better understand how a consensus sequence for a promoter or other element is developed and why identification of these sequences is not as clear-cut as it sounds, you can download an exercise from the *Exploring Bioinformatics* website that will take you through the generation of a prokaryotic promoter consensus sequence using data from sequenced genomes.

Sequence-based methods of gene prediction examine DNA sequences for patterns (often called **motifs**) that provide clues about the existence of transcriptional or translational units. Sequence-based prediction methods rely on **pattern-matching algorithms**: Given a string to search (such as a plasmid or genome sequence) and a pattern to be matched (such as AUG), they can identify whether, how often, and where the pattern occurs. Indeed, content-based and probabilistic methods usually include elements of pattern matching as well.

An ORF-finding program is a good example of pattern matching in gene prediction. This program could begin by **traversing** the searched text—that is, searching through the nucleotide string from beginning to end—examining each group of three nucleotides for the pattern ATG to find a potential start codon. Then, it would have to find an in-frame stop codon. The process of testing three-nucleotide groups for a match to the pattern would stay the same, so a single algorithm could be provided with different **parameters**. Parameters are values set when an algorithm starts that allow it to solve variations of a problem using the same main steps; in this case, our parameters would be the searched text, the pattern, start and stop locations, an increment value, and a threshold value. When looking for the start codon, the start location is the first nucleotide, the stop location is three nucleotides from the end (no point in looking at the last two), the increment value is one in order to search in all three possible reading frames (in the sequence `CCATGGAC`, look first at `CCA`, then `CAT`, then `ATG`, etc.), and the threshold value is 100%, because we need a perfect match to ATG. Once a start codon is found, we would change the increment value to three (after finding `ATG`, look at `GAC` but *not* `TGG` or `GGA`) and the pattern to TAG, TGA or TAA, again requiring a perfect match.

Algorithm

Pattern-Matching Algorithm

1. Initialize parameters of algorithm:
   `pattern` = search pattern
   `searchedText` = text that will be searched for pattern
   `start` = start location of search (assumes first character is position 1)
   `stop` = stop location of search (this represents last location to search from)
   `increment` = incrementing value (negative number for upstream search, positive number for downstream search)
   `threshold` = minimum percentage match required
2. Compare `pattern` to characters of `searchedText` starting at position `start`. If percentage of matching characters is >= `threshold`, output `start` position and end algorithm. If not, add `increment` to `start` and continue to step 3.
3. If `increment` is positive and `start` is <= `stop`, repeat step 2. If `increment` is negative and `start` is >= `stop`, repeat step 2. If neither statement is true, pattern was not found, end algorithm.

You can quickly see, however, that this straightforward algorithm will not make a great ORF finder. ATG is not *just* a start codon but is used every time the amino acid methionine occurs in a protein. That means the simple algorithm would find apparent ORFs that are actually within other ORFs. Furthermore, ORF-like sequences could occur by chance in noncoding DNA: The pattern `ATGGGGTGA` would occur at random once every 4^9 nucleotides, or about 19 times in the *E. coli* genome, but is clearly not an ORF. Thus, ORF-finding programs commonly allow the user to limit results to ORFs of a certain length, perhaps

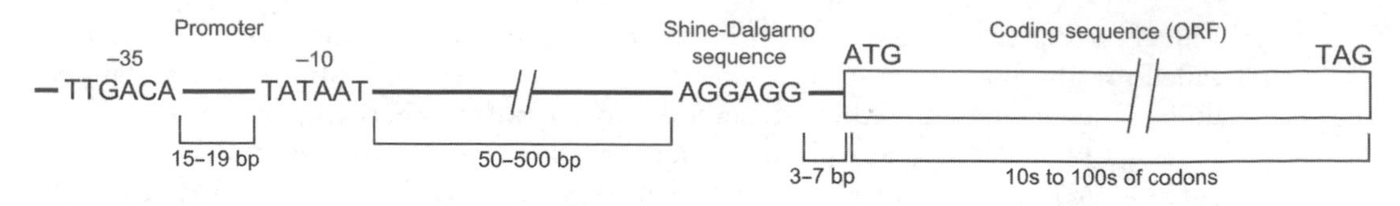

Figure 9.2 Elements of a prokaryotic gene that can be searched by a sequence-based algorithm include the coding sequence or ORF, the Shine-Dalgarno sequence, and the promoter sequence.

100 codons. This would only require setting a start location 300 nucleotides downstream to start looking for the stop codon after finding an ATG—but this modification also brings with it the danger of overlooking small but genuine genes. If you will not be completing the programming projects in this chapter, you may wish to download the sequence of the *Enterococcus* plasmid from the *Exploring Bioinformatics* website and look by hand for some potential ORFs to get an idea of how these parameters would affect the process.

Despite these adjustments, a simple ORF-finding algorithm will not be a very reliable method of gene prediction: Even a reasonably long ORF might not really be a gene, and a short ORF possibly could be a gene. To help distinguish real genes, we can also look for regulatory sequences: In bacteria, genes are preceded by promoter sequences (−10 and −35 sequences) and the start codon is preceded by a Shine-Dalgarno sequence (see **Figure 9.2** and BioBackground). Unfortunately, finding these patterns is less straightforward. In *E. coli*, the Shine-Dalgarno consensus sequence is AGGAGG, but the match to this pattern can be imperfect. The end of this sequence should be approximately five nucleotides upstream of the start codon, give or take two positions (so, -5 ± 2 relative to the ATG). Promoters can also be inexact matches to the consensus −10 (TATAAT) and −35 (TTGACA) sequences; these sequences should be 17 ± 2 nucleotides apart but can occur anywhere from 50 to 500 nucleotides upstream of a start codon. (If you are familiar with prokaryotic molecular biology, you know that even this is a simplified view given the frequent use of operons and alternative sigma factors.) The pattern-matching algorithm can find these sequences given appropriate parameters, such as start and stop locations and threshold values.

Test Your Understanding

1. DNA is double stranded, and one strand may serve as the template (copied) strand for one gene (in one region) but the nontemplate (mRNA-like) strand for another (in another region). The algorithm given could find an ATG start codon in one of three reading frames by reading a sequence entered in the 5′ to 3′ direction, but really we should consider all *six* possible reading frames: three from the DNA as it was entered and three more on the complementary strand. What would we need to do to find ORFs in all six possible reading frames?
2. As noted, the pattern-matching algorithm might find an ORF within another ORF, because within a gene there could be multiple ATG codons. How could your algorithm filter out these undesirable matches?
3. Identify parameters that could be used in the pattern-matching algorithm to search for a Shine-Dalgarno sequence once an ATG is found. Assume an exact match to the consensus sequence.
4. Identify parameters that could be used in the pattern-matching algorithm to search for a promoter once an ATG is found. Assume that five of the six bases in the −10 and −35 sequences must match their consensus.

CHAPTER PROJECT:
Gene Discovery in a Resistance Plasmid

This chapter's project focuses on sequence-based methods of finding genes within DNA sequence data. We consider only prokaryotic genes in the Web Exploration and Guided Programming Project, because the lack of introns and more clearly defined expression signals makes them easier from a practical standpoint. In the On Your Own Project, we consider how these principles apply to eukaryotes. Specifically, we look for genes within the sequence of a plasmid isolated from antibiotic-resistant *Enterococcus* and, in the Web Exploration, annotate those genes by looking for clues to function.

Learning Objectives

- Understand the structure of a gene and which features are useful in developing computational methods for identifying genes
- Appreciate the strengths and limitations of sequence-based methods for gene discovery
- Use Web-based gene discovery tools to annotate a plasmid
- Understand how pattern matching can be used in sequence-based computational solutions
- Apply sequence-based algorithms to the more complex problem of gene discovery in eukaryotes

Suggestions for Using the Project

This project provides an introduction to pattern matching in gene discovery for both programming and nonprogramming courses. The Web Exploration in this project guides students to predict and annotate genes in a plasmid sequence; the Guided Programming Project allows them to implement a pattern-matching algorithm that can be applied to the same problem. The On Your Own Project asks students to implement (in programming courses) or examine (in nonprogramming courses) the application of pattern matching to eukaryotic gene prediction clues. All tools described here could be applied equally well to any other question that the instructor wished to explore.

Programming courses:

- Web Exploration: Use Web-based tools to identify likely genes within a plasmid sequence; complete either Part I or Part II (or both parts, in teams) and Part III. Optionally, annotate genes with BLAST.
- Guided Programming Project: Implement a pattern-matching algorithm and compare its output with the Web-based tools.
- On Your Own Project: Extend the pattern-matching algorithm to eukaryotic gene prediction.

Nonprogramming courses:

- Web Exploration: Use Web-based tools to identify likely genes within a plasmid sequence and annotate the genes with BLAST. Complete either Part I or Part II (or both parts, in teams) and Part III.
- On Your Own Project: Consider how a pattern-matching algorithm could be used to identify sequence-based clues to eukaryotic genes.

Web Exploration: Prokaryotic Gene Prediction and Annotation

In this part of the project, we use Web tools to find genes within an *Enterococcus* resistance plasmid sequence. Sequence-based methods for gene prediction work well for prokaryotes, because they lack exons and have more easily predictable patterns for regulatory elements (see BioBackground). Parts I and II use two different simple ORF finders to accomplish the same task. It is suggested that pairs of students work on these exercises together: Each can use one of the tools and then the results can be compared. Alternatively, an instructor may choose to assign only Part I or Part II. Part III uses a more advanced tool to search for Shine-Dalgarno sequences to better identify actual genes. BLAST can be used to annotate genes with putative functions and potentially to further explore the nature of resistance and the evolution of resistance plasmids (see More to Explore later in the chapter); instructors may skip this part of the exercise if they wish.

Part I: Sequence-Based ORF Identification Using the NCBI ORF Finder

The simplest gene discovery program would simply look for an ORF as described in Understanding the Algorithm: a start codon followed by a coding sequence longer than some length specified by the user and terminating with a stop codon. The ORF could occur in any of the six possible reading frames (three on each strand). Such a program would actually be fairly effective in finding genes in a prokaryotic genome, given the absence of introns. There are many such programs; we use NCBI's ORF Finder to identify ORFs in the *Enterococcus* resistance plasmid.

Start by downloading the sequence of the ***Enterococcus faecium* resistance plasmid** from the *Exploring Bioinformatics* website. Open NCBI's **ORF Finder** and paste the sequence into the input box. There are not a lot of parameters available; note that you could search only a portion of the sequence if desired, or you could change the genetic code used if you were working with something like mitochondrial DNA where a few codons are different. Run the program; you should see a display of ORFs similar to that in **Figure 9.3**.

You may be surprised by the number of ORFs found by this program. How long is the DNA sequence? Click on `View` to find out. Does it seem reasonable to have this many genes in a sequence this long? How are the genes distributed on the two strands of DNA? Notice that the ORFs are listed by size and that some of them are pretty short. By default, ORF Finder shows any ORF longer than 100 nucleotides, or about 33 amino acids. Change the drop-down to view only ORFs that have at least 100 amino acids (300 nucleotides) and see how this changes the list.

Gene prediction is more valuable if we can also annotate the genes with putative functions based on sequence comparison. Click on one of the ORFs either in the list or the graphical view to see its nucleotide and amino-acid sequences. Notice that you can then directly submit the sequence of just this ORF for either a protein (blastp) or nucleotide BLAST search. Try a protein search and try to find a putative function for each ORF. Some should match known antibiotic-resistance genes; for these, find out what antibiotic the gene confers

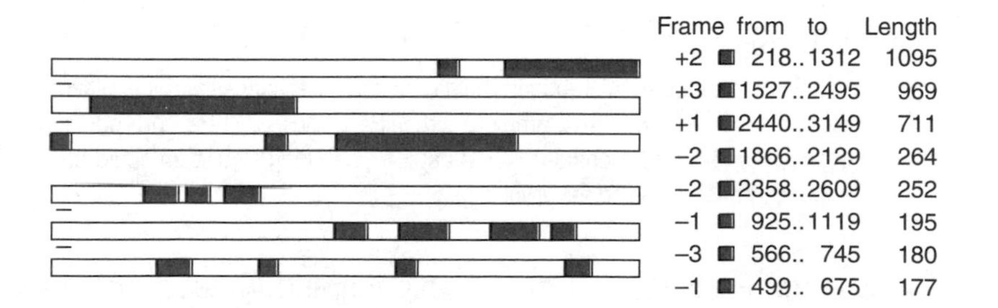

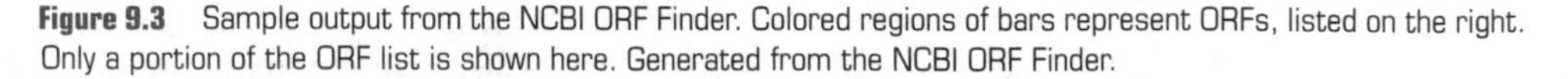
Figure 9.3 Sample output from the NCBI ORF Finder. Colored regions of bars represent ORFs, listed on the right. Only a portion of the ORF list is shown here. Generated from the NCBI ORF Finder.

resistance to and try to find the mechanism of action for the resistance protein (for example, does it inactivate the antibiotic, modify the cellular target of the antibiotic, or perhaps pump the antibiotic out of the cell?). For those that do not appear to be antibiotic-resistance genes, do they have functions that make sense in the context of this resistance plasmid? Remember that some of the ORFs may not be real genes at all. In addition to annotating the genes, notice that this BLAST search step effectively adds an alignment-based gene discovery method to increase the accuracy of our sequence-based predictions.

When you are satisfied with what you have learned about an ORF, use the `Back` button on your browser to return to the ORF Finder view of the gene. If you are convinced that the ORF is a genuine gene, click `Accept` and notice that the program changes the color of the gene in the graphical view and of its symbol in the list. This will help you keep track of the genes you have identified.

Part II: Sequence-Based ORF Identification Using NEBcutter

Another program that is useful for finding and annotating ORFs is New England Biolabs' **NEBcutter**. The primary goal of this program is to identify restriction endonuclease cut sites (useful, for example, in gene cloning; see References and Supplemental Reading), but it also identifies ORFs and places them on a map of the DNA in relation to the restriction sites. Find the NEBcutter page and paste the ***Enterococcus* plasmid sequence** into the input box. This is the complete sequence of a plasmid, so choose the option to show a circular DNA molecule (plasmids are always circular). Notice that you can change the minimum length of the ORF displayed from this page and set other options (you can even customize the colors of the output if you wish). Submit the sequence; you should get a result similar to the sample data shown in **Figure 9.4**.

Like NCBI's ORF Finder, NEBcutter shows the number and size of ORFs that met the specified criteria graphically and in a list. The NEBcutter display, however, does not separate the ORFs by the strand or reading frame in which they were found; notice that this might help you decide which genes might be grouped into operons. You can choose options to see a list of the ORFs with more detailed information. As with ORF Finder, you can change the minimum length of the ORF displayed based on your expectations.

To annotate genes in the plasmid, click on an ORF either in the list or graphical view to see its amino-acid sequence and find a link for a protein BLAST search. Use the BLAST results to find a putative function for each ORF. Some should match known antibiotic-resistance genes; for these, find out what antibiotic the gene confers resistance to and try to find the mechanism of action for the resistance protein (for example, does it inactivate the antibiotic, modify the cellular target of the antibiotic, or perhaps pump the antibiotic out of the cell?). For those that do not appear to be antibiotic-resistance genes, do they have functions that make sense in the context of this resistance plasmid? Remember that some of the ORFs may not be real genes at all. In addition to annotating the genes, notice this BLAST search step effectively adds an alignment-based gene discovery method to increase the accuracy of our sequence-based predictions.

Close the BLAST window when you are satisfied with your investigation of the ORF. If you are convinced it is a real gene, you can click `Edit` to name the gene (you might give it the same name as its orthologs: for example, β-lactamase proteins involved in penicillin resistance are named *bla* in many organisms) and describe its protein product. These data will then show up in the ORF list and in the description when you click on the ORF. When you have finished annotating genes, you can use the `Print` option to save your map to a PDF or GIF file.

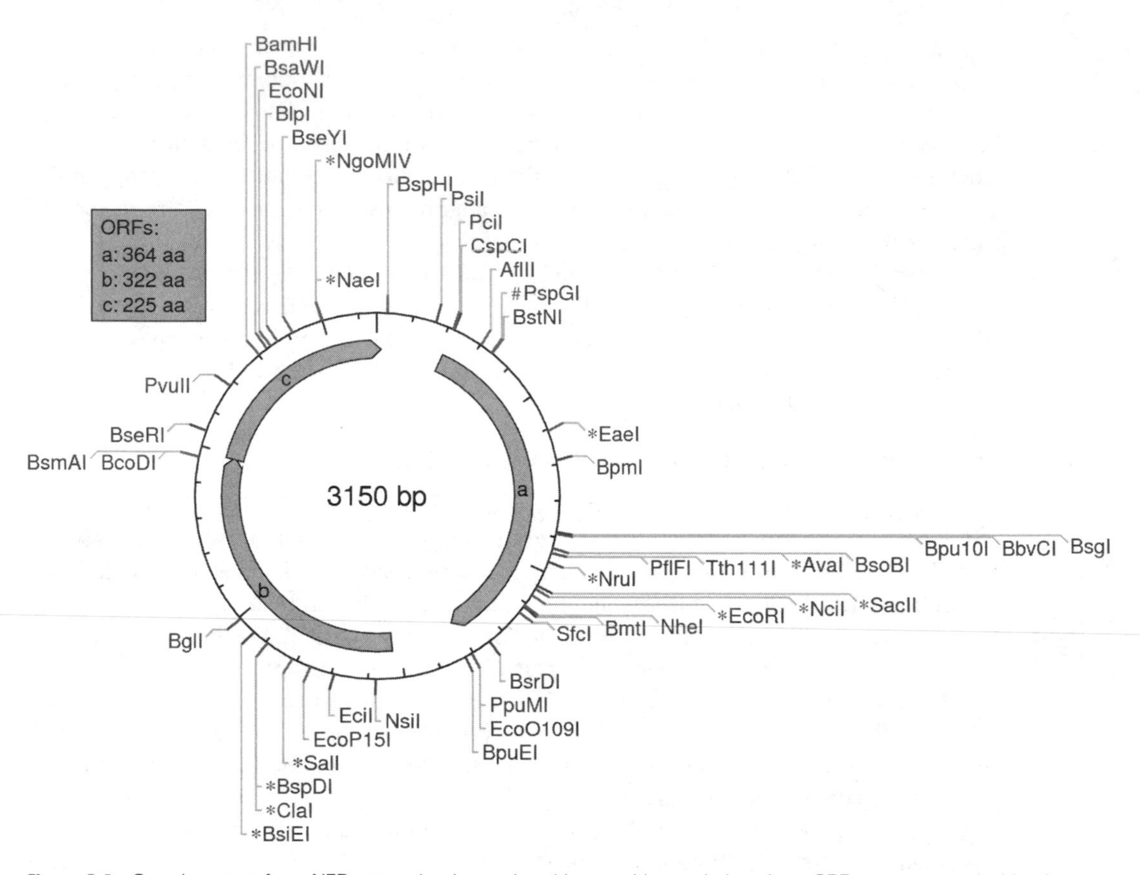

Figure 9.4 Sample output from NEBcutter showing a plasmid map with restriction sites; ORFs are represented by the gray arrows and listed by size in the box at left. Generated from NEBcutter; Vincze, T., Posfai, J. and Roberts, R. J. "NEBcutter: a program to cleave DNA with restriction enzymes." *Nucleic Acids Res.* 31: 3688–3691. (2003).

Part III: Shine-Dalgarno Prediction and Codon Usage Analysis with EasyGene

Using ORF Finder or NEBcutter, we got a long list of potential genes we had to narrow down by hand. We were able to eliminate many ORFs from the list by requiring that the ORFs be at least 100 amino acids long. NEBcutter also ignores overlapping ORFs in its main display. However, we might have missed some genes this way: What if some of the short ORFs also encode functional genes? What if two genes do overlap (rare in eukaryotes but not infrequent in bacteria and common in viruses)? We could improve our ability to find authentic genes by determining whether the start codon is preceded by a Shine-Dalgarno sequence (a sequence similar to 5′ AGGAGG located 5 ± 2 nucleotides before the start codon). This is still a sequence-based method of gene prediction, because we are still looking for a match to a specific sequence pattern; however, to use it effectively, we have to relax the stringency of the search to allow for imperfect matches.

We can use EasyGene (see References and Supplemental Reading) to add this element of sophistication to our prokaryotic gene prediction. EasyGene looks for ORFs and examines the region just before the putative start codon for a possible Shine-Dalgarno sequence. It also adds a content-based method of gene prediction: It asks whether the codons used in the ORF match the typical codon usage for the organism of interest. For example, six different codons all specify the amino acid leucine, but CTG is the codon actually used in *E. coli* genes more than 50% of the time. **Table 9.1** shows the codon usage frequencies for

Table 9.1 Codon usage table for *Escherichia coli*.

Codon (aa)	Freq.[1]	Codon (aa)	Freq.	Codon (aa)	Freq.	Codon (aa)	Freq.
UUU (F)	19.7	UCU (S)	5.7	UAU (Y)	16.8	UGU (C)	5.9
UUC (F)	15	UCC (S)	5.5	UAC (Y)	14.6	UGC (C)	8
UUA (L)	15.2	UCA (S)	7.8	UAA (*)	1.8	UGA (*)	1
UUG (L)	11.9	UCG (S)	8	UAG (*)	0	UGG (W)	10.7
CUU (L)	11.9	CCU (P)	8.4	CAU (H)	15.8	CGU (R)	21.1
CUC (L)	10.5	CCC (P)	6.4	CAC (H)	13.1	CGC (R)	26
CUA (L)	5.3	CCA (P)	6.6	CAA (Q)	12.1	CGA (R)	4.3
CUG (L)	46.9	CCG (P)	26.7	CAG (Q)	27.7	CGG (R)	4.1
AUU (I)	30.5	ACU (T)	8	AAU (N)	21.9	AGU (S)	7.2
AUC (I)	18.2	ACC (T)	22.8	AAC (N)	24.4	AGC (S)	16.6
AUA (I)	3.7	ACA (T)	6.4	AAA (K)	33.2	AGA (R)	1.4
AUG (M)	24.8	ACG (T)	11.5	AAG (K)	12.1	AGG (R)	1.6
GUU (V)	16.8	GCU (A)	10.7	GAU (D)	37.9	GGU (G)	21.3
GUC (V)	11.7	GCC (A)	31.6	GAC (D)	20.5	GGC (G)	33.4
GUA (V)	11.5	GCA (A)	21.1	GAA (E)	43.7	GGA (G)	9.2
GUG (V)	26.4	GCG (A)	38.5	GAG (E)	18.4	GGG (G)	8.6

[1]Frequency of codon per 1,000 codons
Data from: Codon Usage Database.

E. coli. EasyGene compares the codons in each ORF to a training set taken from whichever prokaryotic genome the user selects and calculates a significance score representing the likelihood that it is a genuine gene. Only ORFs scoring above a selected threshold are displayed.

Navigate to the **EasyGene** site. You will want to compare your EasyGene results with what you found with ORF Finder and/or NEB Cutter, so you may want to open this site in a new window or tab. Paste your *Enterococcus* plasmid sequence into the input box. From the list of organisms, choose the most closely related available species; this organism is used to determine what Shine-Dalgarno sequence to search for as well as the codon usage pattern to use for comparison. Note the lack of an option to limit the size of ORFs; given the additional features of EasyGene, it is preferable to limit results by the significance score rather than an arbitrary size cutoff.

Examine the EasyGene results and compare them with your results from ORF Finder. We might expect EasyGene to ignore ORFs that lack Shine-Dalgarno sequences or that do not match codon usage data; does this appear to be the case? How does EasyGene's list compare with ORF Finder's when ORF Finder is limited to 100-amino-acid ORFs? What if ORF Finder is allowed to find 30-amino-acid ORFs? Does EasyGene identify any of the short ORFs excluded by the length limit as actual genes? Does EasyGene fail to identify any genes that you annotated as genuine based on your ORF analysis and BLAST searches? (If so, does lowering the significance score cut-off allow it to find these genes?)

Looking at the EasyGene results, you should see a column showing the initiation codon for each gene it found; do you see any surprises here? In fact, ATG is not the start codon for every gene: tRNA carrying methionine can in some cases bind to a bacterial ribosome positioned at a GTG or TTG codon. Take a look at the ORFs EasyGene identified as having an alternative start codon, and then find the same ORF in ORF Finder. How long was the ORF that ORF Finder identified? What happens if you click `Alternative Initiation Codons`? Why is this result better, biologically? Why did EasyGene's algorithm, even though it is still sequence based, find the longer ORF instead of the shorter one with the more obvious start codon? Does BLAST confirm that this is a better result?

Web Exploration Questions

Report your findings regarding antibiotic resistance in the *E. faecium* strain isolated from the abdominal infection. Discuss whether this strain is multidrug resistant and to what antibiotics it is resistant. Then, provide an annotated list of genes on the plasmid for which you have solid evidence: Name them if possible (refer to them by the starting position of the ORF where you cannot find a suitable name), give their start and stop positions and their length in amino acids, and list their functions briefly but specifically.

More to Explore: Evolution of Antibiotic Resistance and a Resistance Plasmid

If you would like to use your gene prediction data to dig deeper into the nature and evolution of this resistance plasmid, try answering the following questions:

1. Multidrug-resistant bacteria are often capable of transferring resistance to multiple antibiotics on a single plasmid. Such resistance plasmids have frequently evolved when resistance genes become associated with transposons, mobile pieces of DNA able to move from place to place within a genome. If a transposon carries a resistance gene from the chromosome to a plasmid, that gene can now be more easily passed to another strain. As resistance plasmids are passed around among bacteria, there is a good chance they will eventually be in a cell carrying a different transposon-associated resistance gene, so that the resistance plasmid can "collect" additional genes over time. Transposons have repeated sequences at their ends and transposase genes that carry out the reaction of "cutting and pasting" the transposon DNA. Is there evidence to suggest that this resistance plasmid evolved in this manner?
2. Vancomycin is considered a "last resort" antibiotic for infections caused by gram-positive bacteria such as *Enterococcus*. Resistance to this drug is known, but it has developed more slowly than other antibiotic resistances, and most bacteria can still be killed by vancomycin. Physicians therefore do not use it unless it is the only drug that will work in a given situation, so that further spread of resistance is not encouraged. Based on what you have been able to learn about the genes in this resistance plasmid, can you suggest why it is more difficult for bacteria to develop resistance to this antibiotic than to others?

Guided Programming Project: Pattern Matching for Sequence-Based Gene Prediction

Sequence-based gene discovery methods are really quite simple in concept: As you saw previously in the Understanding the Algorithm, they simply search a string (DNA sequence) for a match to a pattern (start codon, stop codon, Shine-Dalgarno sequence, etc.). We can use parameters to limit the range of the search and whether to consider imperfect matches. In this guided project, you are asked to write the code to implement the ORF finder algorithm. We again limit our scope to prokaryotic gene prediction, where we can use sequence-based methods most effectively.

In Understanding the Algorithm and Web Exploration, you saw that a good gene prediction program must be able to search for multiple patterns—for example, to find a start codon and then look upstream in the same sequence for a Shine-Dalgarno sequence. The pattern-matching algorithm described previously can be used repetitively by changing its parameters, so a good programming approach is to modularize your code by implementing a subroutine or function to search the sequence for the pattern. For this chapter's exercise, the focus is on reusing the function to find different kinds of patterns. Therefore, let's review how this might work. To write a function, we need to know the main task of the function, the parameters we need to pass to the function, and the information the function needs to return to the calling routine:

- **Main task:** The main task of our pattern-matching function is to traverse an input sequence searching for a pattern.
- **Parameters passed in**: For a flexible and reusable function, we should use parameters to pass in the distinctive information for a particular search: the pattern, the searched text, the start and stop locations, the increment value, and the threshold value.
- **Information returned**: For a gene prediction program, we need to know the location where the pattern was found. If we use 0 to represent the location of the first character in the sequence, then −1 is an invalid location and we can use this value to represent a failed search. The calling program can determine whether the function returned a positive number (location of a successful match) or a negative number (pattern not found). In some situations, we might also need to return additional information such as the number of matched nucleotides or the number of matches.

The following pseudocode shows a solution for our function.

Algorithm

Pattern-Matching Function

Goal: A function that can be used to find a pattern within a search text.

Parameters: pattern, searched text, start location of search (assumes 0 is the first position in the search text), stop location of search, increment, threshold

Return Value: The location of the pattern in the search text (assumes first character represented by location 0) or −1 if pattern not found.

```
// Function findPattern
findPattern(pattern, searchText, startLoc, stopLoc, increment,
threshold)
    textLen = length of searchText
    patternLen = length of pattern
    for each i from startLoc to stopLoc by increment
        ctr = 0
        j = i
        // count number of matching characters
        for each k from 0 to patternLen
            if searchText[j] == pattern[k]
                ctr ++
            j++
        // compare number of matched characters to threshold
        if ctr/patternLen >= threshold
            return i
    return -1
```

Notice in this example that a `return` statement appears within the loop, so that the loop terminates as soon as a match is found. Some programmers prefer to exit a loop only when the conditional statement of the loop fails, a technique that improves readability in long, complex programs. In this short function, however, the return will not detract from readability and saves unnecessary looping as well as an additional flag variable. If the loop ends (reaches the end of the sequence without finding the pattern), the search has failed and −1 is returned. The function just given can be used to find any of the patterns necessary for gene prediction in prokaryotes and can be called multiple times within a complete gene prediction program. Your main program should carry out the following steps, calling the pattern-matching subroutine to look for each pattern:

1. Search for a start codon. If found, continue; otherwise, end.
2. Search for a stop codon in the same reading frame as you found the start codon. Determine if the number of codons between the start and stop codons is >= a user-defined minimum value. If a large enough ORF is found, continue; otherwise, end.
3. Search for a Shine-Dalgarno sequence no less than three and no more than seven bases upstream of the start codon. The sequence should match at least five nucleotides of the six-nucleotide consensus. If found, the ORF is a possible gene: continue; otherwise, end.
4. Search the 500 nucleotides upstream of the Shine-Dalgarno sequence for a promoter sequence: `TTGACA` located 15–19 nucleotides upstream of `TATAAT`, allowing at most one mismatch in each sequence.

Putting Your Skills Into Practice

1. Implement the pattern-matching algorithm within a complete prokaryotic gene prediction program as described earlier. You may wish to review the Test Your Understanding questions, where you should have already considered parameters that would allow your algorithm to search for these elements. Generate a short test sequence with clearly defined promoter and Shine-Dalgarno sequences to ensure your program works as expected.
2. Modify your program to allow the user to choose the match threshold for the Shine-Dalgarno and promoter sequences. Test the program using the *Enterococcus* plasmid sequence. Because this is a large sequence, you might want to start by testing only the ORF-finding routine on a segment of the plasmid sequence. Use your ORF Finder results for comparison. Then add the Shine-Dalgarno and promoter prediction and compare your results with those obtained using EasyGene. Can the program find the genes and promoters EasyGene found? What happens if the thresholds for the consensus sequences are relaxed?
3. Modify your program so it searches all six reading frames. Did you modify your function or the calling routine?
4. On any sizeable piece of DNA, there will probably be more than one ORF; however, the previous steps stopped searching after any step failed. Modify your program so it continues to search until all possibilities are exhausted.
5. Modify your program to discard an ORF if it has the same stop codon as an ORF already found and is shorter.
6. In prokaryotes, ORFs that are part of operons (see BioBackground) may not be directly preceded by promoters: One promoter is used for the entire operon. However, each ORF will still be preceded by a Shine-Dalgarno sequence. Modify your program to take this information into consideration: For example, you might check for upstream ORFs oriented in the same direction and then look for promoters, or you might look farther upstream for a promoter first.
7. How do the genes identified by your program in the *Enterococcus* sequence compare with those found by EasyGene? Your program uses a sequence-based search for promoters, whereas EasyGene uses a content-based analysis of codon usage; which mechanism seems to have worked best in terms of identifying the genes you classified as genuine in the Web Exploration?

On Your Own Project: Sequence-Based Gene Discovery in Eukaryotes

In the Web Exploration, you used—and in the Guided Programming Project developed—programs to find genes in prokaryotes using ORFs and sequence clues like promoter and Shine-Dalgarno sequences. In this project, you will apply these skills to predicting genes in eukaryotic genome sequences such as the human genome. If you are taking a nonprogramming course, there are exercises dealing with how sequence-based methods can be applied to eukaryotes, and your instructor can make a **completed gene prediction program** available for you from the *Exploring Bioinformatics* website.

Understanding the Problem: Sequence-Based Pattern Matching in Eukaryotes

Clearly, our gene prediction program from the guided project does not care whether the input sequence is from a prokaryote or a eukaryote; it can just as well find eukaryotic patterns. Unfortunately, it is more difficult to determine what patterns to search for in eukaryotes. Although the start and stop codons are identical, there is no Shine-Dalgarno sequence to identify the correct start codon, nor is there a single, clear promoter sequence (see BioBackground). Worse, the ORFs are usually interrupted by introns, so we cannot start with simple ORF finding. However, we do have some options.

In eukaryotes, the start codon is almost always the *first* AUG from the 5′ end of the mRNA and thus the first one after the transcriptional start site. Furthermore, in about 75% of cases, the transcriptional start can be identified by the presence of a **core promoter** pattern. Thus, you should be able to modify your solution to the Guided Programing Project to look for a start codon (not an entire ORF, because of the intron problem) preceded by the core promoter pattern. Subsequent analysis could then identify the ORF by looking for splice sites and predicting where the exons are (much more on this topic in Chapter 10).

The core promoter can be recognized by a consensus sequence called the **TATA box**, a sequence similar to `5´ TATA(A or T)A(A or T)` followed by three additional nucleotides that are rarely cytosine or guanine. The TATA box is usually found within about 150 nucleotides upstream of the start codon and at about the –25 position relative to the +1 nucleotide (first nucleotide transcribed into mRNA). The transcriptional start site itself (+1) commonly lies within an additional consensus sequence, the **initiator sequence (*Inr*)**. *Inr* consists of six nucleotides: The first two are usually `C` or `T`, the last two are usually `G` or `A`, and the middle two are `CA`, where the `C` is usually the +1 nucleotide. We can write this sequence more easily by using code letters to represent combinations of nucleotides (so-called **ambiguous** nucleotides): Y (for pYrimidine) to represent C or T and R (for puRine) to represent G or A. The *Inr* sequence is then `YYCARR`. Similarly, in the TATA sequence, W is used to represent A or T, and the sequence is written `TATAWAW`. **Table 9.2** shows the complete set of ambiguous nucleotide codes.

We can think of the core promoter as the minimal requirement for eukaryotic transcription. Unlike prokaryotic RNA polymerase, which binds directly to the –10 and –35 promoter sequences, eukaryotic RNA polymerase II (the form of RNA polymerase that transcribes mRNA) binds to **transcription factors**: proteins that in turn bind to the DNA sequences. The transcription factors that bind the core promoter (e.g., TFIID, which binds the TATA box) direct RNA polymerase to the correct location for transcription, but a gene with *only* these promoter elements is only very weakly transcribed. Higher-level transcription requires additional transcription factors bound to additional sequences. Some transcription factors bind to sequences common at many different promoters, such as the CAT box (`5´ CAAT`) and the GC box (`5´ GGGCGG`), both of which usually occur within about 100 bp of the +1 site. Other transcription factors promote the transcription of genes only in a specific cell type or in response to some particular condition; their binding sites may be hundreds or even thousands of bp upstream. Examples include the estrogen response

Table 9.2 One-letter code for ambiguous nucleotides.

Code	Meaning
N	A, T, C or G (aNy base)
R	A or G (puRine)
Y	C or T (pYrimidine)
W	T or A (Weak)
M	C or A (aMino)
K	T or G (Keto)
S	C or G (Strong)
B	C, T, or G ("not A")
D	A, T, or G ("not C")
H	A, C, or T ("not G")
V	A, C, or G ("not T")

element (ERE; 5´ AGGTCANNNTGACCT) bound by the estrogen receptor in response to the hormone estrogen, the NF-κB site (5´ GGGRNNYYCC) used to activate growth and genes of the immune system, and the heat-shock element (5´ AGAAN repeats) activated in response to elevated temperature. Finding binding sites like these in a putative promoter region not only strengthens the case that a transcribed region has been identified but also provides clues about how the gene is regulated.

Solving the Problem

The questions in this section should help students in programming courses develop their implementation of a eukaryotic gene prediction program. Students in nonprogramming courses may wish to use these questions as exercises to test their understanding of the algorithms involved in sequence-based gene prediction.

The pattern-matching algorithm discussed earlier uses a threshold parameter to decide how closely a sequence must match the pattern. How is this different from matching ambiguous nucleotides? If the eukaryotic gene prediction algorithm can match ambiguous nucleotides, does it still need the threshold parameter? Which of the sequence patterns discussed previously would you want to require a program to find to identify a gene, and which would be optional or perhaps user-selected?

ATG codons used as start sites occur most commonly within a consensus sequence known as the Kozac sequence: 5´ gccRccATGG. In this sequence, capital letters represent highly conserved bases and lower case letters represent bases that are common but not as highly conserved. How could the algorithm be modified to account for the Kozac sequence? A short distance past the stop codon, eukaryotic genes have a polyadenylation site where the mRNA is cleaved and the poly(A) tail added. Although this sequence, 5´ AAUAAA, is known, why would it probably not be worthwhile to search for this sequence as a marker for the end of a predicted gene?

Programming the Solution

Your eukaryotic gene prediction program should search for start codons preceded by a core promoter sequence and allow users the opportunity to select other regulatory patterns from a list or read them in from a file (for example, one user might want to find estrogen-regulated genes but someone else might be interested in heat-shock genes).

Your program will need to recognize the codes for ambiguous nucleotides such as Y and W. Suppose you are searching for `YYCARR` (the *Inr* sequence). One approach is to search for the unambiguous bases `CA` and then search backward and forward for valid nucleotides. Or, you could create a list of all possible values (`CCCAGG, CTCAGG, TCCAGG, TTCAGG, CCCAGG, CTCAGG`, etc.) and then search for an exact match with any one of those values. Regular expressions or character classes could be used to help with this search if appropriate for your language. Your program will also need to allow for some variation from the consensus sequence.

Running the Program

Create some short sample sequences to test your program; include ATGs that are and are not preceded by core promoter sequences or other promoter elements. Once you have a working version of your program, download a **test sequence** containing a eukaryotic chromosome region with one predicted gene from the *Exploring Bioinformatics* website. How does your program fare with this complex sequence? After completing the Web Exploration in Chapter 10, you may wish to compare your program's output to that of a program with more sophisticated prediction methods.

More to Explore: Transcription Factor Binding Sites

Although most currently popular eukaryotic gene prediction programs incorporate content-based or probabilistic methods (Chapter 10), sequence-based methods remain important for exploring how predicted genes might be regulated by identifying binding sites for known transcription factors. If you would like to explore this idea further, you may want to look at the **Jaspar** database of transcription factor binding sites or the **TFSEARCH** or **MAST** search tools that can look for binding sites in a sequence you provide.

Connections: Ongoing Need for Gene Discovery

With the human genome "finished" since 2003, you might wonder if the need for gene discovery is fading. On the contrary, gene prediction remains a thriving part of bioinformatics for a number of reasons. Next-generation sequencing offers more sequences faster and cheaper than ever before, and new genomes—from viruses and bacteria to vertebrates—are being sequenced at the rate of dozens per month. Although there are often related genomes that allow annotation by alignment, each genome is unique and has genes never previously sequenced. Sequencing of metagenomes (see Chapter 8) of completely unknown organisms is resulting in the identification of many genes unlike anything in the databases. Even within sequenced genomes, gene discovery is an ongoing process; as discussed in Chapter 10, no one yet knows with certainty the actual number of genes in the human genome—let alone how many total proteins (considering alternative splicing and other complications) they encode.

The study of small RNAs has become a key area of molecular genetics within the past few years, with the increasing recognition that short functional RNA molecules play important roles in the lives of cells. In addition to tRNAs, small RNAs are found as components of the ribosome, the spliceosome, and some enzymes ("ribozymes" such as telomerase, the enzyme that constructs the ends of chromosomes). Genes encoding the extremely small (20–25 nucleotide) short-interfering RNAs (siRNAs) and micro RNAs (miRNAs) recently shown to regulate gene expression and contribute to antiviral defenses are especially difficult to predict, and some estimates suggest there may be tens of thousands of such genes in the human genome. It is certain that the need for gene discovery will not soon disappear. New kinds of genes require the development of new computational algorithms and bioinformatic techniques, and similarity and structure analyses will continue to be needed to uncover the functions of newly discovered genes.

BioBackground: ORFs, Consensus Sequences, and Gene Structure

There are many ways to define a gene. One that covers most bases is that a gene is a **transcription unit**: a segment of DNA that can be transcribed into RNA. Although we most often think about genes encoding proteins, this definition also covers genes that encode functional RNAs, such as tRNAs and rRNAs used in the process of translation, as well as small regulatory RNAs and components of various enzymes. A transcription unit must have a **promoter**: DNA sequences allowing RNA polymerase to identify and transcribe the gene. If it is a protein coding gene, then within the transcribed region, there must be an **open reading frame**: a start codon (ATG, or AUG in RNA), a set of codons encoding various amino acids, and a stop codon (TGA, TAG or TAA).

For a protein coding gene, the eukaryotic ribosome begins translating at the first start codon of an mRNA. Thus, the eukaryotic transcription unit can contain only a single ORF. However, this ORF may occur in segments called **exons** broken up by noncoding regions called **introns**. In a prokaryotic cell, the ribosome finds the correct start codon by binding to a sequence known as the **Shine-Dalgarno sequence** or **ribosome binding site** that precedes the start codon by a few bases. Thus, a prokaryotic transcriptional unit may contain multiple ORFs, each encoding a distinct protein and each preceded by a Shine-Dalgarno sequence. A transcription unit containing two or more ORFs is known as an **operon**, and the proteins encoded by genes in an operon often function together in some cellular process. **Figure 9.5** compares eukaryotic (A) and prokaryotic (B) transcription units.

Because prokaryotes lack introns, we can readily identify unbroken ORFs by looking for start and stop codons; the amino acids encoded by the codons between the two can be identified by reading the nontemplate (mRNA-like) strand and applying the genetic code (Chapter 2). Certainly, long ORFs are likely to be genes, but it is harder to tell if a short ORF might encode a short protein. An ORF preceded by a Shine-Dalgarno sequence and (farther upstream) a promoter sequence can be identified as a gene with more confidence, although the possibility that an ORF may be separated from its promoter by one or more other genes in the same operon must be considered.

Eukaryotic DNA lacks Shine-Dalgarno sequences to conveniently mark start codons, and an intron-interrupted ORF may be spread over tens or hundred of thousands of nucleotides. Promoter

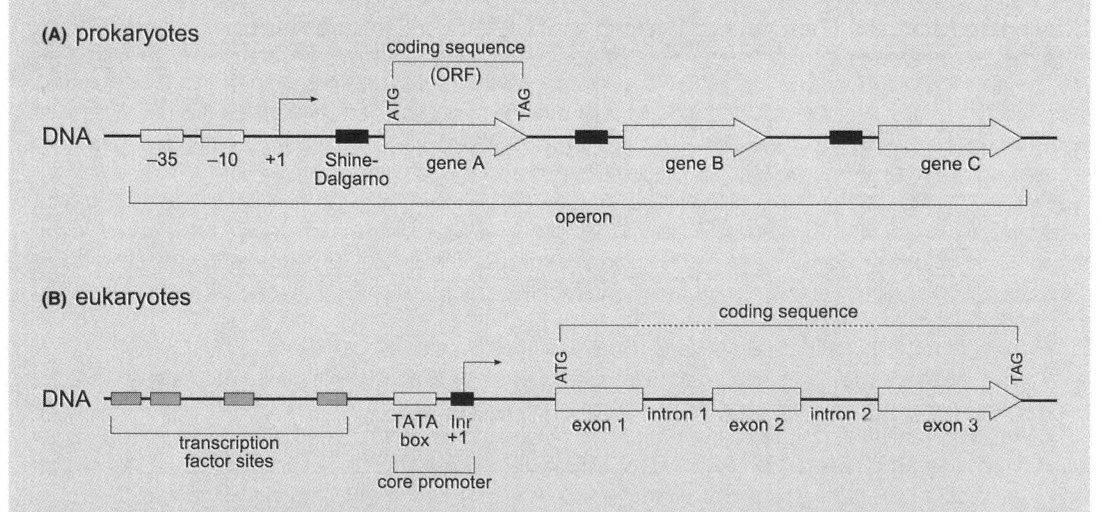

Figure 9.5 Comparison of (A) a prokaryotic transcription unit, showing a three-gene operon with a single promoter and individual Shine-Dalgarno sequences marking the start codon for each ORF; and (B) a eukaryotic transcription unit, showing a single gene interrupted by introns and preceded by a core promoter region and additional transcription factor binding sites.

regions still serve as useful clues, but whereas prokaryotes have clear consensus sequences for promoters, eukaryotic RNA polymerase looks not for a specific sequence but rather for an assembly of transcription factors bound to sites that may be near the transcriptional start site or hundreds of base pairs away. Some transcription factors bind most promoters, whereas others are specific to a particular cell type or condition. Furthermore, in both prokaryotes and eukaryotes, variation among species can be seen in the binding proteins and thus the sequences they bind. Promoters are also used to initiate transcription of genes for noncoding RNAs, but in eukaryotes, there are three distinct RNA polymerases (I, II, and III) that transcribe different kinds of genes (rRNA, mRNA, and tRNA/small RNAs, respectively), each with its own distinct promoter structure.

Table 9.3 shows some DNA sequences that are important in gene expression in prokaryotes and eukaryotes. These are referred to as **consensus sequences**, because they are not as precise as might be imagined. The prokaryotic promoter, for example, is defined by two six-nucleotide sequences. One, the **−10 sequence**, is centered at about 10 bp upstream of the transcriptional start site and is similar to 5′ TATAAT. The other, the **−35 sequence** is centered at about 35 bp upstream of the start site and is similar to 5′ TTGACA. However, few if any natural promoters contain exactly these two sequences. Genes expressed at a high level tend to have closely matching promoter sequences, whereas weaker promoters are farther from the consensus sequence, but even strongly expressed promoters typically vary from these "canonical" sequences by a nucleotide or two. The consensus sequences were developed by sequencing and aligning the promoter regions (determined by biochemical and molecular experiments) of multiple genes and looking for the sequences that are conserved among them (**Figure 9.6A**). The nucleotides in the consensus are those that occur most frequently; ambiguous nucleotide codes (**Table 9.2**) are used when two or more occur with nearly equal frequency. A graphical representation called a **sequence logo** (**Figure 9.6B**) gives a better idea of the relative occurrences of the four nucleotides at each position. The sequences given in this chapter for the Shine-Dalgarno site, TATA box, *Inr* site, transcription factor binding sites, and so on are all consensus sequences derived from studying the sequences found in many genes.

Table 9.3 Consensus sequences for gene expression in prokaryotes and eukaryotes.

Sequence	Consensus (5′ → 3′)	Function
	Prokaryotes	
−10 sequence	TATAAT	RNA polymerase binds to start transcription
−35 sequence	TTGACA 17±2 from −10	RNA polymerase binds to start transcription
Shine-Dalgarno	AGGAGG 5±2 from ATG	Ribosome binds to find start codon
	Eukaryotes	
TATA box	TATAWAW	Core promoter; binds TFIID
Inr sequence	YYCARR	Core promoter; contains +1 sequence (C)
GC box	GGGCGG	Transcription factor binding site
CAT box	CAAT	Transcription factor binding site
Kozak consensus	gccRccATGG	Context of start codon
5′ splice site	MAG | GTragt	Bound by spliceosome to remove introns
3′ splice site	cAG | G	Bound by spliceosome to remove introns
intron branch site	CTRAY	3′ end of intron binds to mark for degradation
polyadenylation site	AAUAAA	Cleavage of mRNA for poly(A) tail

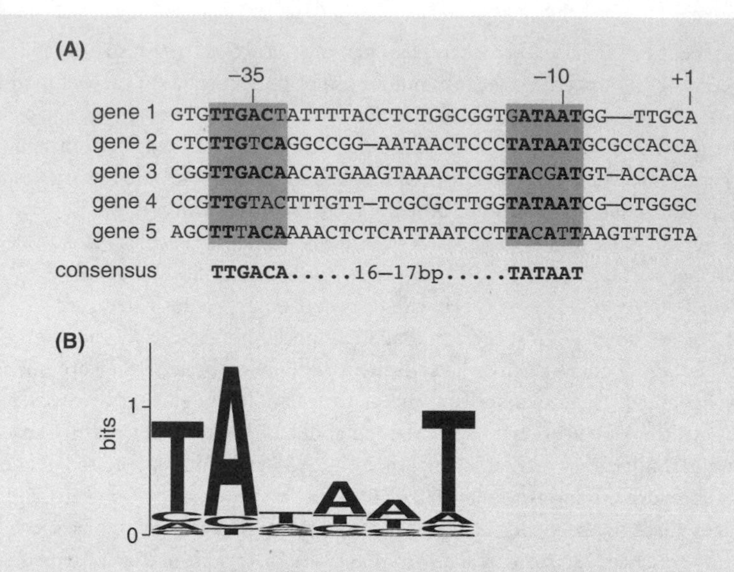

Figure 9.6 Generation of a consensus sequence. (A) The prokaryotic promoter consensus sequence derived from sequences of individual promoters. Conserved regions are shaded, with individual nucleotides that match the consensus in bold. (B) Sequence logo showing the occurrence of the four nucleotides at each position in the −10 promoter consensus, generated with WebLogo from a subset of the data published by Harley and Reynolds (see References and Supplemental Reading). Sequence logo generated from WebLogo: Crooks et al., *Genome Res.* 14:1188 (2004).

References and Supplemental Reading

Introduction to Gene Prediction Methods

Burge, C. B., and S. Karlin. 1998. Finding the genes in genomic DNA. *Curr. Opin. Struct. Biol.* **8**:346–354.

NEBcutter

Vincze, T., J. Posfai, and R. J. Roberts. 2003. NEBcutter: a program to cleave DNA with restriction enzymes. *Nucleic Acids Res.* **31**:3688–3691.

EasyGene

Larsen, T. S., and A. Krogh. 2003. EasyGene—a prokaryotic gene finder that ranks ORFs by statistical significance. *BMC Bioinform.* **4**:21–35.

Consensus Sequences for E. coli *Promoters and Sequence Logos*

Crooks, G. E., G. Hon, J. M. Chandonia, and S. E. Brenner. 2004. WebLogo: a sequence logo generator. *Genome Res.* **14**:1188–1190.

Harley, C. B., and R. P. Reynolds. 1987. Analysis of *E. coli* promoter sequences. *Nucleic Acids Res.* **15**:2343–2361.

Schneider, T. D., and R. M. Stephens. 1990. Sequence logos: a new way to display consensus sequences. *Nucleic Acids Res.* **18**:6097–6100.

Chapter 10

Advanced Gene Prediction:
Identification of an Influenza Resistance Gene

Chapter Overview

This chapter builds on the sequence-based gene prediction methods discussed in Chapter 9 and examines content-based and probabilistic methods of gene discovery. These methods are of particular importance in eukaryotic gene prediction: The division of eukaryotic coding sequences into multiple exons separated by variable-length introns with poor consensus sequences at their boundaries greatly increases the difficulty of identifying coding sequences computationally. Codon usage and CpG island identification are introduced as content-based algorithms contributing to gene prediction, and neural networks and hidden Markov models are presented as examples of probabilistic gene prediction. The Web Exploration gives students the opportunity to use some of these prediction methods, whereas the Guided Programming Project enables programming students to experiment with prediction of CpG islands. In the On Your Own Project, students explore the design of a gene prediction method based on a hidden Markov model.

Biological problem: Identification of an influenza-resistance gene

Bioinformatics skills: Exon–intron prediction, neural networks, hidden Markov models

Bioinformatics software: GENSCAN, AUGUSTUS, Sequence Manipulation Suite (CpG island prediction), Neural Network Promoter Prediction

Programming skills: Frequency matching and sliding windows, hidden Markov modeling

Understanding the Problem:
Exon Prediction

Among the priorities for influenza research laid out by the World Health Organization in 2009 is the investigation of genetic factors affecting susceptibility of individuals to influenza virus infection. Understanding how individual genetic variation might result in either increased susceptibility to influenza or increased resistance to the disease could lead to new preventative or therapeutic measures, either conventional or genetic. To be useful, however, recognition of heritable factors altering resistance must be followed by identification of specific genes and alleles. Methods such as genome-wide association studies (GWAS; see Chapter 1) can identify general areas of the genome connected to a phenotype, but gene prediction methods may be needed to identify specific genes located in the identified region.

One of the surprises in the "rough draft" of the human genome announced in June 2000 was the small number of protein coding genes: Whereas many researchers had predicted 80,000 to 100,000 genes in the human genome, the actual number appeared to be less than 30,000. Indeed, by the time a "finished" genome sequence was announced in April 2003, the estimate of protein coding genes had been further revised downward to between 20,000 and 25,000. Even today, the exact number of genes in the genome remains uncertain. Annotation of the genome, the identification of genome elements and their functions (**Figure 10.1**), is an ongoing effort. A 2012 report from the ENCODE consortium, whose goal is to definitively catalog the human genomes, identified 20,687 protein coding genes, but further studies are likely to change that number. Gene discovery software plays an important role in this continuing process.

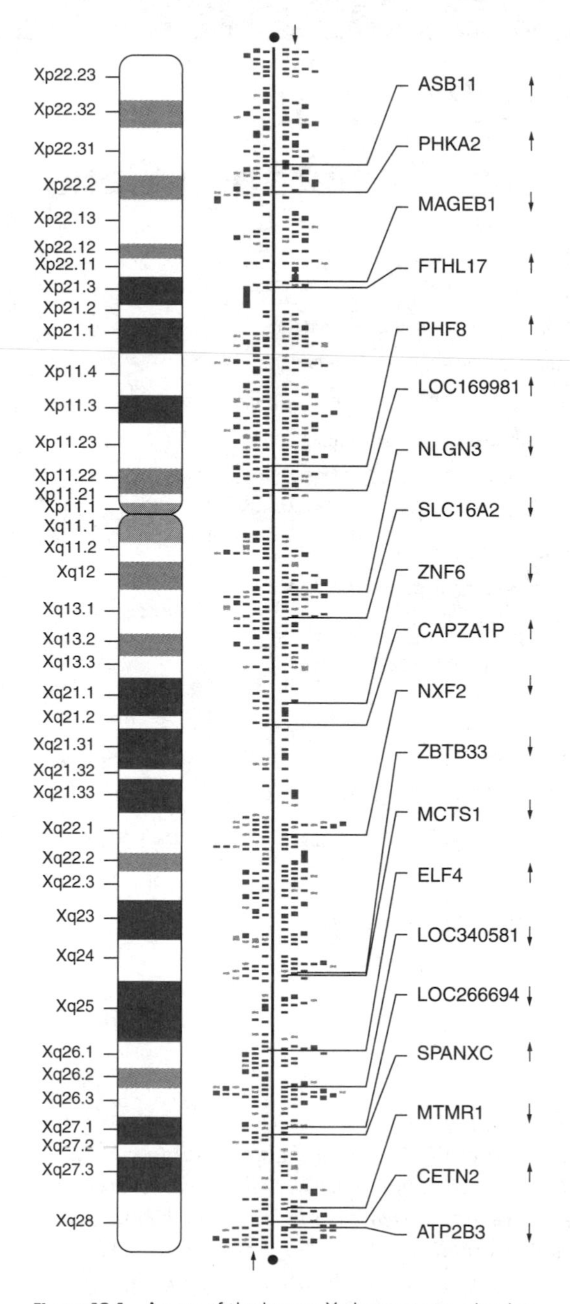

Figure 10.1 A map of the human X-chromosome, showing locations and identities of some of its genes.

As discussed in Chapter 9, sequence-based methods of gene prediction are the most straightforward and are quite reliable in prokaryotes. In eukaryotes, however, a number of problems arise. First, there is no Shine-Dalgarno sequence to mark the start codon; eukaryotic translation begins at the *first* start codon in the mRNA, and unambiguous identification of the transcriptional start site is difficult. Second, eukaryotic promoters are a collection of transcription factor binding sites rather than the more consistent −10 and −35 sequences of prokaryotes; many include the TATA box and *Inr* sequences of the core promoter, but this is not universally the case. Third, most importantly, there are very few unbroken ORFs: Nearly all genes in eukaryotes, especially higher eukaryotes, are split into multiple exons separated by introns. Finally, the sequence patterns at the intron–exon boundaries lack the clarity needed for reliable sequence-based prediction; it is clear from the sequence logos for the 5′ (or splice donor; **Figure 10.2A**) and 3′ (or splice acceptor; **Figure 10.2B**) sites that only a dinucleotide is truly conserved at each boundary, surrounded by a weak consensus. Thus, we need to consider additional methods of gene discovery in annotation of eukaryotic genomes.

Gene prediction is used to identify genes within a newly sequenced genome but is also valuable in identifying genes when a particular genome region has been associated with a disease or phenotype of interest. In this chapter, we see how gene discovery

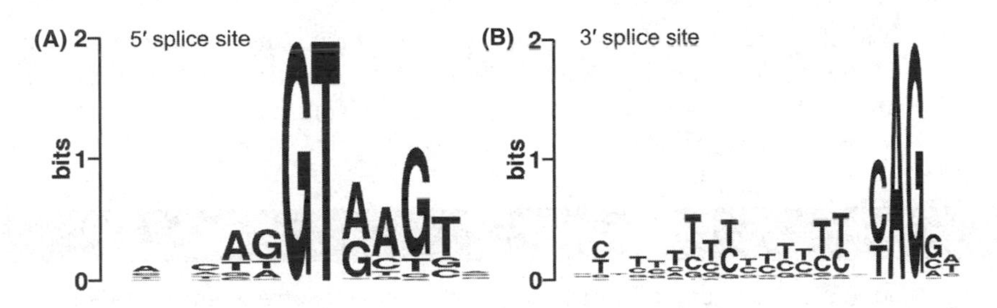

Figure 10.2 Sequence logos showing the poor consensus sequences found at the (A) 5′ (splice-donor) and (B) 3′ (splice-acceptor) sites between introns and exons. Sequence logo generated from WebLogo: Crooks et al., *Genome Res.* 14:1188 (2004).

algorithms designed to distinguish exons from introns can lead to the identification of a potential influenza resistance gene within a large DNA region correlated with inherited resistance. The gene examined in this chapter is known to interact with the influenza virus and has been suggested by Wolff et al. as a possible resistance gene (see References and Supplemental Reading); however, the identification of its chromosomal region with resistance is hypothetical.

Bioinformatics Solutions:
Content- and Probability-Based Gene Prediction

If we cannot rely on ORFs and consensus binding sites to clearly define the set of genes in a eukaryotic genome, how else can we approach this problem? In Chapter 9's Web Exploration, we used one method that did not depend on identifying particular sequence patterns: EasyGene combines sequence-based searches for ORFs and Shine-Dalgarno sequences with an examination of codon-usage patterns. Codon usage is an example of a content-based method of gene prediction: A putative sequence is examined to see if the frequency of usage of different codons matches that observed for the organism as a whole. In reality, there could be reasons why some genes have a different codon bias than others (for example, some genuine genes may have been acquired by horizontal gene transfer), but, in general, authentic genes all show similar codon usage within one organism. This technique can also be applied to prediction of introns and exons within a presumed transcription unit: Where codon usage changes noticeably from the norm, a boundary between an exon and an intron has probably been crossed. Another content-based method is looking for **CpG islands** (see BioBackground), structures associated with transcribed regions.

A problem with content-based methods is that they are not very precise. We may be able to find regions where codon usage matches the expected frequency well or poorly, for example, but this is unlikely to tell us exactly where an exon–intron boundary lies. Combining two methods, such as looking for a consensus exon–intron boundary sequence within the region where codon usage changes, can yield a better prediction than either the sequence- or content-based method alone.

Better predictions still can be achieved by probabilistic methods such as **hidden Markov models** (**HMMs**). These are not truly distinct methods, but rather they use sequence and content data to calculate probabilities, such as the probability that any given nucleotide lies within an exon. Points where that probability declines sharply are likely to mark the boundaries between exon and intron, whereas points where it increases sharply mark intron–exon boundaries. This chapter considers some content- and probability-based methods to see how they are used to identify the segments of a coding sequence within a larger sequence. In the Guided Programming Project, you will experiment with how to identify CpG islands, and in the On Your Own Project, you will try your hand at designing an HMM to identify introns and exons.

BioConcept Questions

1. Why is codon usage a poor predictor of the point where an exon and intron are joined? Why is the 5′ splice site consensus also a poor predictor?
2. How much of a typical human gene is usually coding sequence, versus intron sequences that are spliced out (you may wish to recall the gene displays you saw in the UCSC Genome Browser in Chapter 1)? How does this pattern affect the difficulty of predicting introns and exons?
3. Why are CpG islands considered valuable for gene prediction? Where would you expect to find one with respect to a eukaryotic transcription unit? What other elements might you look for in connection with the CpG island to increase the strength of a gene prediction?
4. How could alignment of a sequence with orthologous sequences contribute to the prediction of exons and introns? How could expression data (e.g., cDNA sequences) contribute?

Understanding the Algorithm:
Codon Usage, Frequency Matching, HMMs, and Neural Networks

Learning Tools

The *Exploring Bioinformatics* website has a link for an online hidden Markov model demo that you can use to get a better idea of how this model chooses the most likely hidden states given a set of probabilities.

This section briefly considers algorithms for two content-based methods of gene prediction: codon usage and identification of CpG islands. We then spend some time on understanding hidden Markov models and briefly cover neural networks, two probabilistic methods that can be applied to gene prediction but are also used in many other areas of bioinformatics, including protein structure prediction and sequence alignment.

Using Codon Frequencies in Gene Prediction

In a protein coding sequence, codons are not used with equal frequency. Some amino acids are much more common in proteins than others: Serine is the most common amino acid in vertebrate protein sequences (about 8% of all amino acids), whereas tryptophan is the least common (only 1%). Additionally, the genetic code is redundant, and where there are multiple synonymous codons for one amino acid, they are not used with equal frequency. This idea was discussed briefly in Chapter 9, with the codon frequency table for *E. coli* given in Table 9.1.

How might we apply the idea of codon frequency to predicting which sequences are exons and which are introns? An exon–intron boundary would be expected to separate a region where the codon frequency closely matches the expected frequency for the organism from a region where the frequency matches poorly, and an intron–exon boundary would do the reverse. As shown in **Figure 10.3**, we could examine a range or "window" of nucleotides, perhaps 75 nt (illustrated with a short sequence as window 1A in Figure 10.3), break it into codons (25 codons, in this case), and measure codon usage. Several codon usage measures

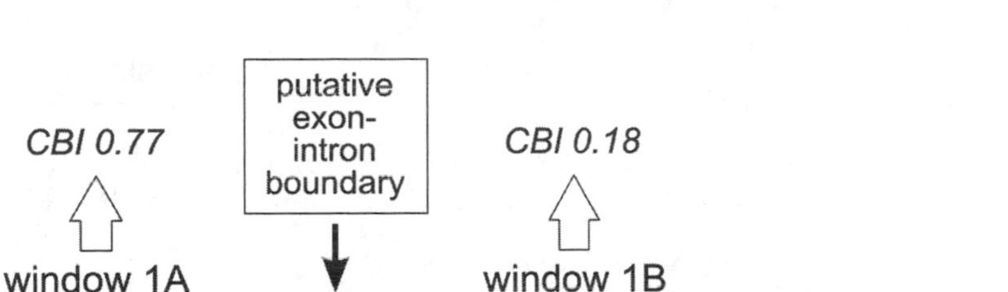

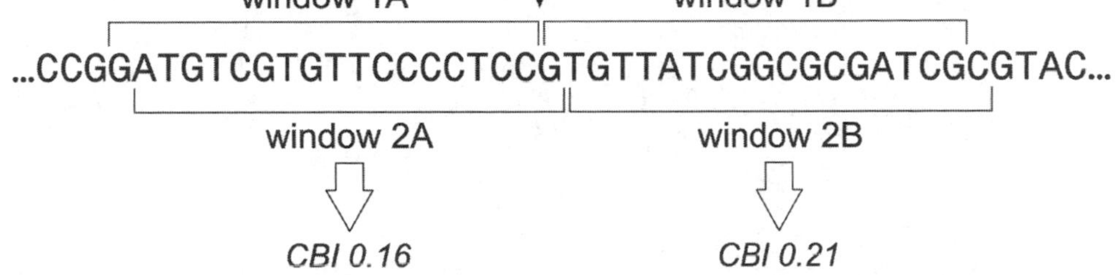

Figure 10.3 Sliding-window approach to exon prediction by codon-usage bias. Codon usage is compared for two adjacent same-length sequence windows (1A and 1B); a large difference suggests an exon–intron boundary. The windows slide along the sequence (2A and 2B) to identify potential boundaries in different reading frames along the length of the sequence.

are in common use; one is the **codon bias index** (**CBI**) proposed by Bennetzen and Hall (see References and Supplemental Reading) that compares the usage of "preferred" (most common codons) to the random occurrence of those codons, giving a number between 0 (random codon usage) and 1 (exclusive usage of preferred codons). The same procedure is then repeated for the 75 nucleotides immediately downstream (window 1B in Figure 10.3) and the difference between the two is determined. The two windows are then shifted by one nucleotide (windows 2A and 2B in Figure 10.3), and the difference in CBI is computed again; note that the codons examined here are in a different reading frame.

Continuing through the sequence with this **sliding window** approach, we expect to find points at which the boundary between the "A" and "B" windows corresponds to a drop in CBI to near zero (exon–intron boundary) or a sudden increase in CBI from near zero to a larger number (intron exon boundary). Additional constraints can be added to the algorithm based on our understanding of gene structure. For example, the putative boundaries can be rejected if the conserved GT and AG pairs are not present. Additionally, the first exon must start with ATG and should not be preceded by a splice consensus, and the last exon ends with a stop codon and is not followed by a splice consensus.

Prediction of CpG Islands

Given the difficulty of unambiguously recognizing a eukaryotic promoter region based on consensus sequences, identification of CpG islands (see BioBackground) adds valuable corroboration and can be used in combination with sequence-based methods and exon prediction techniques to help identify the first exon of a gene. We can find CpG islands with a **frequency matching** algorithm. This algorithm uses a sliding window approach like the one just discussed (except that only one sliding window is needed) combined with elements of a pattern-matching algorithm (Chapter 9), counting up CG pairs within each window and computing a CpG ratio. The steps of this algorithm are outlined next. Notice that the CpG ratio is really an odds ratio: The result is 1.0 if the number of CpG pairs found in a window is the same as the number that would be expected by chance. **Figure 10.4** shows the result of graphing the CpG ratio as the window slides through a DNA sequence.

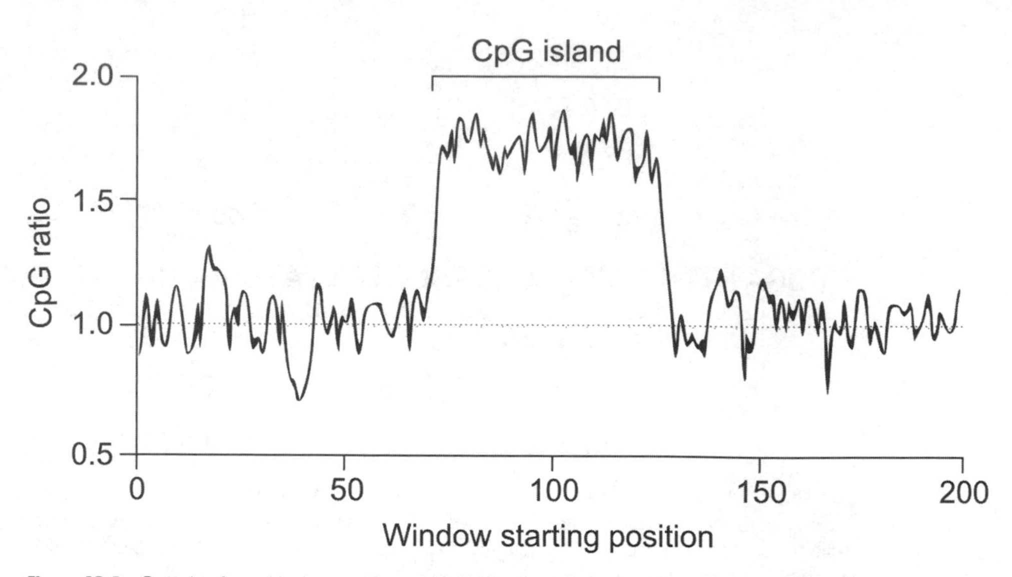

Figure 10.4 Sample of graphical output from a CpG island prediction program, with the CpG ratio (1.0 if the CpG frequency is the same as expected by chance) measured for each window as a sliding window moves across a sequence. A region of consistently high CpG ratio values represents a CpG island.

Algorithm

Frequency-Matching Algorithm

1. Determine window size and set start position to the first nucleotide in the sequence.
2. Count the number of CG pairs, C nucleotides, and G nucleotides in the window.
3. Calculate the ratio of observed to expected CpGs for the window:

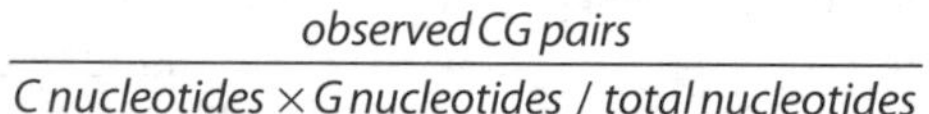

$$\frac{\textit{observed CG pairs}}{\textit{C nucleotides} \times \textit{G nucleotides} \,/\, \textit{total nucleotides}}$$

4. Increment the start position by 1. If the window is not longer than the remaining sequence, repeat step 3; otherwise, continue.
5. Examine the CpG ratios for all the windows and identify areas of CpG islands where the ratio is higher than a threshold.

HMMs for Gene Prediction

The difficulties with eukaryotic exon prediction discussed previously in combination with the explosion of genomic information available (especially with the advent of faster, cheaper next-generation sequencing) have driven the development of gene discovery algorithms to be more powerful even than combinations of sequence- and content-based methods. Many popular gene prediction programs are now based on implementations of hidden Markov modeling, probability-based algorithms that use sequence and content data to inform a calculation of the likelihood that a given sequence is part of an intron or exon.

Simply put, an HMM seeks to draw a conclusion about something that cannot be directly observed ("hidden") based on a set of observations and a set of known probabilities. A commonly given example is someone who wants to determine the weather in a certain city based on an observation such as umbrella sales or the activities a friend chooses. Given these observations and some basic data, such as the overall frequency of sunny and rainy days in that city, an HMM can compute the highest probability for the actual weather, which is the hidden state.

Applying an HMM to gene prediction, the nucleotides of a DNA sequence would represent the input observations. In a simple model for an exon–intron boundary (**Figure 10.5**), the nucleotides could exist in one of three hidden states: exon (E), intron (I), or splice site (S). For each state, we have an **alphabet** of possible symbols that could be output. A position in an exon, for example, could be any nucleotide from the alphabet A, C, G, and T. We then use the data we have about genes in the organism being studied to determine **emission probabilities** (*e*): the likelihood of each output. For example, we might assume that As, Cs, Gs, and Ts occur with equal frequency within an exon and assign each one an emission probability of 0.25. Given more information, however, we could refine the probabilities further: It turns out that in human exons, codon bias and other factors increase the likelihood of a G or C (see References and Supplemental Reading), so a better set of emission probabilities for the exon or E state would be 0.3 for G and C and 0.2 for A and T (Figure 10.5).

We also know there is a nucleotide bias at the splice-donor site (Figure 10.2A). A more realistic HMM could take into account all the data depicted in this sequence logo, but to keep our example simple, let's only use the data for the first two nucleotides of the intron, which are almost always G and T. Knowing the frequencies with which each nucleotide is found at these two positions (**Table 10.1**), we can construct a list of emission probabilities for each nucleotide of a two-nucleotide splice-donor site (the S_1 and S_2 states in Figure 10.5). Finally, we need emission probabilities for the intron or I state. Human introns tend to be slightly AT rich, with T (0.3) favored over A (0.27) and G (0.23) favored over C (0.2).

The last parameters needed for our model are the **transition probabilities** (*t*): the likelihood of changing from one state to the next versus the likelihood of remaining in the same state. A genuine splice-donor site is always followed by an intron, never an exon or another splice site, so we can assign $S_2 \rightarrow I$ a transition probability of 1.0. The probabilities for $S_2 \rightarrow E$ and $S_2 \rightarrow S_1$ are zero and therefore not shown in Figure 10.5. Similarly, we require a two-nucleotide splice-donor site, so $S_1 \rightarrow S_2$ would also have a transition probability of 1.0. For this example, we set the transition probability for $E \rightarrow S_1$ at 0.1, with $E \rightarrow E$ (remaining in the exon state) at 0.9. E cannot go to I without going to S_1 first, so $E \rightarrow I$ is zero. Finally, we set the probability of continuing in an intron, $I \rightarrow I$ at 0.9 as well, with the probability of ending the intron at 0.1. These transition probabilities are shown as arrows in Figure 10.5. Notice that the entire model can be easily represented with a picture; many authors have commented that the ability to make a statistical model for anything you can represent visually is a strength of hidden Markov modeling.

Now, our HMM can examine all possible states for each nucleotide in our input nucleotide sequence and then determine the overall probability of each outcome, or path through the states. Suppose we have a sequence that represents a two-codon exon followed by a GT splice site and four more intron nucleotides: `ATGCGCGTATTC`. In our simple model, because we have to start in an exon, end in an intron, and the splice site is a dinucleotide pattern, there are nine possible paths for this short sequence, as shown in Figure 10.5. For each, we can determine the probability at each position and then multiply to get the total probability. For example, for $ES_1S_2IIIIIIIII$, the emission probability of A in an exon position is 0.2, and the transition probability for $E \rightarrow S_1$ is 0.1. Then, the emission probability for T as the first nucleotide of a splice-donor site is

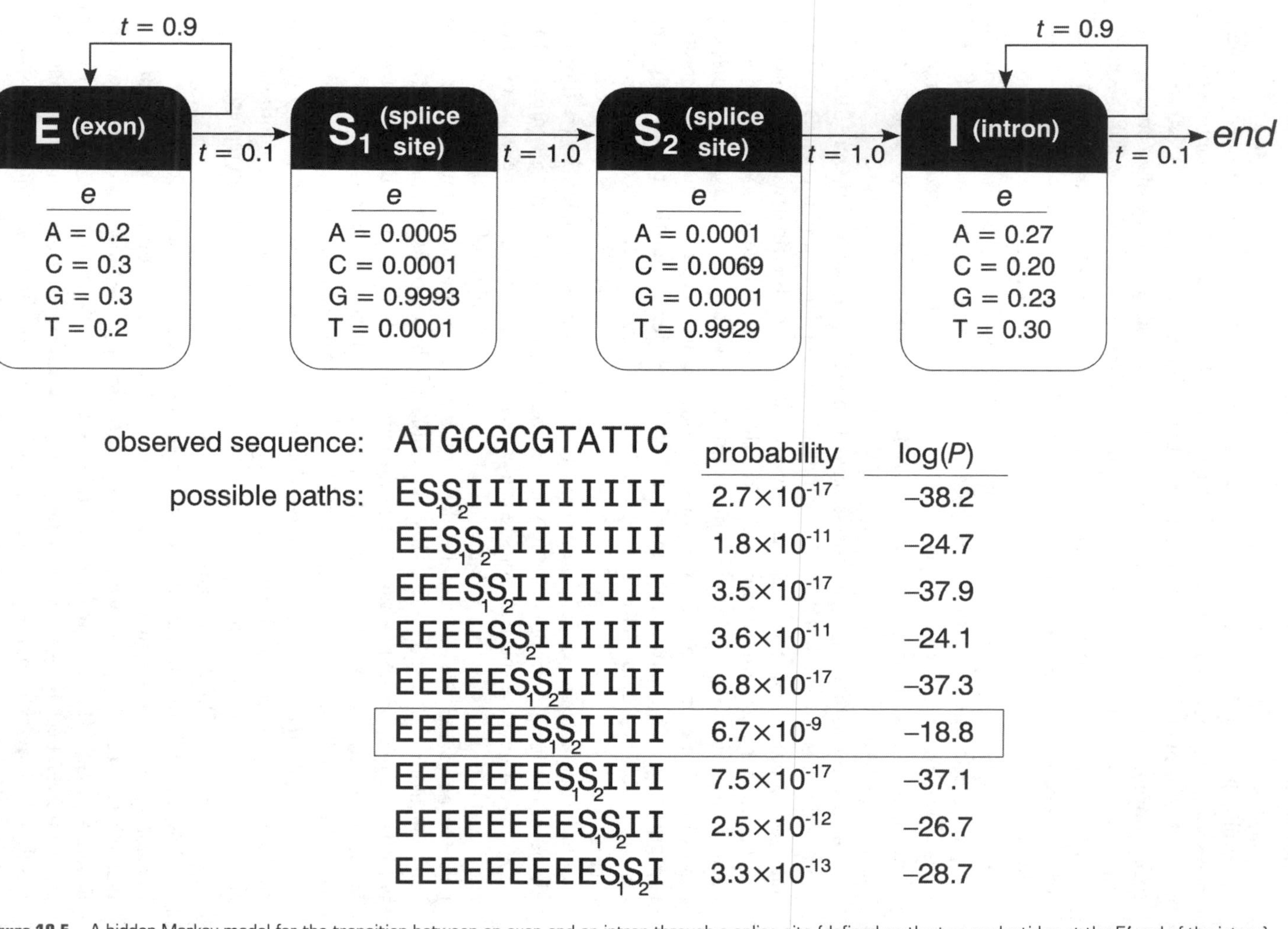

Figure 10.5 A hidden Markov model for the transition between an exon and an intron through a splice site (defined as the two nucleotides at the 5′ end of the intron). Black boxes show the four possible states in this model, with emission probabilities (*e*) in the white boxes below each state and transition probabilities (*t*) shown by the arrows between states. Below the model are the nine possible paths for a short DNA sequence and the probability of each; the highest probability is boxed and corresponds to a GT splice site.

Table 10.1 Nucleotide frequencies for the first two intron positions.

Nucleotide	Position 1	Position 2
A	0.0005	0.0001
C	0.0001	0.0069
G	0.9993	0.0001
T	0.0001	0.9929

0.0001, the transition probability for $S_1 \rightarrow S_2$ is 1.0, and the emission probability for G at S_2 is 0.0001. Next, the transition probability from $S_2 \rightarrow I$ is 1.0, the emission probability for C in an intron position is 0.2, the transition probability for $I \rightarrow I$ is 0.9 and so on. The total probability, *P*, is the product of all these individual probabilities: $0.2 \times 0.1 \times 0.001 \times 1.0 \times 0.001 \times 1.0 \times 0.2 \times 0.9 \ldots$, which works out to 2.7×10^{-17}. Taking the natural log of *P* gives a log probability value of −38.2.

After computing the probability for each of the nine possibilities (see Figure 10.5), it is easy to determine which probability is the greatest (largest log *P*). In this example, the result matches the design of our test data, with a splice site following the two-codon exon.

In our simple model, we are not considering what happens downstream of the intron. In reality, there would be another transition to another splice site and then to another exon—which we could similarly model by adding additional states with corresponding transition and emission probabilities. We also used somewhat arbitrary transition probabilities; a better model would base these on the typical length of exons and introns in the organism. We also have not yet accounted for the fact that the first exon begins with an ATG and is not preceded by an intron, whereas the last exon ends with a stop codon and is not followed by an intron. We can further strengthen the model by explicitly including the probabilities of other nucleotides surrounding the two splice sites. Additional sophistication could be built into the model in many ways: The CG bias in the promoter region could also be taken into account; for example, our codon bias data could be calculated into the exon emission probabilities. Some HMMs even include advanced statistical methods such as Bayesian statistics to calculate the emission and transition probabilities at each step. You will use existing HMM-based gene prediction software in this chapter's Web Exploration, and the On Your Own Project will give you an opportunity to design an HMM that is a little more complex than our initial example.

Neural Network Modeling

The **neural network** (**NN**) algorithm is one more important gene prediction method that we touch on briefly here. It takes its name from the network of neurons in the brain, which clearly recognizes patterns better and faster than a computer can. You immediately recognize a friend's face regardless of its setting, for example, whereas face-recognition software can readily be fooled by a hat or sunglasses. Although no one knows exactly how neural processing works, we know that each of your neurons is connected to many other neurons and fires when the sum of its many inputs, positive and negative, exceeds some threshold. It is this behavior that neural network algorithms attempt to mimic.

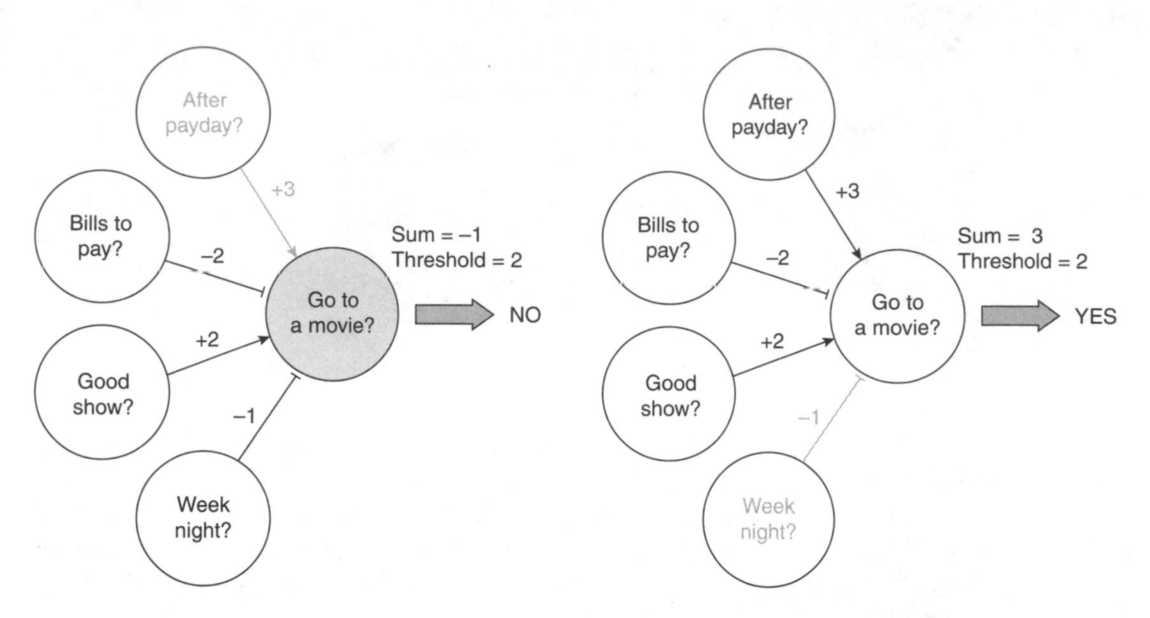

Figure 10.6 Decision making with a neural network. Four inputs, each weighted differently, contribute to deciding whether or not to see a movie. The sum of the inputs must exceed 2 in order to see the movie; this is not true in the left diagram but is true in the right diagram.

The decision-making process illustrated in **Figure 10.6** is a simple example of a neural network: We decide whether to go to a movie based on the sum of four inputs. Each input is given a different weight, and the sum must exceed a threshold (2) to make the choice to see the movie. Similarly, inputs for a neural network to predict exons might include codon bias, CG content, consensus sequences, length, and so on.

The hardest part of developing a neural network algorithm is deciding how to weight the inputs and set the threshold. Often, this is accomplished by adding a machine learning algorithm. An initial model is developed and used to classify a training set of known sequences as intron or exon sequences; the algorithm "learns" by adjusting weights and threshold until it can classify the training set with minimal errors. You will use a neural network algorithm in this chapter's Web Exploration.

Test Your Understanding

1. Suppose you use the sliding window algorithm described to analyze codon bias. At several points in a DNA sequence, you see a high score in your first window and a low score in your second window. But, when you slide the window by one or two nucleotides, you get low scores in both windows. How would you explain this pattern? How might you want to account for it in deciding where your exon–intron boundaries are?
2. Explain why the codon-usage method is likely to be imprecise in defining exon–intron boundaries.
3. CpG island prediction algorithms generally require not only a higher-than-expected frequency of CG pairs but also that the region under examination has an overall higher percentage of G+C than the average in the genome. What is the value of this constraint?
4. CpG islands are associated with promoter regions. How can this help with exon prediction?

5. Draw an HMM that requires an ATG followed by some exon nucleotides, a splice-donor site, and then some intron nucleotides.
6. How might the first exon be distinguished from internal exons in an HMM?
7. Suggest some qualities of a DNA sequence that you would weight positively and some that you would weight negatively in developing a neural network model to identify an exon.

CHAPTER PROJECT:

Identifying an Influenza Resistance Gene

Often, the study of a genetic disease or another genetic trait leads to a general region of the genome but does not immediately identify a particular gene. Chapter 1 dealt with how SNPs can be identified in GWAS experiments; as you saw in that chapter, the extensive human genome data now available often allows us to simply browse a genome region to look for genes of potential interest. But what happens when there is less information with which to work? This chapter's projects focus on a hypothetical but realistic scenario involving a chromosome region suspected of including an influenza resistance gene.

Learning Objectives

- Understand how eukaryotic genes introduce additional complexity into the problem of gene prediction and recognize the limitations of sequence-based methods
- Know some content-based methods of gene prediction and appreciate their strengths and limitations
- Be able to combine content-based and probabilistic methods of gene discovery to identify the most probable locations of introns and exons in a eukaryotic DNA sequence
- Know how to design an HMM to integrate sequence and content data for a more precise and accurate determination of exon–intron boundaries

Suggestions for Using the Project

In the Web Exploration for this project, students analyze a large DNA sequence to look for potential genes using several different gene prediction techniques. The different methods have different strengths, and the value of combining multiple methods will be recognized. If time is limited, the first part of the Web Exploration gives the most comprehensive look at gene prediction. In the Guided Programming Project, students implement a sliding window algorithm for a content-based gene prediction method, identifying CpG islands. In the On Your Own Project, students design (and, in programming courses, implement) an HMM that builds on the discussion in Understanding the Algorithm and includes a splice-acceptor site.

Programming courses:

- Web Exploration: Use existing tools including CpG island prediction, HMMs, and neural networks to identify exons, introns, and transcriptional units within a 90-kb segment of human DNA sequence. Part I could be used alone if needed.
- Guided Programming Project: Implement an algorithm to identify CpG islands using a sliding window algorithm.

- On Your Own Project: Design an HMM that incorporates both splice-donor and splice-acceptor sites and implement the HMM in a desired programming language. Optionally, increase the sophistication of the model by incorporating start codons and the potential for multiple exons.

Nonprogramming courses:

- Web Exploration: Use existing tools including CpG island prediction, HMMs, and neural networks to identify exons, introns, and transcriptional units within a 90-kb segment of human DNA sequence. Part I could be used alone if needed.
- On Your Own Project: Design an HMM that incorporates both splice-donor and splice-acceptor sites and then increase the sophistication of the model by incorporating start codons and the potential for multiple exons.

Web Exploration: Finding Genes in a Eukaryotic Genome Sequence

As an influenza researcher, you have become interested in a small number of individuals you know were unvaccinated and repeatedly exposed to the 2009 H_1N_1 influenza virus but did not become ill. When immunological testing showed they were not actually immune to the virus, you began to seek a genetic link that might explain their resistance to this disease. Using next-generation sequencing, you were able to identify common transcripts from respiratory epithelial cells that are missing in your resistant patients. This leads you to sequence a particular genome region in one patient, some 90,000 bp (90 kb) from the 1q25.3 region of chromosome 1. You would now like to analyze that genome fragment to identify genes within it that might be involved in susceptibility or resistance to influenza.

Part I: Gene Prediction with GENSCAN and AUGUSTUS

The number of available tools for gene prediction is somewhat mind-boggling. Several popular gene prediction programs are comprehensive in nature, bringing together several kinds of analysis in one piece of software; these would be a good place to start the analysis of a genome sequence or segment. We initially work with two such gene prediction programs, GENSCAN and AUGUSTUS.

GENSCAN (see References and Supplemental Reading) combines HMM-based models for coding-region and splice-site prediction with models that attempt to account for additional factors that affect splice-site choice as well as observed changes in splice sites and gene density in low-GC versus high-GC regions of human DNA. GENSCAN claims to correctly identify 70–80% of known exons. This comprehensive program produces clear and compact graphical output, making it easy to compare other programs' results.

Start by downloading **1q25.txt**, containing 90 kb of DNA sequence from human chromosome 1, from the *Exploring Bioinformatics* website. Navigate to a Web-based implementation of **GENSCAN** (there are several available) and input or upload your sequence in FASTA format. Choose a training set appropriate to analyzing human DNA from the drop-down menu: the GENSCAN implementation at the Pasteur Institute provides `HumanIso`, suitable for humans and other vertebrates and *Drosophila* (click the help icon to see this information), whereas other implementations provide a vertebrate training set. The Pasteur implementation includes additional options for `Verbose output`, providing some additional information in the output file and to `Create Postscript`

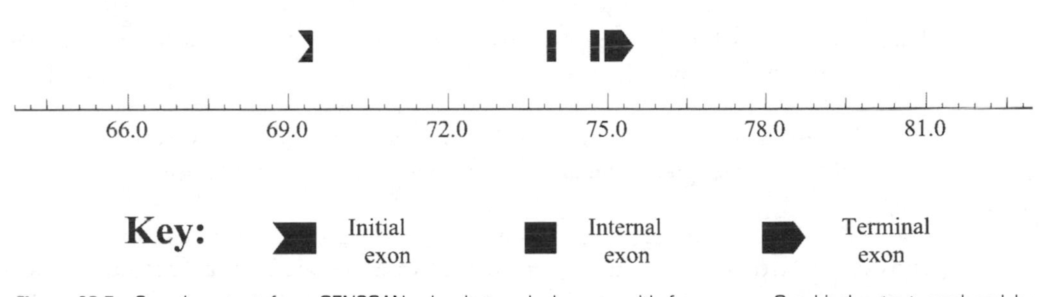

Figure 10.7 Sample output from GENSCAN, showing a single gene with four exons. Graphical output produced by GENSCAN. *J. Mol. Biol.* 268:78, 1997.

`output`, giving a graphical representation of the results; set these options to get the most useful output. Other parameters can be left at their defaults for now; if needed, they could be set to reduce the stringency of the criteria for exons or to scale the output. Run the program.

When the results appear, you will see a window containing text output (make this window full screen to make it easier to see). The output includes the specific locations of the predicted introns and exons, information on reading frames and splice sites, and translations for the putative coding regions. There is also useful information about the reliability of the predictions. You may want to save this output to a text file (from which you could copy protein sequences for later alignment, for example) and/or print it for later reference.

There will also be a window for graphical output. If you are using a Macintosh, you can simply right-click the small visible region of the graphical output and choose `Open with Preview`. PCs unfortunately lack built-in software to deal with PostScript files; alternatives include uploading the file to Google Drive, downloading the free Ghostscript viewer, opening with a graphics program such as Inkscape or Photoshop, installing a utility that makes PostScript files viewable with Adobe Reader, or finding an online conversion program. Choose one of these options as appropriate to view your graphical results. **Figure 10.7** shows an example of the kind of output expected from GENSCAN.

Web Exploration Questions

1. List the genes that GENSCAN found within the sequenced region, along with their lengths and the approximate length of the processed mRNAs. Why do the gene arrows point in different directions?

2. What is the difference between an exon marked `Init` and an exon marked `Intr` (in the text output)? Why is this difference significant in predicting genes?

3. Look at how the predicted proteins begin. Does this information strengthen or weaken the case for any of the genes?

4. What other features did GENSCAN identify (look in the text output)? Do these provide additional support for any of the predicted genes?

Unfortunately, there is no perfect gene prediction algorithm. Not only will most prediction programs return some potential genes that aren't "real," but they may place introns and exons at different positions. However, we might imagine that "real" genes should be detected by a variety of algorithms while false positives might tend to be more program specific. So, it is useful to run other prediction programs on the same sequence and see how their results compare.

AUGUSTUS is another popular gene prediction program that combines multiple kinds of prediction into a single piece of software (see References and Supplemental Reading). The core of AUGUSTUS is an *ab initio* prediction algorithm that uses HMMs to find the most likely sequence of hidden states (i.e., exon or intron for each nucleotide) that accounts for the sequence as a whole. The program can be "trained" by uploading sets of data (e.g., known genes from the organism being studied) and can incorporate user-defined information (such as locations of known expressed sequences) to improve its accuracy.

Navigate to the Web-based implementation of **AUGUSTUS**. Choose the Web interface, then upload the sequence from **1q25.txt**. Choose the correct organism from the drop-down menu; this will change the dataset used to "train" AUGUSTUS, so the training set should match the organism from which the sequence being analyzed originates. Note that you have some options for where AUGUSTUS will look for genes, as well as some "expert" options you can leave alone for now. Run the program to look for genes in your sequence.

AUGUSTUS will initially show text output that is quite similar to the output from GENSCAN: lists of predicted initial, internal, and terminal exons and translations of the predicted genes. Use the link provided to get to a list of available files containing graphical and text output, then choose `graphical browsable results`, which will show the results in a genome browser format similar to the UCSC Genome Browser (see **Figure 10.8A**). Exons are shown in color, with darker colors representing greater confidence in the predictions. Hovering over or clicking on regions of predicted genes will display details such as the coding sequence or predicted amino-acid sequence.

Will two different gene prediction programs give the same results? By now, you should realize there are no perfect criteria to identify exons, so you can probably guess that different programs using different algorithms will not necessarily identify the same sequences as exons. Indeed, if an exon is identified as such by more than one method, it would strengthen the evidence that it's a genuine exon. Thus, it's useful to compare the results of GENSCAN and AUGUSTUS. You could do this by examining the text output (importing it into a spreadsheet could make it easier to line up the exons identified by each program) or by using the graphical output. One approach would be to print the graphical output of one of the two programs and draw in the exons found by the other.

AUGUSTUS has a feature that makes this comparison easy: because its output is in genome browser format, custom tracks can be added. For example, you could add a track listing exons found by GENSCAN and compare them side by side with the AUGUSTUS results. **Figure 10.8B** shows the format of a text file listing a series of exons that could be added to AUGUSTUS as a track; you can easily manipulate your GENSCAN text output into this format. Unfortunately, the Web implementation of AUGUSTUS does not support the addition of custom tracks, so to use it would require that you install AUGUSTUS locally. If you choose this option, you could also upload tracks with your data on CpG islands or predicted promoters (see part II) as shown in Figure 10.8B.

(A)

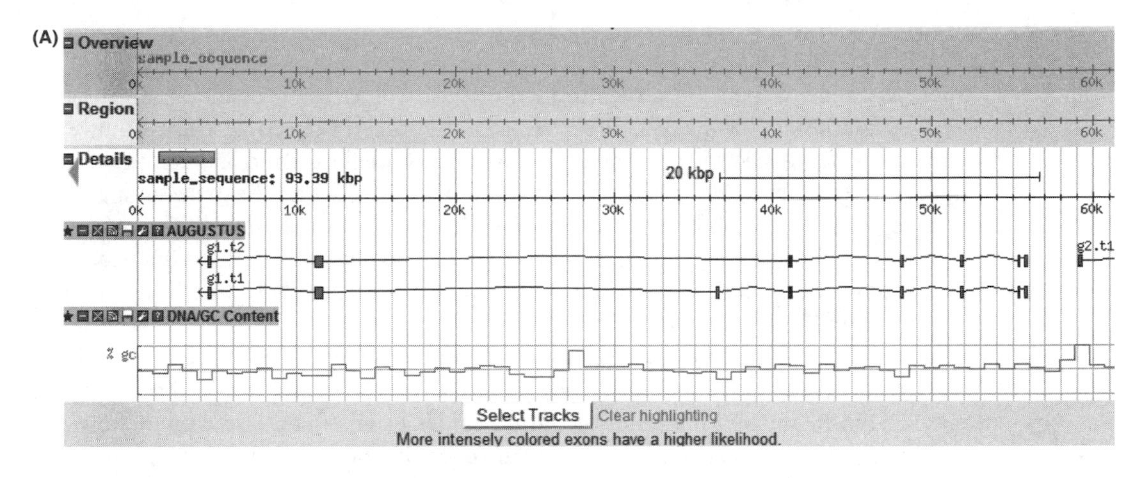

(B)

```
[GENSCAN]
key = Genes predicted by GENSCAN
GENSCAN "Gene 1" influenza_resistance_region:7236..7231
GENSCAN "Gene 1" influenza_resistance_region:13868..13615
GENSCAN "Gene 1" influenza_resistance_region:15463..15290
GENSCAN "Gene 2" influenza_resistance_region:48686..48538
GENSCAN "Gene 2" influenza_resistance_region:56933..56894
[CPG]
key = Predicted CpG Islands
CPG "CpG1" influenza_resistance_region:6000..6500
CPG "CpG2" influenza_resistance_region:4600..4700
```

Figure 10.8 (A) Sample output from AUGUSTUS. Exons (rectangles) and introns (thin lines connecting rectangles) are shown in a format similar to a typical genome browser. Potential splice variants are identified, and the overall G+C content of the DNA is shown in the bottom track. (B) Format of a text file in Feature File Format (FFF) to add two custom tracks to AUGUSTUS. Bracketed text is the name of the track; each line requires the name of the track, gene, or feature name and its location in the sequence. Graphical output produced by AUGUSTUS (Bioinformatics 19S2:215, 2003).

After comparing the two programs' output (by any method), you should be able to identify one major gene on which the two programs agree to a significant degree (though not perfectly). This would represent a gene on which further efforts to understand influenza resistance should be concentrated.

Web Exploration Questions

5. How does the number of genes predicted by AUGUSTUS compare to the results from GENSCAN?
6. How does the structure (i.e., length, number of introns and exons, position in the DNA) of the genes predicted by AUGUSTUS compare to GENSCAN?
7. How do the predicted proteins compare? Clearly, they're not identical, but do they appear related? For example, are they basically the same protein with perhaps some different splicing choices, or do they come from entirely different reading frames or even regions of the DNA? (You can of course use EMBOSS or BLAST to directly compare the proteins or their exons if you wish.)
8. Describe the gene that you conclude may be important in influenza resistance: total length, number of exons, processed length, number of amino acids, etc.

Part II: Evidence of Gene Expression

GENSCAN and AUGUSTUS served to identify at least a candidate gene of interest that might be responsible for the observed resistance to influenza infection. Clearly, however, the matter is not settled. At this point, the investigator might turn to less comprehensive programs to look for some specific features that might support the existence of a gene in this region and hopefully clarify its specific location. Indeed, we do not yet know for sure whether any gene expression occurs in this region: The putative coding sequence could turn out to be a pseudogene. Therefore, let's look for evidence that something could be expressed from this region of interest.

CpG islands are commonly found in the promoter regions of expressed genes, so let's start with a content-based method to see if there are CpG islands within the sequenced fragment. **The Sequence Manipulation Suite** includes a simple CpG island prediction program. Navigate there and paste or upload your sequence and submit it. At first, the resulting long list of CpG islands may seem daunting. However, notice that many of the results overlap: As discussed earlier in the chapter, CpG prediction uses a sliding window, and SMS shows results for *each* 200-bp window that meets the criteria. Therefore, consider how many nonoverlapping islands the program found. Given a set of overlapping sequences, one island would extend from the first nucleotide of the first sequence found to the last nucleotide of the last sequence in that set. It may also be useful to apply more stringent criteria; although the definition of a CpG island is operational, islands at least 500 bp in length with an overall GC content of at least 55% and a ratio of observed to expected CpG pairs exceeding 0.65 are considered most likely to genuinely function in gene expression.

Next, we might look at whether programs specifically designed to identify promoters would find any transcriptional signals in reasonable locations relative to the putative genes in our sequenced region. **Neural Network Promoter Prediction (NNPP)** looks for core promoter features using a neural network algorithm based on training sets containing known promoters. Promoter prediction, however, often returns too many putative promoters to be useful from any large region of DNA. It is thus desirable to cut down the size of the DNA sequence to be examined. Using your GENSCAN and/or AUGUSTUS map, decide how much sequence to use. Include the first exon and all upstream sequences for the putative gene on which you are focusing. To avoid having to count nucleotides, use the `Group DNA` option in the Sequence Manipulation Suite to number the sequence. Then cut the numbered sequence down to the nucleotides you decided on and use the `Filter DNA` option to get rid of the numbers again. Save your cut-down DNA, now the potential promoter region, to a new file.

Finally, submit your potential promoter region to NNPP for processing and view the results. Remember to consider whether you need to look at both strands or can focus on just one. You may be surprised by the number of potential promoters predicted; this should give you some insight into the complexity of eukaryotic genome data.

Again, we can increase our confidence in the results by comparing them with the results from other programs using different algorithms. **TSSG** claims to be the most accurate mammalian promoter prediction program; it uses a combination of sequence motifs and nucleotide composition analysis to identify promoters. Submit your putative promoter region to this program for analysis. You may wish to print the results for easy comparison with NNPP. If you have time and are interested, you may also wish to try analyzing your sequence with **TSSW**, which is very similar to TSSG but is based on a different database of protein sequence motifs.

Web Exploration Questions

9. Do the CpG islands within the sequenced region support your hypothesis about the genes that are found here? Do they provide any information that might help distinguish between the GENSCAN and AUGUSTUS results?
10. Higher scores in the NNPP results mean putative promoters that better match the criteria. Note on your map where the strongest predicted promoters are. The large letters represent the predicted transcriptional start sites. Can you see good matches to the consensus TATA box sequence (tATAWAW) upstream of potential translational starts?
11. How does the number of promoters returned by TSSG compare with the NNPP results? What else is different about the TSSG results, and how might this difference be useful?
12. Higher scores from TSSG again represent better promoter predictions. Do any of the high-scoring promoters match up (at least approximately) with high-scoring promoters from NNPP?
13. Does your expression analysis help to reconcile the differences between the GENSCAN and AUGUSTUS predictions?
14. Choose the gene you believe is founded on the most solid evidence, obtain its coding sequence, and use BLAST and OMIM to find out what is known about this gene. Have you actually identified a gene that makes sense in the context of influenza resistance?

More to Explore: Further Analysis

You could further pursue the discrepancies in identification of introns and exons between GENSCAN and AUGUSTUS by using additional analyses. Two programs in common use that focus more specifically on splice site identification are **HMMgene** and the neural network-based **NetGene2**. NetGene2 integrates a variety of rules that affect identification of exons, including nucleotide and codon bias, splice site consensus sequences, reading frame predictions, and lengths of introns and exons. This program claims to detect 95% of donor and acceptors sites with less than 0.4% false positives. HMMgene, as its name suggests, uses an HMM algorithm to predict gene structure. It only finds splice sites that make sense in the context of a whole gene, leading to fewer predicted genes but better predictions.

Once a putative gene has been identified and we have a hypothesis about the locations of its exons, promoter, and other features, we still need confirming data, which usually come from "wet lab" experiments. We might, for example, obtain complementary DNA from cells of interest and carry out a microarray or deep sequencing experiment to identify all the expressed genes and determine whether any match our putative gene. Given the wealth of available information about the human genome, we can also take advantage of experiments done by others. One way to find out if our putative gene is actually expressed is to compare it with the Expressed Sequence Tag (EST) database to see if a unique expressed sequence has been identified within our gene. Another method is to use a BLAST search with output limited to sequences that include "mRNA" in their titles to look for DNAs from this region and compare them with our predicted exons.

Guided Programming Project: Predicting CpG Islands

Rather than searching DNA for a particular site or sequence, content-based gene prediction methods look at the DNA sequence more broadly for clues to which sequences are genes (or, more precisely, which are within exons). Here, we work with one specific example of a content-based algorithm to search sequences for CpG islands (see BioBackground) that may indicate a nearby promoter. An increase in the frequency of CG pairs has been observed between nucleotides −1,500 and +500 relative to a transcriptional start site; finding such a CpG island appropriately positioned upstream of a putative gene would strengthen the case that it is an actual gene.

In a random DNA sequence, we would expect CG dinucleotides to occur once in every 16 nucleotides (1 of every 4 nucleotides should be a C, and the next nucleotide will be a G one-fourth of the time). To identify CpG islands, we will not merely search for the sequence pattern (CG) but will also need to determine how frequently it occurs. As described in Understanding the Algorithm, a frequency-matching algorithm is a variation on the pattern-matching algorithm (Chapter 9) that can accomplish this. We use a sliding window to traverse our sequence, counting up CG pairs within each window and looking for higher than average CpG ratios. The following pseudocode shows how this could be done. In this example, all CpG ratios are stored and displayed; however, if a CpG ratio is >1.5 (strong indicator), stars (***) print next to the value to highlight the ratio. Of course, another alternative is to only print the windows where the ratio is >1.5. In the skills exercises, we explore other options.

Algorithm

CpG Island Prediction Algorithm

Goal: To identify regions of CpG islands

Input: A FASTA formatted input file containing a sequence

Output: Window start positions, CpG ratios, and text indicating high ratios.

```
// Initialization—Read in sequence data
Open input file containing sequence: infile
Input window size from user: window
read and discard first line (fasta comment) from infile
for each remaining line of data in infile
    seq = seq + line

// Step 1: Determine CpG ratios
lenSeq = length of seq
ratios = array of size lenSeq-window+1 (holds CpG ratio of each
window)
for each i from 0 to lenSeq-window+1
    cCtr = gCtr = cgCtr = 0
    for each j from 0 to window-1
        if seq[j+i] == 'C'
            cCtr++
            if seq[j+i+1] == 'G'
                cgCtr++
        else if seq[j+i] == 'G'
            gCtr++
    if cCtr*gCtr != 0
        ratios[i] = cgCtr/((cCtr*gCtr)/window)
    else
        ratios[i] = 0
```

```
// Step 2: Print window start position and CpG ratios
for each i from 0 to length of ratios
    if ratios[i] > 1.5
        output i+1, ratios[i], '***'
    else
        output i+1, ratios[i]
```

Putting Your Skills Into Practice

1. Write a program to implement the CpG island prediction algorithm in the language of your choice as outlined in the given pseudocode. You should read in a sequence from a file and produce a tabular list of high-CpG regions with their scores. Devise some simple test sequences to test your program, and then try it on the long sequence (**1q25.txt**) used in the Web Exploration.
2. Compare the output of your program with the output of the CpG island prediction program from the Sequence Manipulation Suite. How similar are the predictions of the two programs? Can you suggest an explanation for any discrepancies? You may also want to look for additional CpG island prediction programs for comparison, such as **CpGProD**.
3. The initial program as described here has the same problem we saw when we used the CpG island prediction program from the Sequence Manipulation Suite (Web Exploration, earlier): because it shows each window where the CpG ratio exceeds a threshold value, it produces a long list of overlapping CpG islands. Make the output of your program more user-friendly by merging overlapping CpG islands into single entries in the results table.
4. To make your program even more effective, you might apply additional criteria. CpG islands associated with actual promoters are usually at least 500 bp in length and have an overall G+C content greater than 55% and a ratio of observed to expected CpG pairs exceeding 65%. Implement these additional criteria as part of your program.

On Your Own Project: Hidden Markov Modeling in Gene Prediction

Understanding the Algorithm introduced HMMs as a very flexible means of identifying coding segments by calculating the most probable match between an observed sequence and an exon–intron pattern based on our understanding of content and sequence cues. A fairly simple model accounting only for an exon–intron junction was presented there (Figure 10.5). This On Your Own Project asks you to design (and, for programming courses, implement) an HMM that also considers the 3′ splice-acceptor site.

Understanding the Problem

Our original HMM example included four states: exon nucleotides, a two-nucleotide splice site (the GT nucleotide pair occurring at nearly all 5′ intron boundaries), and intron nucleotides. We determined emission probabilities based on observed nucleotide frequencies in human introns and exons and established the probability of a transition from exon to splice site at 10%. Clearly, there are many more parameters that should be considered for a program to accurately identify exons and introns.

Solving the Problem

Although an HMM could become very complex indeed, let's add only a moderate level of complexity to our model. First, let's consider the difference between the first exon and an internal exon. The first exon begins with the ATG start codon, and in eukaryotes this is essentially the only possible start codon. Therefore, we could require an invariant ATG as the states of the first nucleotides of our model. The next states could be exon nucleotides, a splice-donor GT site, and intron nucleotides as described in Understanding the Algorithm.

The splice-acceptor site can be defined for the purposes of this model as a near invariant AG occurring as the last two nucleotides of the intron. To determine the emission frequencies, use the following data: A occurs with a frequency of 99.98% at the first position, with all other nucleotides occurring at equal frequency. G occurs with 99.93% frequency at the second position, C with 0.05% frequency, and A or T with equal frequency. This leaves the transition probabilities to be considered. For this exercise, allow an intron to transition to a splice-acceptor site with a 10% probability, similar to the original model. The splice-acceptor site always transitions to an exon—but not to the start codon, which is only in the first exon. Exons should have a 10% probability of transitioning to a splice-donor site but also a 10% probability of being the last exon and terminating the gene. Based on these parameters, design an HMM using a diagram similar to Figure 10.5 that will find a multiple-exon gene.

Programming the Solution

Once you have developed an appropriate design for your HMM, it should be relatively easy to implement in a programming language, if you are in a programming course. The first task is to generate the list of possible paths for the observed sequence. A recursive approach is appropriate because a state may be able to transition to any number of possible states, including itself. You should consider how you will deal with the start codon, because it is not expected to be the first three nucleotides of the input sequence. The end of the gene is also a problem. For this project, we assume any exon could be the last exon, and thus we need to assign a low transition probability from E→end, such as 0.001.

Then, for each path, the emission and transition probabilities are calculated for each nucleotide and multiplied to give an overall probability, P. The natural log of P is then stored for each possibility, and the maximum value for log(P) is chosen as the best way to classify the observed sequence into exons, splice sites, and introns.

You certainly do not want to turn your program loose on the entire 90-kb sequence from the Web Exploration without testing it carefully first. Develop some short test sequences with obvious start codons and splice sites (similar to the very short sequence used as the example in Figure 10.5) to test the program. Then, test it with longer sequences—perhaps a single gene as predicted by GENSCAN or AUGUSTUS. If your program proves capable of handling these longer sequences, you may then want to try it on the full-length sequence and compare its results with those of the programs you used in the Web Exploration.

More to Explore

To make your HMM even more realistic, you could incorporate the observed frequencies of nucleotides at other positions within the splice site (Figure 10.2). If you would like to try this, **Table 10.2** gives the nucleotide frequencies for the dataset used to make the sequence logos.

Table 10.2 Nucleotide frequencies for the 5′ and 3′ splice sites.

	Splice-donor (5′) Site				Splice-acceptor (3′) Site			
Position	A	C	G	T	A	C	G	T
−21					0.22	0.31	0.10	0.37
−20					0.28	0.15	0.25	0.32
−19					0.13	0.37	0.29	0.21
−18					0.08	0.44	0.11	0.37
−17					0.16	0.22	0.22	0.40
−16					0.08	0.26	0.16	0.50
−15					0.08	0.31	0.20	0.41
−14					0.16	0.20	0.11	0.53
−13					0.03	0.24	0.13	0.60
−12					0.07	0.26	0.12	0.55
−11	0.30	0.25	0.27	0.17	0.04	0.41	0.09	0.46
−10	0.36	0.27	0.28	0.08	0.05	0.37	0.20	0.38
−9	0.16	0.23	0.29	0.31	0.11	0.32	0.11	0.46
−8	0.16	0.31	0.36	0.16	0.04	0.35	0.17	0.44
−7	0.34	0.23	0.25	0.17	0.08	0.36	0.15	0.41
−6	0.30	0.22	0.22	0.25	0.03	0.31	0.08	0.58
−5	0.45	0.23	0.13	0.18	0.07	0.36	0.04	0.53
−4	0.29	0.28	0.25	0.17	0.27	0.20	0.17	0.36
−3	0.22	0.45	0.11	0.21	0.04	0.69	0.00	0.27
−2	0.61	0.09	0.10	0.20	0.98	0.00	0.02	0.00
−1	0.17	0.05	0.60	0.18	0.00	0.02	0.98	0.00
+1	0.0005	0.0001	0.9993	0.0001	0.16	0.21	0.56	0.07
+2	0.0001	0.0069	0.0001	0.9929	0.38	0.19	0.07	0.36
+3	0.59	0.02	0.38	0.01	0.32	0.15	0.25	0.28
+4	0.68	0.18	0.06	0.08	0.18	0.25	0.20	0.37
+5	0.02	0.04	0.83	0.11	0.14	0.29	0.29	0.28
+6	0.03	0.15	0.19	0.63	0.15	0.22	0.37	0.26
+7	0.31	0.31	0.27	0.10	0.17	0.28	0.23	0.32
+8	0.25	0.24	0.30	0.20	0.33	0.17	0.30	0.20
+9	0.16	0.34	0.23	0.26	+0.19	0.40	0.13	0.28
+10					0.09	0.21	0.45	0.25
+11					0.32	0.34	0.18	0.16
+12					0.20	0.36	0.26	0.18
+13					0.27	0.30	0.14	0.29
+14					0.21	0.14	0.42	0.23
+15					0.14	0.35	0.33	0.18
+16					0.25	0.20	0.30	0.25
+17					0.34	0.20	0.23	0.23
+18					0.18	0.43	0.26	0.13
+19					0.22	0.22	0.36	0.19

BioBackground: Splicing and CpG Islands

mRNA Splicing in Eukaryotes

When a gene is expressed, it is transcribed in the nucleus to make a single-stranded RNA complementary to the entire template strand of the DNA for that gene: the **pre-mRNA**. A methylated G nucleotide is added to the 5′ end of the mRNA by an unusual 5′-to-5′ linkage; this **5′ cap** is the structure by which a ribosome recognizes the mRNA. At the 3′ end, cleavage occurs at a **polyadenylation site** (consensus sequence 5′ AAUAAA), and a **poly(A) tail** of 200–300 A nucleotides is added to protect the mRNA from rapid degradation.

Splicing is carried out by the **spliceosome**, a large complex made up of several **small nuclear ribonucleoproteins** (snRNPs, pronounced "snurps"): functional units composed of both RNA and protein. The snRNPs direct the binding of the spliceosome to sites at the beginning and end of an intron to cut an mRNA, remove the intron, and rejoin the ends (see **Figure 10.9**). At the 5′ end of an intron (5′ splice site), the exon usually ends with a consensus sequence close to MAG, and the intron almost invariably begins with GU, usually followed by RAGU. On the other end, the 3′ splice site is defined by an AG sequence, most often CAG, at the end of the intron, with G as the first base of the next exon. Within the intron itself is a **branch site** with the consensus sequence CURAY 20–50 bases from the 3′ end of the intron; after cutting the mRNA, the 3′ end of the intron is joined to this site, forming a "lariat" structure that marks the intron for degradation rather than transport to the cytoplasm. The exons are joined together, and when splicing is complete, the **mature mRNA** moves to the cytoplasm for translation.

CpG Islands

Although each species has a characteristic ratio of G and C nucleotides in its DNA to A and T nucleotides, the frequencies of these nucleotides are not constant across the genome. A pattern noted in the study of genomes is that the promoter regions of known genes tend to be higher in G and C nucleotides than A and T nucleotides. Furthermore, the dinucleotide CG—which molecular biologists call **CpG**, with the letter *p* representing the phosphate in the sugar-phosphate DNA backbone—occurs in these regions much more frequently than would be expected by chance. Because the C in a CG pair is a target for methylating enzymes,

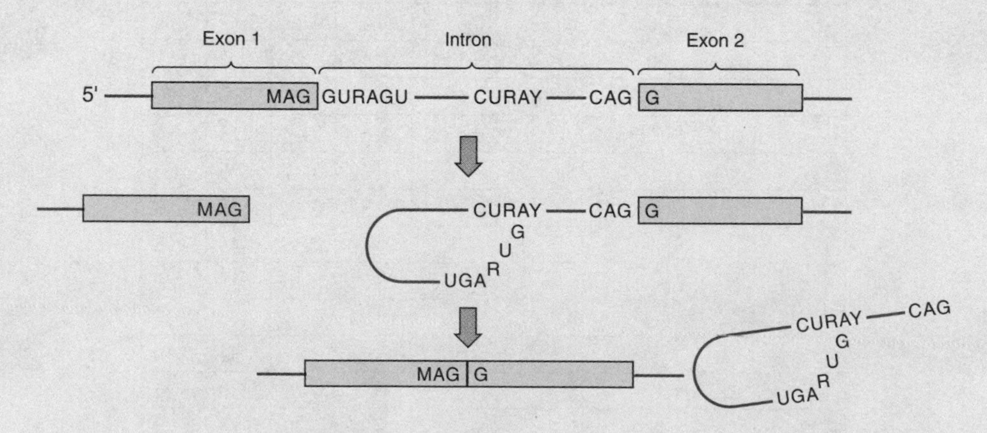

Figure 10.9 The process of mRNA splicing in a eukaryotic cell, showing the consensus sequences occurring at the two splice junctions and the internal branch site.

the concentration of methylated nucleotides is higher in promoter regions that overlap CpG islands, altering gene expression patterns. The identification of CpG islands is therefore one marker for a promoter region. Remember that the promoter region for a eukaryotic gene can be long, and it is not precisely defined with regard to the translational start site. Similarly, CpG island(s) are not precisely aligned with a particular promoter element but can occur anywhere within the broadly defined promoter region. However, CpG-rich regions in an area where there is other evidence of gene expression can add credibility to the prediction of a promoter and a downstream first exon.

References and Supplemental Reading

ENCODE Project's Report on Human Genome Elements

The ENCODE Project Consortium. 2012. An integrated encyclopedia of DNA elements in the human genome. *Nature* **489**:57–74.

Influenza Resistance Genes

Wolff, T., R. E. O'Neill, and P. Palese. 1998. NS1-binding protein (NS1-BP): a novel human protein that interacts with the influenza A virus nonstructural NS1 protein is relocalized in the nuclei of infected cells. *J. Virol.* **72**:7170–7180.

World Health Organization. 2009. *WHO Public Health Research Agenda for Influenza.* WHO Press, Geneva.

Zhang, L., J. M. Katz, M. Gwinn, N. F. Dowling, and M. J. Khoury. 2009. Systems-based candidate genes for human response to influenza infection. *Infect. Genet. Evol.* **9**:1148–1157.

Gene Prediction and Annotation in Eukaryotes

Brent, M. R. 2007. How does eukaryotic gene prediction work? *Nat. Biotechnol.* **25**:883–885.

Do, J. H., and D. K. Choi. 2006. Computational approaches to gene prediction. *J. Microbiol.* **44**:137–144.

Yandell, M., and D. Ence. 2012. A beginner's guide to eukaryotic genome annotation. *Nat. Rev. Genet.* **13**:329–342.

Codon Usage Measurement

Bennetzen, J. L., and B. D. Hall. 1982. Codon selection in yeast. *J. Biol. Chem.* **257**:3026–3031.

Hidden Markov Models

Eddy, S. R. 2004. What is a hidden Markov model? *Nat. Biotechnol.* **22**:1315–1316.

Henderson, J., S. Salzberg, and K. H. Fasman. Finding genes in DNA with a hidden Markov model. *J. Computat. Biol.* **4**:127–141.

Neural Networks

Krogh, A. 2008. What are artificial neural networks? *Nat. Biotechnol.* **26**:195–197.

Nucleotide Bias in Human Genes

Louie, E., J. Ott, and J. Majewski. 2003. Nucleotide frequency variation across human genes. *Genome Res.* **13**:2594–2601.

GENSCAN and AUGUSTUS:

Stanke, M. and S. Waack. 2003. Gene prediction with a hidden Markov model and a new intron submodel. *Bioinformatics* **19S2**:215–221.

Burge, C. and S. Karlin. 1997. Prediction of complete gene structures in human genomic DNA. *J. Mol. Biol.* **268**:78–94.

Chapter 11

Protein Structure Prediction and Analysis:

Rational Drug Design

Chapter Overview

Thus far, we have worked with the *sequences* of proteins: we have viewed them as simple chains of amino acids. But, a protein is actually a folded, three-dimensional structure (see BioBackground at the end of the chapter), and this structure is crucial to the protein's function. In this chapter, we use Web-based software to model protein structure and see how such molecular modeling can aid in drug design. We learn to "align" protein structures and observe that even when sequence similarity is limited, proteins can be very similar in structure and thus function. In the Web Exploration we also examine how a protein's structure might be predicted from its sequence, and in the Guided Programming Project and On Your Own Project, we implement one algorithmic solution to this complex problem.

Biological problem: Designing an HIV protease inhibitor

Bioinformatics skills: Protein structure modeling and structural comparison, structure prediction

Bioinformatics software: Jmol, SWISS-MODEL, PDBeFold, PSIPRED

Programming skills: Chou-Fasman algorithm, sliding windows, hash tables

Understanding the Problem:

Structure Prediction

*When HIV-1, the virus that causes AIDS, was discovered in 1984, it was commonly assumed a vaccine, effective antiviral drugs, or both would be found within a few years. However, 2012 marked the 25th World AIDS Day, and the pandemic is still going strong, with an estimated 34 million living with HIV or AIDS worldwide and nearly 2 million annual deaths (*__Figure 11.1__*). Despite two and a half decades of intensive research, we still have no vaccine and no drugs that can cure the infection. Perhaps this is less surprising when we realize no antiviral drug exists that can cure any viral disease, and indeed there are few effective antivirals on the market. Part of the reason for this is that unlike bacteria, viruses replicate within our own cells and use our own cellular machinery to copy their genomes and synthesize their proteins, leaving us few virus-specific targets to attack with pharmaceuticals.*

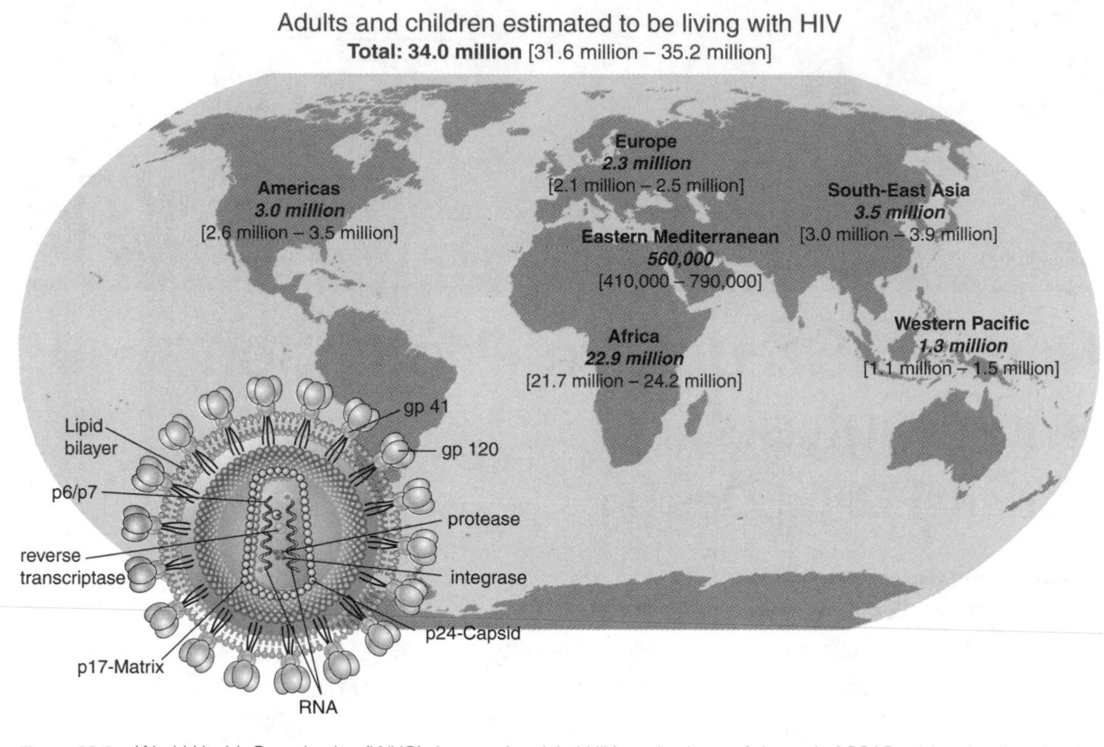

Figure 11.1 World Health Organization (WHO) data on the global HIV pandemic as of the end of 2010, with a drawing showing the structure of the HIV virus. Data from: WHO.

A detailed understanding of the three-dimensional structure of virus proteins may be one route to new breakthroughs in antiviral research. The two key goals of any antimicrobial drug are (1) to be effective against the disease-causing agent and (2) to be **selectively toxic**: able to kill or inhibit the microbe without causing harm to the patient. Viruses have no metabolism outside host cells and few proteins of their own; this makes it difficult to identify effective and selective antiviral drugs by the standard approach of testing libraries of potentially bioactive molecules. **Rational drug design** provides an alternative: By examining the three-dimensional structure of a viral protein, one should be able to design a molecule to precisely fit some part of that protein and block its function. Two of the first examples of commercially available antiviral agents designed this way are anti-HIV drugs: raltegravir (Isentress), an inhibitor of the HIV integrase enzyme, and enfuvirtide (Fuzeon), which blocks entry of HIV into cells.

Unfortunately, rational drug design poses its own difficulties. Determination of the detailed three-dimensional structure of a protein requires crystallizing that protein and then measuring how the crystal scatters x-rays, a process called **x-ray crystallography**. Many proteins are difficult to crystallize, particularly if they have hydrophobic regions that insert into membranes, and this process is slow and labor intensive. Once a crystal structure is known, there remains the problem of accurately determining the shape of a molecule that fits into some part of the structure, synthesizing that molecule, and then testing it to see if it has the desired biological effect. Furthermore, although our skills in these areas are improving, it still remains difficult to predict potential toxicity of a prospective therapeutic molecule as well as how quickly it will be metabolized by the patient and lose its effect. Fortunately, today's bioinformatic techniques are improving our ability to predict and model protein structure.

In addition to its application to drug development, we can use protein structure in many other ways. For example, a key functional region of a protein may actually be made up of amino acids scattered throughout its primary sequence but brought together by folding and thus not recognized in ordinary alignments. Furthermore, we are becoming increasingly aware that changes in macromolecular structure are important components of many diseases, both genetic and infectious: For example, the ΔF508 mutation causes cystic fibrosis (Chapter 2) by interfering with the folding of the CFTR protein, and prion diseases such as "mad cow disease" result from "contagious" misfolding of a specific protein (see References and Supplemental Reading for more on protein folding in human disease).

Bioinformatics Solutions: Predicting and Modeling Protein Structure

Molecular biology and bioinformatics have worked together to make great strides in sequencing genes and even entire genomes, identifying genes within genomes, predicting amino-acid sequences of proteins, and comparing sequences to obtain clues to function and evolutionary relatedness. However, determining the nucleotide sequence of a gene allows us to predict only the **primary structure** (amino-acid sequence; see BioBackground) of the protein it encodes. An actual cell is a three-dimensional arena where molecules with specific structures interact, and the three-dimensional structure of a protein (**Figure 11.2**) determines what interactions it can have with other molecules. An enzyme must have the

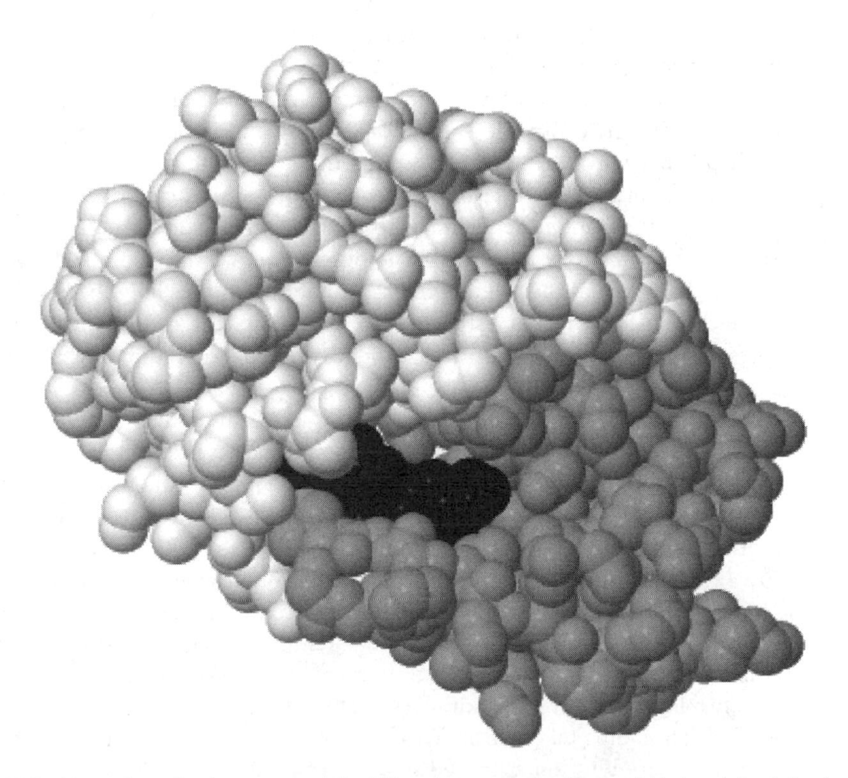

Figure 11.2 Three-dimensional structure of the HIV protease, showing its two folded protein chains (gray and white) and a protease inhibitor in its active site (black). Structure from the RCSB PDB (www.pdb.org): PDB ID 1AID E. Rutenber et al. (1993) Structure of a non-peptide inhibitor complexed with HIV-1 protease: Developing a cycle of structure-based drug design. *J. Biol. Chem.* 268:15343–15346.

correct shape to bind a specific substrate and exclude nonsubstrate molecules, for example, whereas a transport protein on the surface of a cell must have a specific structure to selectively allow specific molecules to enter or exit.

To date, no experimental methods for determining the structure of either proteins or nucleic acids can keep up with the tremendous rate at which their primary sequences are being determined. Although we have successfully determined tens of thousands of protein structures, genome sequencing projects have given us tens of *millions* of primary sequences of nucleic acids and proteins. One goal of computational structural biology is to solve this problem by predicting the structure of a protein given only its primary sequence. The possible conformations any protein can assume are determined by its amino-acid sequence, and its final, folded state is thus determined to a large degree by its primary structure (see BioBackground). Thus, given sufficient understanding of individual amino acids and the conditions under which they are folding, this should be possible. However, it is a big problem: We might know that a particular amino acid has an –OH group that can form a hydrogen bond with an amino group on another amino acid, but how do we know which two amino acids to pair up in a protein hundreds or thousands of amino acids long?

The number of possible folded structures for a protein is enormous, so algorithms that predict folding from sequence rely on structural rules to arrive at a likely folded structure. Many of these rules originated with Linus Pauling's pioneering work on protein structure (see References and Supplemental Readings), which defined the nature of the chemical bonds between amino acids and how bond angles, rotation of atoms, and flexibility of chains limit the structures that can be formed. Pauling predicted the structure of the **α-helix** (see BioBackground) as a major component of folded proteins, later confirmed by x-ray crystallography. In an α-helix, the C=O group of one amino acid must be able to form a hydrogen bond with the amino group of an amino acid located four residues farther down the chain. However, not just any amino acid can be included in the helix; proline, for example, introduces a turn into the protein backbone and disrupts helical structure. Similar rules can be worked out for amino acids likely to form β-sheets (see BioBackground) and other elements of protein secondary structure. Anfinsen (see References and Supplemental Reading) and others then went a step further, explaining that the thermodynamics of the cellular environment determines how these structures fold into a three-dimensional tertiary structure. Bioinformatic algorithms use secondary structure rules and thermodynamic optimization algorithms to predict how a protein folds into an overall stable structure.

Our ability to effectively predict tertiary structure from sequence alone (**ab initio** or **de novo prediction**) is unfortunately quite limited at present. However, the combination of increasing numbers of experimentally determined protein crystal structures with the enormous explosion in genomic data has given rise to two additional bioinformatic techniques that are very important in modeling protein structure. **Homology modeling** (**Figure 11.3A**) is used to find the structure of a protein when an ortholog or paralogs with a known structure can be identified. To construct a homology model, the protein of interest is aligned with the sequence of a similar **template** protein, and the alignment is used to map its amino acids onto a structural model based on the template structure. If there is no closely related protein with a known structure, **threading** (**Figure 11.3B** and **C**) can be used instead. Threading takes advantage of the observation that most proteins whose structures are known are built on a limited number of basic folded units. For example, the immunoglobulin fold shown in Figure 11.3B is a basic structural unit found one or more times in dozens of different proteins; although many of these proteins function in the immune system, their molecular functions are very diverse. As shown in Figure 11.3C, new protein

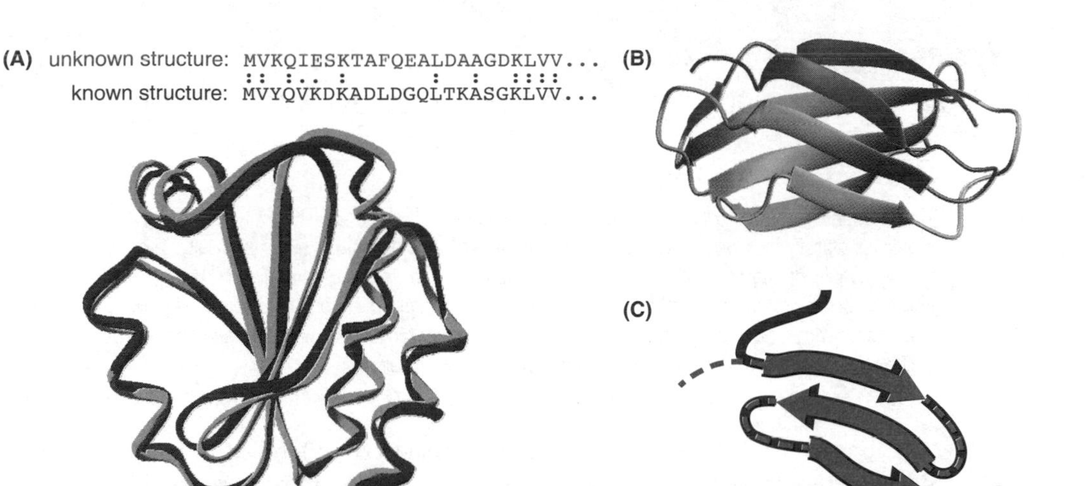

Figure 11.3 Predicting protein structure based on similarity to known structures. (A) Homology modeling: protein of unknown structure (blue) is an ortholog or paralog of a protein of known structure (black), allowing structure to be modeled from a sequence alignment. Courtesy of Tim Vickers. (B) The immunoglobulin fold, a common protein structural domain. (C) Threading: sequence comparison allows part of a protein of unknown structure (blue) to be threaded onto a protein of known structure (black), showing that it contains an immunoglobulin domain. Structures created from MOLMOL.

sequences can be "threaded" onto common structural units, allowing at least a partial structural model to be constructed.

This chapter's projects explore protein structure prediction and modeling in the context of rational drug design. In the Web Exploration, we use modeling software to examine the structure of the HIV protease, examine how its structure relates to function, and then construct a homology model of a drug-resistant protease mutant. In the Guided Programming Project, we examine de novo structure prediction and compare predicted secondary structure with experimentally verified protein conformation, implementing a more complete solution in the On Your Own Project.

BioConcept Questions

1. Why is it valuable to know the three-dimensional structure of a protein?
2. Both secondary and tertiary structures of proteins are three-dimensional structures; what is the difference between the two?
3. What characteristics of amino acids help determine how they will participate in the folding of the protein?
4. Sickle-cell anemia results from changing a single hydrophilic amino acid (glycine) found on the surface of the folded protein to a hydrophobic amino acid (valine). Discuss how the hydrophobicity of the amino acid could be so important in this disease.
5. The amino-acid sequence of a protein clearly must determine what folded structures are possible for that protein. What other factors contribute to the structure that is actually chosen? What complications arise in trying to predict a folded structure from an amino-acid sequence?

Understanding the Algorithm:
The Chou-Fasman Algorithm for Secondary Structure Prediction

Learning Tools

The Protein Data Bank (PDB), managed by the Research Collaboratory for Structural Bioinformatics, is the major repository for proteins whose structures have been determined experimentally. The PDB's long-standing "Molecule of the Month" series is an excellent way to improve your understanding of the relationship between protein structure and function. Every month, a protein important to some key biological process is discussed from a structural perspective and illustrated by molecular models made from structures available in the PDB; the site's archives now include hundreds of proteins.

The ab initio prediction of the three-dimensional (tertiary) folded structure of a polypeptide structure from its amino-acid sequence is a "holy grail" of structural biology. Because of the enormous complexity of proteins and the many factors that could affect amino-acid interactions, this is a very difficult problem to solve. Indeed, even accurately predicting the folding of the amino-acid chain into the secondary structures (e.g., α-helices and β-sheets) that underlie tertiary structure remains an open problem in bioinformatics.

Many of our ideas about secondary structure prediction stem from an algorithm proposed by Peter Chou and Gerald Fasman in 1974 (see References and Supplemental Reading). At that time, a handful of protein crystal structures were known, and Chou and Fasman developed the idea of examining these known structures to determine which specific amino acids within the proteins contributed to each secondary structure. Using this information, they developed **propensity values** (the likelihood that an amino acid would appear within a particular secondary structure) and **frequency values** (the frequency with which an amino acid is found in a hairpin turn) for each amino acid (**Table 11.1**). These values were updated in 1978 (see References and Supplemental Reading) using new training data and became known as the Chou-Fasman parameters.

Chou and Fasman calculated three different propensity (*P*) values for each amino acid: *P*(a), *P*(b), and *P*(turn), representing the likelihood of finding the amino acid within an α-helix, β-strand, and β-turn, respectively. These values are log-odds ratios, where $P > 1.0$ indicates the amino acid has a greater than average chance of contributing to that particular structure, $P < 1.0$ means it has a less than average chance, and $P = 1.0$ means it is no more likely to contribute to that structure than any randomly chosen amino acid. Each amino acid also has four frequency (*f*) values: *f*(i), *f*(i+1), *f*(i+2), and *f*(i+3), the frequencies with which it is found at each of the four positions of a hairpin turn (β-turn). From these parameters, Chou and Fasman developed rules to predict the locations of α-helices, β-strands, and β-turns. Different implementations of this algorithm vary in the threshold values for the parameters or the criteria for designating a region an α-helix or a β-sheet. One implementation is presented here.

Algorithm

Chou-Fasman Algorithm

1. Identify α-helices
   a. Find a region of six contiguous residues where at least four have *P*(a) > 103.
   b. Extend the region until a set of four contiguous residues with *P*(a) < 100 is found.
   c. If the region's average *P*(a) > 103 and Σ*P*(a) > Σ*P*(b) for the region, then that region is predicted to be an α-helix.

2. Identify β-strands
   a. Find a region of five contiguous residues where at least three have $P(b) > 105$.
   b. Extend the region until a set of four contiguous residues with $P(b) < 100$ is found.
   c. If the region's average $P(b) > 105$ and $\Sigma P(b) > \Sigma P(a)$ for the region, then that region is predicted to be a β-strand.
3. Determine β-turns
   a. For each residue j, determine the turn propensity or $P(t)$ for j as follows:

$$P(t)_j = f(i)_j \times f(i+1)_{j+1} \times f(i+2)_{j+2} \times f(i+3)_{j+3}$$

   b. A turn is predicted at position j if $P(t) > 0.000075$, and the average $P(\text{turn})$ for residues j to $j+3 > 100$, and $\Sigma P(a) < \Sigma P(\text{turn}) > \Sigma P(b)$.
4. Handling overlaps
   If an α-helix region overlaps with a β-sheet region, the region's summed values for $P(a)$ and $P(b)$ are used to determine the overlapping region's most likely structure. If $\Sigma P(a) > \Sigma P(b)$ for the overlapping region, then it is considered an α-helix. If $\Sigma P(b) > \Sigma P(a)$, then the overlapping region is considered a β-sheet, and if $\Sigma P(b) = \Sigma P(a)$, then no valid determination can be made.

Table 11.1 The Chou-Fasman parameters.

Amino Acid	*P*(a)	*P*(b)	*P*(turn)	*f*(i)	*f*(i + 1)	*f*(i + 2)	*f*(i + 3)
Alanine	142	83	66	0.060	0.076	0.035	0.058
Arginine	98	93	95	0.070	0.106	0.099	0.085
Asparagine	67	89	156	0.161	0.083	0.191	0.091
Aspartic acid	101	54	146	0.147	0.110	0.179	0.081
Cysteine	70	119	119	0.149	0.050	0.117	0.128
Glutamic acid	151	37	74	0.056	0.060	0.077	0.064
Glutamine	111	110	98	0.074	0.098	0.037	0.098
Glycine	57	75	156	0.102	0.085	0.190	0.152
Histidine	100	87	95	0.140	0.047	0.093	0.054
Isoleucine	108	160	47	0.043	0.034	0.013	0.056
Leucine	121	130	59	0.061	0.025	0.036	0.070
Lysine	114	74	101	0.055	0.115	0.072	0.095
Methionine	145	105	60	0.068	0.082	0.014	0.055
Phenylalanine	113	138	60	0.059	0.041	0.065	0.065
Proline	57	55	152	0.102	0.301	0.034	0.068
Serine	77	75	143	0.120	0.139	0.125	0.106
Threonine	83	119	96	0.086	0.108	0.065	0.079
Tryptophan	108	137	96	0.077	0.013	0.064	0.167
Tyrosine	69	147	114	0.082	0.065	0.114	0.125
Valine	106	170	50	0.062	0.048	0.028	0.053

Data from: Chou & Fasman, *Adv. Enzymol. Relat. Areas Mol. Biol.* 47:45–148 (1978).

Neural network methods (see Chapter 10) are common in secondary structure prediction programs such as PSIPRED, which we will use in the Web Exploration Project. However, although the Chou-Fasman algorithm is sometimes denigrated for its accuracy of only 50–60%, the ideas behind it underlie many of these newer methods. Indeed, some methods in current use are much more complicated yet only slightly more accurate. The Chou-Fasman algorithm remains very valuable for understanding the principles of protein structure prediction.

Test Your Understanding

1. Find an α-helix in the short sequence `N-MDGPDFWEAMKRISTQTYSNGHKMPS-C` using the Chou-Fasman rules.
2. Examine the Chou-Fasman rules carefully, and look at the *P*(a) and *P*(b) values for various amino acids in **Table 11.1**. What can you see that might reduce the ability of this algorithm to clearly distinguish between α-helices and β-sheets?
3. How do we define a β-turn in a protein structure? Given this definition, can you think of a simple rule you could add to the algorithm for identification of β-turns that might increase its accuracy?
4. Would it improve the predictive ability of the algorithm to specify that a region should be identified as a β-strand only if it is either preceded or followed by a β-turn? Why or why not?
5. Proteins that are part of the cell membrane or an organelle membrane typically have one or several α-helical domains about 20 amino acids long that pass through the membrane. These membrane-spanning helices consist almost entirely of very hydrophobic amino acids such as L, I, V, F, and W and are anchored in place by hydrophilic amino acids on their two ends. If you applied the Chou-Fasman algorithm to a membrane protein, why would it likely fail to predict the membrane-spanning helices?

CHAPTER PROJECT:

Protein Structure Prediction

This chapter's projects address the problem of identifying potential anti-HIV drugs that block the action of the viral protease and of overcoming the rapid development of drug resistance. We examine both ab initio and homology-based methods of predicting protein structure and examine how changes to the structure of a protein may affect its function.

Learning Objectives

- Understand how protein structure and function are related and why structure prediction is important
- Know how to use available tools to examine the experimentally determined structures of proteins and visualize structural and functional features
- Use homology-based tools to compare a novel protein sequence with a well-studied one and identify potentially significant differences
- Appreciate the value and limitations of ab initio approaches to protein structure prediction
- Understand how protein structure prediction and analysis can inform drug design

Suggestions for Using the Project

In the Web Exploration for this chapter, students start by using Web-based structure visualization tools to explore protein structure and understand the value of different ways of showing protein structure. They then use homology-based methods to compare an HIV protease mutant to the unmutated protein and see how mutation can affect drug effectiveness. They then experiment with ab initio structure prediction, comparing these results with the known structure of the protein. In the Guided Programming Project, they develop a solution for part of the Chou-Fasman algorithm and then completely implement this algorithm in the On Your Own Project.

Programming courses:

- Web Exploration: Use Web-based tools to become familiar with protein structure, model a mutant protein, and test ab initio structure prediction. If time is limiting, we recommend completing at least Part I to become familiar with protein structure and Part III to generate comparison data for the programming projects.
- Guided Programming Project: Implement the Chou-Fasman algorithm to find α-helices in an amino-acid sequence and compare results with known sequences and predictions from other ab initio tools.
- On Your Own Project: Fully implement the Chou-Fasman algorithm to find α-helices, β-strands, and β-sheets in an amino-acid sequence and compare results.

Nonprogramming courses:

- Web Exploration: Use Web-based tools to explore protein structure, homology modeling to examine the structure of a mutant protein, and ab initio methods to predict secondary structure from amino-acid sequence. Parts I, II, and III are independent enough to be used separately to match the focus of a particular course.
- On Your Own Project: Download an implementation of the Chou-Fasman algorithm for ab initio secondary structure prediction. Compare its results with those of prediction programs used in Part III of the Web Exploration and to experimentally determined structures.

Web Exploration: Protein Structure Modeling and Drug Design

Traditionally, new drugs have been discovered by performing initial testing of a huge number of molecules that might possibly affect some process of interest (for example, inhibiting bacterial growth, blocking pain receptors, or halting allergic responses). Pharmaceutical companies maintain large libraries of potentially useful chemicals for this reason; once a candidate molecule is found, it can then be chemically modified to increase its activity, reduce its toxicity, and so on. In many cases, the new drug needs to interact with an enzyme or other protein, and this is where rational drug design could drastically improve the selectivity and effectiveness of our pharmaceuticals and the speed with which we can identify new candidate drugs. If we were able to easily and quickly determine the structure of the protein and connect structural domains with protein functions, we could design a drug to "fit" precisely in an appropriate spot.

HIV and AIDS have been a major focus of pharmaceutical discovery for more than 25 years, and indeed we have developed an unprecedented number of new antivirals, some of which resulted from the study of protein structure and rational design. In this project, we focus on the HIV **protease**. When HIV infects a cell (**Figure 11.4**), one of the earliest steps

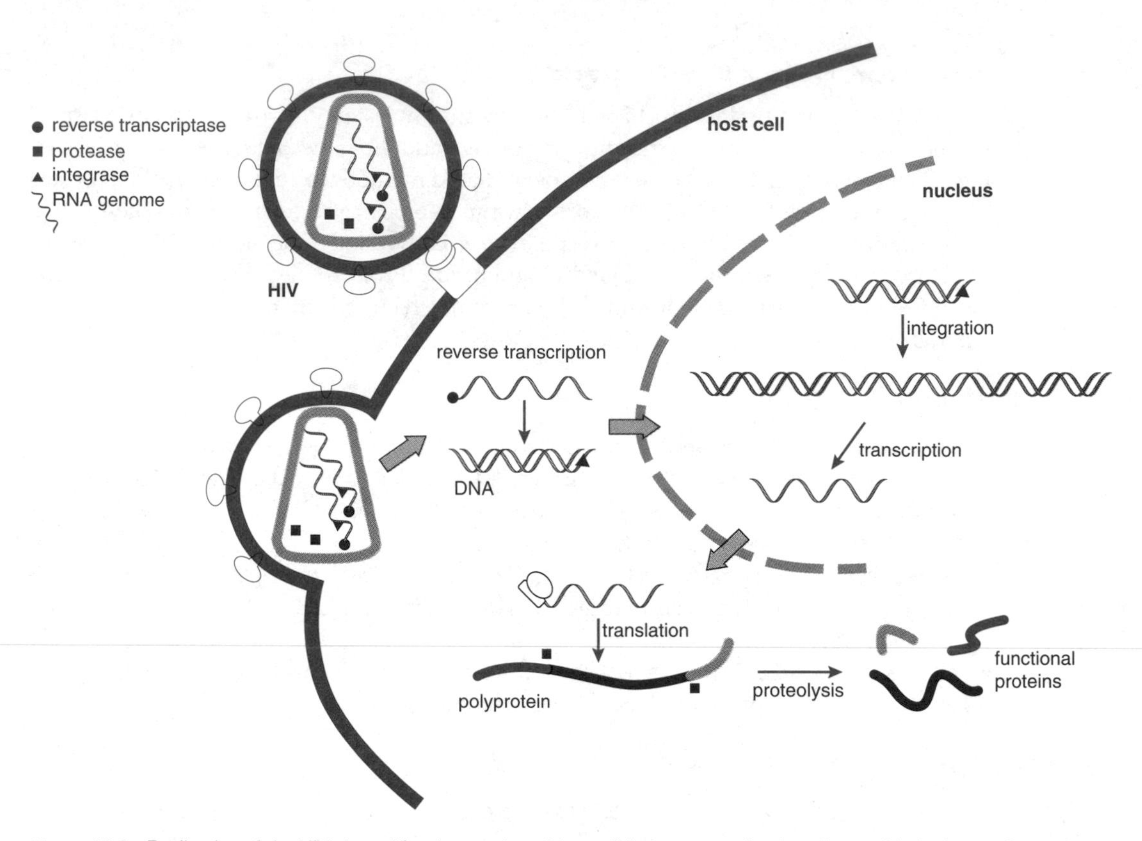

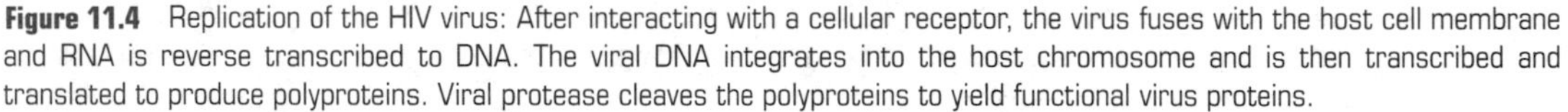

Figure 11.4 Replication of the HIV virus: After interacting with a cellular receptor, the virus fuses with the host cell membrane and RNA is reverse transcribed to DNA. The viral DNA integrates into the host chromosome and is then transcribed and translated to produce polyproteins. Viral protease cleaves the polyproteins to yield functional virus proteins.

is to make a DNA copy of the virus' RNA genome, a process called reverse transcription that does not occur in uninfected cells. To accomplish this, the virus must carry the enzyme reverse transcriptase (also a target of drug therapy). The HIV integrase protein then inserts the DNA into one of the host cell's chromosomes, where the viral genome behaves just like any ordinary gene. There is only one promoter within the HIV genome, so a single mRNA is made by transcription (although it can be spliced in more than one way to produce a few different mature mRNAs for translation). Because eukaryotic ribosomes begin translation with the *first* AUG on an mRNA, only one protein can be made from any particular mRNA, so to produce all the proteins HIV needs, the **polyprotein** product of translation is cleaved by the HIV protease into individual functional protein units (see References and Supplemental Reading). For example, it cleaves a single polypeptide to become the functional reverse transcriptase, integrase, and protease proteins required for viral replication. Blocking the function of the HIV protease therefore inhibits the replication of the virus. The first protease inhibitor was approved for use in treating HIV and AIDS in 1996, and today 10 such drugs are on the market.

Part I: Exploring the Structure of the HIV Protease

When the structure of a protein is "solved," we know where the atoms that make up its amino acids are found in space, allowing us to generate representations that show the locations of the various amino-acid side chains and how they interact to form secondary and tertiary structures. X-ray crystallography is the current gold standard for protein structure and can under the best conditions distinguish the positions of atoms less than 1 Å (10^{-10} m) apart. More flexible proteins may form less perfect crystals and generate structures with lower resolutions of 3 Å or more. Other techniques, such as nuclear magnetic resonance (NMR), can also be used to determine the structures of proteins; they typically generate lower-resolution structures but may have other advantages. NMR, for example, can be applied to uncrystallized proteins in solution. Structural data are deposited in public databases, most notably the **Protein Data Bank (PDB)**, in a standardized format that can be read by various kinds of software to visualize and work with the structure.

A text search of PDB for the HIV-1 protease (note that HIV-1 is the proper name of HIV and will return the best search results) will return a large number of results, mostly variations in which the protein is bound to various inhibitors. We want to see the protease interacting with more natural substrates, so search instead for a specific accession number, `1KJF`, to see a structure where a peptide substrate is used. You may wish to explore some of the features of the PDB entry for a protein; like many of the DNA and protein databases, many resources are brought together at this site, and you can find the sequence of the protein, information about the methods used to produce the structure, biochemical information about the enzyme, references, and more. On the right side of the page, you can see a graphical representation of the protein structure (discussed in more detail later in the chapter). However, the actual PDB data are not graphical at all: Take a look at what is actually stored in the PDB database by using the `Display Files` drop-down menu to examine the `PDB file` for the protease. As you can see, this is purely a text file. If you scroll down, you will realize that the heart of the file is simply a list of atoms, the amino acids to which they belong, and coordinates describing their spatial position (**Figure 11.5**). This is all the information required to minimally describe the protein's structure. Additional information in the file includes the

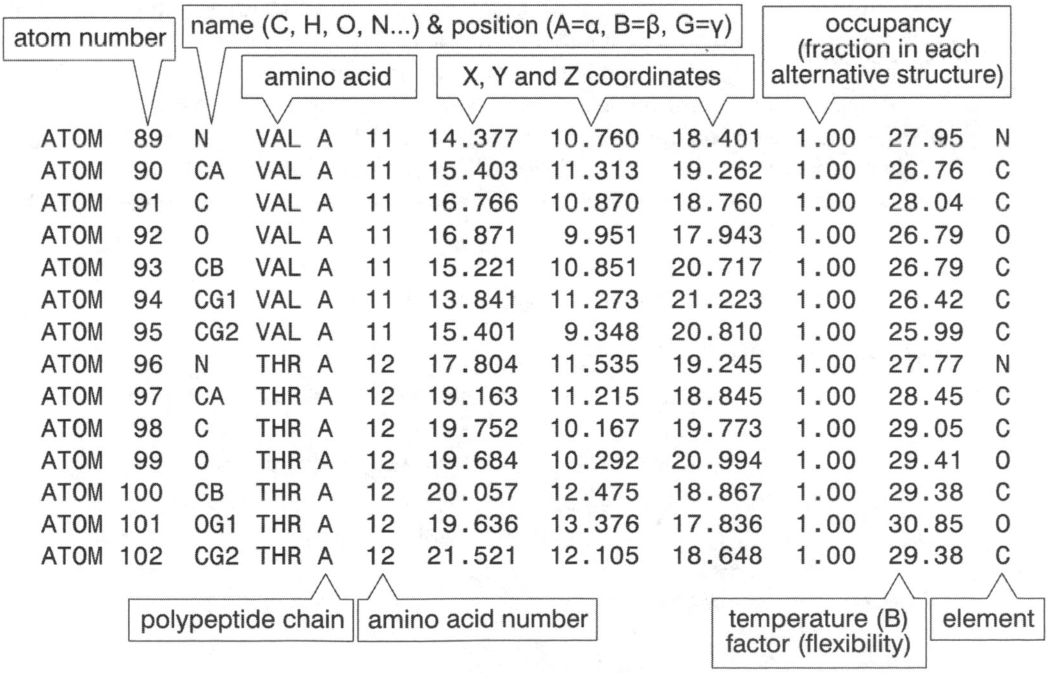

atom number | name (C, H, O, N...) & position (A=α, B=β, G=γ) | amino acid | X, Y and Z coordinates | occupancy (fraction in each alternative structure)

```
ATOM  89  N   VAL A 11 14.377 10.760 18.401 1.00 27.95 N
ATOM  90  CA  VAL A 11 15.403 11.313 19.262 1.00 26.76 C
ATOM  91  C   VAL A 11 16.766 10.870 18.760 1.00 28.04 C
ATOM  92  O   VAL A 11 16.871  9.951 17.943 1.00 26.79 O
ATOM  93  CB  VAL A 11 15.221 10.851 20.717 1.00 26.79 C
ATOM  94  CG1 VAL A 11 13.841 11.273 21.223 1.00 26.42 C
ATOM  95  CG2 VAL A 11 15.401  9.348 20.810 1.00 25.99 C
ATOM  96  N   THR A 12 17.804 11.535 19.245 1.00 27.77 N
ATOM  97  CA  THR A 12 19.163 11.215 18.845 1.00 28.45 C
ATOM  98  C   THR A 12 19.752 10.167 19.773 1.00 29.05 C
ATOM  99  O   THR A 12 19.684 10.292 20.994 1.00 29.41 O
ATOM 100  CB  THR A 12 20.057 12.475 18.867 1.00 29.38 C
ATOM 101  OG1 THR A 12 19.636 13.376 17.836 1.00 30.85 O
ATOM 102  CG2 THR A 12 21.521 12.105 18.648 1.00 29.38 C
```

polypeptide chain | amino acid number | temperature (B) factor (flexibility) | element

Figure 11.5 A segment of the PDB file for the HIV protease describing the locations of the atoms in the protein. Data from: PDB.

amino-acid sequence of each polypeptide chain (look for SEQRES), locations of secondary structures (HELIX, SHEET, etc.), comments (REMARK), and references (JRNL).

Many programs can produce interactive three-dimensional visualizations based on PDB files. Web-based software is usually based on Jmol (see References and Supplemental Reading), a scriptable open-source viewer that runs within a browser as a Java applet. Indeed, a Jmol viewer can be invoked directly from the PDB entry page by clicking on the `View in 3D` link. For this exercise, we use FirstGlance in Jmol, which includes both a full featured Jmol viewer and scripts to facilitate viewing of key structural features. Alternatively, you may wish to use one of the more powerful viewers listed in **Table 11.2**, which can be downloaded to run from a desktop computer; the activities in this section could equally well be completed with one of these programs.

From the **FirstGlance in Jmol** start page, enter `1KJF` to see the HIV protease model you found at PDB. When the applet loads, you should see the protease structure in a "cartoon" view similar to **Figure 11.6**, where α-helices are shown by spiral ribbons (arrows point toward the C-terminus of the protein) and β-sheets by parallel flat ribbons. Unstructured (**random coil**) areas of the protein look like thin ropes. When the program starts, the protein is rotating to show you the three-dimensional view; click on the menu at left to halt it. Notice that three different colors are used. The HIV protease functions as a **homodimer**, that is, the functional protease is composed of two identical polypeptides (quaternary structure). You should see that two colors represent two polypeptides with the same structure joined together. The third color shows a short peptide that represents a segment of a protein substrate in the active site of the enzyme.

Jmol is an interactive program that allows the user to control how the protein is visualized. Notice that by clicking and dragging on the structure, you can rotate it to any desired position. Try rotating the molecule so you get a clear view of the substrate peptide. Can you see the distinct cleft where the substrate binds? This is where the active site of the enzyme is located. You can zoom in and out by clicking on the molecule and rotating the scroll wheel on your mouse or by holding shift while you click and drag. Holding shift also constrains the rotation of the molecule so it moves around a fixed point instead of in three dimensions. You can identify any amino acid in the protein by hovering over it.

Notice the links on the menu at the left. These run preset scripts to show you the kinds of information a typical user would want. Start by clicking on `Secondary Structure` to change the color scheme. Now, the α-helices, β-sheets, and random coils have distinct colors. Likewise, `Hydrophobic/Polar` allows you to see the hydrophobicity of the

Table 11.2 Desktop software for protein structure visualization.

Program	Description
Cn3D	NCBI's protein structure viewer; structures can be downloaded from NCBI databases in Cn3D format. Free.
DeepView	Viewer comparable to Cn3D maintained by the Swiss Institute of Bioinformatics. Free.
PyMOL	Powerful Python-based visualization tool known for creation of publication-quality images. Source code and a limited prebuilt educational version are free; fully supported prebuilt versions require a paid subscription.
Chimera	Developed by a molecular visualization group at the University of California San Francisco. Free for academic and nonprofit use.
RasMol	One of the first popular visualization tools. Requires use of command-line commands. Free open-source and user-supported versions available.

Figure 11.6 Cartoon structure of the HIV protease (monomers shown in dark and light gray) with a short peptide ligand (white) in its active site. Structure from the RCSB PDB (www.pdb.org): PDB ID 1KJF: M. Prabu-Jeyabalan et al., Substrate shape determines specificity of recognition for HIV-1 protease: analysis of crystal structures of six substrate complexes. *Structure* 10:369–381 (2002).

amino acids that make up the protein (you can click on Water to see where water molecules have access to the protein) and `Charge` lets you see amino acids colored by their charge. Notice these last two options change the view of the molecule to a **space-filling** model, which helps demonstrate that the protein is really not just a ribbon of amino acids but a three-dimensional structure. However, now it is hard to see the two chains and the substrate. Click on `Contacts` to see these highlighted in color again; does this change your understanding of how the peptide fits in the active-site cleft?

In addition to these preset shortcut links (unique to FirstGlance), there are two other ways to interact with Jmol (in any implementation): by menu or by using a command-line console. Right-click on the structure window to access the menus. Suppose, for example, you want to see only the peptide backbone. Open the menu and choose `Style | Structures | Backbone` (if nothing happens, choose `Select | All` and try again). But now you cannot see the individual chains, so choose `Color | Structures | Backbone | By Scheme | Chain` to change this. Many options here will allow you to look even at individual atoms and amino acids. For example, choose `Style | Scheme | CPK Spacefill` to show the space-filling model and `Color | Atoms | By Scheme | Chain` to highlight the individual chains again. Now, click on some of the atoms that seem like they are in close contact with the substrate and watch the display at the bottom to see which amino acids you have chosen and where they are on which chain.

The HIV protease is a member of the aspartyl protease family: The catalytic mechanism for these proteases involves an aspartate in the active site that can be recognized by the three-amino-acid motif Asp-Thr-Gly. Normally, HIV protease contains this motif, but to obtain a crystal structure with a peptide in the active site, a mutation changing the Asp to structurally similar asparagine (Asn) was used for the 1KJF structure. This mutation does not change the structure of the protein but prevents it from cleaving the substrate. Use `Select | Protein | By Residue Name` followed by `Color | Structures | Cartoon` (if you are in cartoon mode) or `Color | Atoms` (if you are in spacefill mode) to highlight asparagines. Then explore the adjacent amino acids by mousing over them (this is easier in cartoon or backbone view) or by selecting and coloring them and see if you can identify the Asn-Thr-Gly combination at the 1KJF active site.

It might be easier to see how the protease and substrate interact if we could get one of the chains out of the way. It is tricky to select a whole chain from the menus but easy from the command line. Show your protein in spacefill mode and choose `Console` from the menu to open the command-line interface. The two protein subunits and the peptide substrate are labeled A, B, and P, respectively (you could find this out by looking at the first few lines of the PDB file). Select all the atoms in the A subunit and color them blue by typing `select *:A; color atoms blue`. Then, color the B subunit red and the substrate yellow. Now hide the A subunit by simply typing `hide *:A` and rotate the molecule to get a good view of how the substrate fits in the cleft. Select and color your three active-site amino acids with commands similar to `select 10:B; color atoms white` and see how they interact with the substrate; hide the substrate if needed to see them better.

Web Exploration Questions

1. The HIV protease functions as a dimer. Some enzymes that form dimers then have two active sites. Is this the case for the HIV protease? Briefly describe the relationship of the active site and peptide-binding cleft to the subunits of the enzyme.
2. What kinds of amino acids do you find in the areas of the protein exposed to the water around it (e.g., when the protein is in solution in the cytoplasm)?
3. If you were to design an inhibitor of the HIV protease, where would you want it to bind? What kind of molecule might you use as the prototype to develop the structure of a good inhibitor?
4. Using the cartoon or ribbon view, you should be able to identify where a long β-strand on each subunit of the protease makes a hairpin turn, forming flexible flaps that cover the active site cleft. These flaps control access of the substrate to the active site. Which amino acids form the flaps (just give the range of numbers)? Although this region is very important to protease function, why are the flaps not likely to make a good target for rational drug design?
5. What are the numbers of the amino acids on each chain that form the Asp-Thr-Gly (Asn-Thr-Gly in this mutant) aspartate protease motif in 1KJF?

Part II: Homology Modeling of a Mutant HIV Protease

One of the major obstacles to pharmaceutical control of HIV is the virus' rapid rate of mutation. The DNA polymerases that replicate DNA in our cells "proofread" during synthesis, reducing their error rate to about one nucleotide in a billion. Reverse transcriptase, however, does not proofread and in addition appears to be much less accurate than other nonproofreading polymerases, producing one mutation for approximately every 10,000 nucleotides of DNA it synthesizes. Combined with its long-term residence in a single host and rapid rate of replication (up to 10^{10} new viruses per infected patient per day), this gives HIV extraordinary genetic variability and many strains can be in competition within a single patient, leading to the rapid evolution of variants that can escape from immune system controls as well as drug-resistant strains. Current drug therapies combine three or more individual antivirals in an attempt to stave off resistance, but even so, patients must be closely monitored and their drug regimens altered in response to the inevitable rise of resistance.

How do changes in HIV proteins lead to drug resistance? From the *Exploring Bioinformatics* website, you can download the **amino-acid sequence of a drug-resistant mutant HIV protease.** Because this protease variant has not been crystallized, its exact structure is not known. We expect, however, that its structure will vary only in specific locations and probably in minor ways (especially because this variant does function as a protease) from the protease we have already examined. Homology modeling is therefore an appropriate method of structure prediction: The sequence of the mutant can be aligned with the original sequence (**template**) and a structure generated that follows the template wherever the amino acids are identical. Where the two sequences are different, the program attempts to predict the effect of the substituted amino acids on the structure based on their properties.

SWISS-MODEL is a Web-based homology modeling program suitable for analysis of the mutant protease; its automated mode provides an easy way to model a protein expected to closely match the template. From the **SWISS-MODEL home page**, choose `Automated Mode` and enter the mutant protease sequence. Although the program can search the entire PDB to find a suitable template by similarity, in this case we know the identity of our protein. A suitable template would be an HIV protease structure that also does not include a substrate (because there is no substrate in our mutant sequence); we can use PDB structure 1ODW. Enter this accession number at the bottom of the page to be used as the template; you can enter either chain A or B, because both are the same. You can wait for the results (usually only a few minutes) or provide an email address to be notified when the analysis is complete.

The output of SWISS-MODEL is a PDB file for the mutant protein—a model structure, because it is based not on crystallography but on homology. This structure can be visualized with a Jmol-based viewer, such as the basic AstexViewer linked on the results page (go ahead and try this; the result should look very familiar). However, it would be more instructive to directly compare the mutant structure with the unmutated protease. Download the PDB model for the mutant using the appropriate link and save it as a local file. Then navigate to **PDBeFold**, a Web interface to a program capable of constructing a pairwise *structure* alignment.

Use the mutant protease PDB file you just downloaded as the query sequence (choose `Coordinate file` from the drop-down menu to upload it) and enter the 1KJF accession number as the target. `Chains` can be set to `* (all)`. Uncheck `match individual chains`—because our two chains are identical, there is no point in doing an A versus B and B versus A comparison. Leave the rest of the options at their defaults. Submit the alignment for processing. A single match should be returned; click on its number to see a detail page. Scroll down the page to see how the two proteins' amino acids matched up: In red are query amino acids matched with the same amino acid in the template, whereas blue shows those that aligned with a different amino acid. Back near the top of the page, you should see a button to superpose the two structures (there are two; use the top one); be sure `superpose whole entries` is checked (so we see both chains) and click on the button to see the structures in a Jmol viewer.

The default view is in cartoon format, with the two chains of the unmutated protease shown in cyan and the two chains of the mutant shown in gray. Set `Screen` to 80% or 90% to see the molecule better and then explore the structure. As you rotate the model, in most places the two structures are so similar that you see a single ribbon or rope, but you should be able to recognize some places where they are quite distinct. Let's focus on how the mutations affect the area of the active site. PDBeFold has essentially produced a composite PDB file in which the two chains of the mutant protease are A and B and the two chains of the original protease are D and E, with the substrate as chain F. To make it easier to see the overall outline of the structure, set `Rendering` to Backbone. Now let's use the console to highlight a couple of specific areas near the active site: Try `select 48-53:A; color backbone blue`, `select 48-53:B; color backbone blue` and `select 76-83:B; color backbone blue` to highlight regions of the mutant protein in blue, and then color amino acids 48–53 red on chains D and E and 76–83 red on chain E to show the unmutated protein. Explore the model to see how these regions relate to the location of the substrate; how might the mutations affect the fit of an inhibitory drug in the active site? To make this clearer, try `select *:F; spacefill` to make only the substrate chain spacefilling and `color atoms white` to make the colors less distracting. Of course, you are free to explore further with different views and color schemes.

Web Exploration Questions

6. How many mutations are there in the mutant protease sequence, as compared with the sequence of the protease you examined in Part I? Use pairwise alignment to find out.
7. In the regions you highlighted, how would you characterize the effect of the mutations on the structure of the protein, in general?
8. How would these structural changes affect the binding of a small inhibitor molecule to the protease active site? Why would they have less effect on the binding of the natural substrate?
9. If you wanted to design a drug that would inhibit this mutant protease, what characteristics would you want it to have?
10. Change the colors of your model so that everything is white except the three amino acids of the aspartyl protease motif (make the substrate gray for contrast). Make these three amino acids blue on the mutant chains, and then see what happens when you color them red on the nonmutant chains. Does their position change in the mutant relative to the unmutated protein? Is this what you expected? Certainly changes in the sequence or structure at these positions could lead to drug resistance; why then do we not observe them among drug-resistant HIV isolates?

More to Explore: Binding of the Mutant Protease to Inhibitors

The previous exercise allowed you to formulate a hypothesis about why this mutant protease is drug resistant. As you saw, PDB has many examples of protease structures with various inhibitors bound to the protease. You could use PDBeFold to make alignments of the mutant protease with some of these structures to see the structural changes in the mutant relative to actual inhibitor binding.

Part III: Predicting Secondary Structure from Amino-Acid Sequence

Finally, let's look at the ability of bioinformatic software to predict secondary structure ab initio—from an amino-acid sequence unassisted by a known structure. Because we know the crystal structure of the HIV protease, we can try predicting secondary structures using its sequence and then compare the results with the known locations of α-helices and β-sheets; use the 1KJF sequence, which you can download from its PDB page. For the structure prediction, we use **PSIPRED** to look for regions of the protein likely to form α-helices, β-sheets, or random coils. PSIPRED uses a neural network algorithm and integrates both a Chou-Fasman–like prediction algorithm and comparative data obtained by searching for orthologous sequences with PSI-BLAST (see References and Supplemental Reading).

From the PSIPRED page, choose `Predict Secondary Structure`. (Notice that the same server offers two other structure prediction options.) Enter the protease sequence and your email and submit your request. You should get an email within half an hour or less indicating the job is complete. You can examine the results either in text form in the email or graphically by clicking the emailed link. Either way, you should see that each amino acid in the protein has been assigned a letter indicating whether it is predicted to be in an alpha (H)elix, a strand of a beta sh(E)et, or a random (C)oil. Each also has a number indicating the statistical level of confidence in the prediction (nine is highest). In the graphical version (**Figure 11.7**; the PDF file provides the nicest view), the confidence value is replaced by a bar whose height shows the level of confidence, and the α-helices and β-strands are shown graphically with cylinders and arrows, respectively. Save or print your results for easy comparison.

Now, return to FirstGlance in Jmol to visualize the HIV protease structure 1KJF. Color the structure by secondary structure so you can see the α-helices and β-strands clearly. You may want to hide one of the chains and the substrate for convenience. Now, identify the start and end points of the α-helices and β-strands in the crystal structure and note them on the PSIPRED results. How does PSIPRED's prediction compare with the actual structure?

Web Exploration Questions

11. How well did PSIPRED predict the secondary structures in the HIV protease? Give specific examples of structures predicted accurately by PSIPRED, predicted structures not found in the actual structure, and actual structures that were not predicted.
12. PSIPRED uses a prediction algorithm not unlike the Chou-Fasman algorithm we will use in the Guided Programming Project. However, instead of applying its algorithm directly to your input sequence, it first does a PSI-BLAST search to get a collection of sequences related to your input. It then applies its prediction algorithm to the results. Why might this method be advantageous in improving the program's ability to identify genuine secondary structure?

More to Explore: More Structure Tools

We have barely scratched the surface of protein structure prediction and analysis tools. **Table 11.3** lists a number of additional tools you may wish to apply to these or to other protein structure questions.

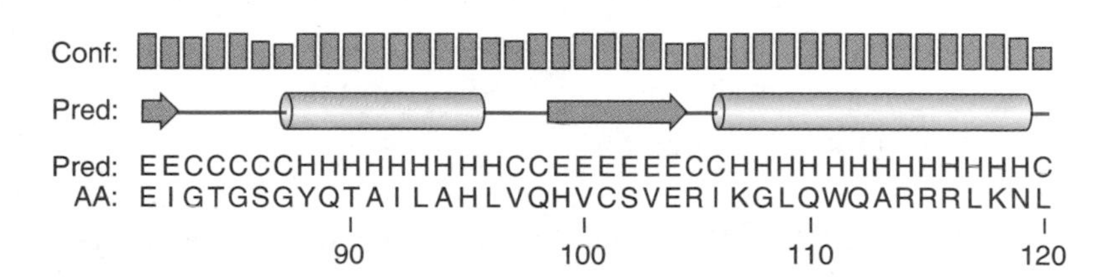

Figure 11.7 Sample output from the PSIPRED server. The bars at the top represent the confidence level of each prediction. Arrows and cylinders in the next line represent predicted β-strands and α-helices, respectively, followed by text showing whether each amino acid is within a predicted β-strand (E), α-helix (H), or random coil (C). Data from PSIPRED server: McGuffin et al., *Bioinformatics* 16:404 (2000).

Table 11.3 Additional recommended protein structure analysis software.

Program	Description
Ab Initio Protein Structure Prediction	
Jpred3	Secondary structure prediction, multiple neural network methods
PEP-FOLD	Tertiary structure prediction based on hidden Markov modeling
ROBETTA	Tertiary structure prediction: structure generation for short fragments followed by energy minimization
Membrane Protein Prediction	
MEMSTAT	Neural network–based prediction of transmembrane domains
HMMTOP	Hidden Markov model-based prediction of transmembrane domains
Homology Modeling	
ESyPred3D	Alignment and model generation; uses MODELLER algorithm to examine a probability density function for each atom
FoldX	Homology modeling and prediction of effects of mutations; useful to design protein variants with desired effects on structure
Threading	
GenTHREADER	Threading based on secondary structure prediction
HHpred	Based on multiple sequence alignment of related sequences identified by PSI-BLAST

Guided Programming Project: Structure Prediction with the Chou-Fasman Algorithm

As described in Understanding the Algorithm, the Chou-Fasman algorithm looks at the likelihood that each amino acid in a protein sequence occurs within an α-helix, β-strand, or β-turn. In this project, you will develop a program that implements the first step of this algorithm: finding α-helices. The complete Chou-Fasman algorithm will be implemented in the On Your Own Project.

Before you begin to write code, think about the data structures you need to store the Chou-Fasman parameters. You may want to consider hash table structures for easy and quick access using amino-acid names as keys. The following pseudocode presents a solution for finding α-helices.

Algorithm

Chou-Fasman Algorithm for Predicting Protein Structure

Goal: To predict the location of α-helices.

Input: An amino-acid sequence in FASTA format

Output: The location of α-helices.

```
// Step 1: Initialization and Read in Sequence

open input file 1: infile1
aminoSeq = ""
read and ignore first line of data in infile1
```

```
for each line of data in infile1
    concatenate line of data to aminoSeq

// Step 2: Find Alpha Helices
// find region of six (step 1a)
lenSeq = length of aminoSeq
window = 6
pScore = 103
minWindow = 4
paHash = map of all amino acids to P(a) values
pbHash = map of all amino acids to P(b) values

for each i from 0 to (lenSeq-window)
    ctr = paSum = pbSum = 0
    // find possible alpha helices
    for each j from 0 to window-1
        paSum = paSum + paHash[aminoSeq[i+j]]
        pbSum = pbSum + pbHash[aminoSeq[i+j]]
        if paHash[aminoSeq[i+j]] > pScore
            ctr++
    if ctr >= minWindow
        output "Possible alpha helix region found at" + (i+1)

        // extend region left (step 1b)
        extend = i-1
        done = false
        while extend >= 0 and !done
            if extend >= 3
              && paHash[aminoSeq[extend]] < 100
              && paHash[aminoSeq[extend-1]] < 100
              && paHash[aminoSeq[extend-2]] < 100
              && paHash[aminoSeq[extend-3]] < 100
                done = true
            else
                paSum = paSum + paHash[aminoSeq[extend]]
                pbSum = pbSum + pbHash[aminoSeq[extend]]
                extend--

        left = extend + 1
        // extend region right (step 1b continued)
        extend = i + window
        done = false
        while extend < lenSeq and !done
            if extend <= lenSeq - 3
              && paHash[aminoSeq[extend]] < 100
```

```
                && paHash[aminoSeq[extend+1]] < 100
                && paHash[aminoSeq[extend+2]] < 100
                && paHash[aminoSeq[extend+3]] < 100
                  done = true
              else
                  paSum = paSum + paHash[aminoSeq[extend]]
                  pbSum = pbSum + pbHash[aminoSeq[extend]]
                  extend++
          right = extend - 1
          // see if step 1c fulfilled
          lenRegion = right - left
          if paSum/lenRegion > pScore and paSum > pbSum
              output "Alpha Region:" + (leftStart+1) + "to"
                  + (rightStart+1)
```

Putting Your Skills Into Practice

1. Write a program to implement the given pseudocode in the programming language used in your course. Short amino-acid sequences can be downloaded from the *Exploring Bioinformatics* website and used to test your program.
2. The PSIPRED secondary structure prediction program gives text output showing the predicted secondary structure for each position in the amino-acid sequence (Figure 11.7). Modify your program to produce output similar to PSIPRED, using H to represent helices and a dash (–) to indicate amino acids that are not in an α-helix.
3. Each chain of the HIV protease contains one α-helix. Identify the amino acids in one chain of the 1KJF structure that are within the α-helix, and then run your program on this sequence and compare its prediction with the actual crystal structure and to the PSIPRED prediction.

On Your Own Project: A Complete Chou-Fasman Program

In this project, you will complete the implementation of the Chou-Fasman algorithm that you started in the Guided Programming Project. If your course does not involve programming, you can download a **completed Chou-Fasman program** from the *Exploring Bioinformatics* website and use it to answer the questions that follow.

Understanding the Problem

The Guided Programming Project showed how to implement step 1 of the Chou-Fasman algorithm, finding all possible α-helices. Understanding the Algorithm introduced the remaining steps of the algorithm: predicting β-strands and β-turns, as well as dealing with overlaps where the same amino acid is within two structures. Amino acids not within any of these structures are considered to be within random coils.

Solving the Problem

A straightforward approach to code the entire algorithm is to traverse the sequence three times, each time searching for a particular structure (steps 1–3). You could then compare the results to handle overlaps (step 4). However, storing all the information from steps 1–3 before tackling step 4 may not be the most efficient approach, because many overlapping areas would require more storage than necessary. Additionally, making a separate pass through the sequence to find each structure adds unnecessary complexity.

Alternatively, your program could find all possible α-helices and then look for β-sheets, checking for overlaps as each is found before continuing. It could then continue on to find β-turns. To accomplish this, you would need to change your guided project solution so that each α-helix is stored rather than simply printed. Think carefully about what data you need to store as you find each α-helix.

Programming the Solution

Extend your solution to incorporate steps 2–4 of the Chou-Fasman algorithm. Your program should display text output similar to that of PSIPRED (**Figure 11.7**), showing the predicted structure for each amino acid: H for α-helices, E for β-strands, T for β-turns, and C for random coil.

Test your program with the **short test sequence** you can download from the *Exploring Bioinformatics* website. Then, run it on the 1KJF protease sequence and see if it finds the known locations of the α-helix and the β-strands.

1. How did your Chou-Fasman prediction compare with the actual structure of the HIV protease?
2. How did your prediction compare with that of PSIPRED? PSIPRED is a much more sophisticated program; does it give significantly better results?
3. You may also want to test your program on other proteins to better evaluate its capabilities. Try, for example, the HIV reverse transcriptase or the HIV capsid protein. For a bigger challenge, try it on the HIV envelope protein, which is a transmembrane protein.
4. It is possible that where Chou-Fasman fails to make an accurate prediction, it may be making the wrong choice between α-helix and β-strand in overlap regions. If you are in a programming course, you could modify your program so it reports overlaps and shows the decision it made, allowing you to see if the opposite decision would have led to a better prediction.

Connections: Distributed Computing to Improve Ab Initio Protein Structure Prediction

By now you have an appreciation for the complexity of protein folding and how hard it is to predict the final three-dimensional conformation of a protein based on its primary structure. Even our best computational algorithms for predicting secondary structure can do so with only moderate confidence. The enormous number of possible ways in which these secondary structures might fold into a tertiary structure compounds the problem. Furthermore, folding occurs differently in different environments—such as for a membrane protein, which is typically inserted into the membrane as it is being synthesized. Computational power is one limiting factor in coping with this complexity: Protein folding algorithms can be refined by comparing predicted structures with the increasing number of known protein structures, but a great deal of computer time is necessary to process the huge numbers of possible models.

Distributed computing offers an intriguing approach to this problem. At least two current projects, **Folding@home** and **Rosetta@home**, use software that can be downloaded freely by anyone and used like a screensaver, working on folding models when the computer is idle. A central server parcels out pieces of the problem to individual computers that process data and return the results to the server, thus harnessing the unused capacity of hundreds of thousands of individual computers. This yields total computing power much greater than any single computer and at very low cost. Both projects focus on structures important to understanding human disease, particularly diseases such as Huntington disease, Alzheimer disease, and prion diseases, which involve misfolded proteins.

BioBackground: Protein Structure

A protein's function depends on both its amino-acid sequence and its conformation, or folded structure. The three-dimensional shape of a protein determines the interactions it can have with other molecules. For example, a DNA-binding protein such as a transcription factor (**Figure 11.8A**) needs structural regions (**domains**), allowing it to fit into the grooves of a DNA molecule. In these binding domains, positively charged amino acids are needed to interact with the negatively charged DNA backbone, and additional amino acids interact with specific DNA bases to determine the DNA sequence to which the transcription factor binds. A channel protein (**Figure 11.8B**) has long helices that pass through the membrane; the exterior of these helices consist of amino acids with hydrophobic side chains to interact with the hydrophobic membrane lipids, but the interior contains hydrophilic amino acids that can interact with some molecule to be transported across the membrane.

How a protein can fold depends on its amino-acid sequence, known as its **primary (1°) structure** (**Figure 11.9A**). Folding results from the interaction of amino-acid side chains, mostly weak noncovalent interactions such as hydrogen bonds (the attraction of a hydrogen attached to an oxygen or nitrogen atom for a nearby oxygen or nitrogen), ionic bonds (attraction between positively and negatively charged side chains), or hydrophobic interactions. Where two cysteine amino acids are close together, a covalent disulfide bond can be formed, as well. Thus, we can

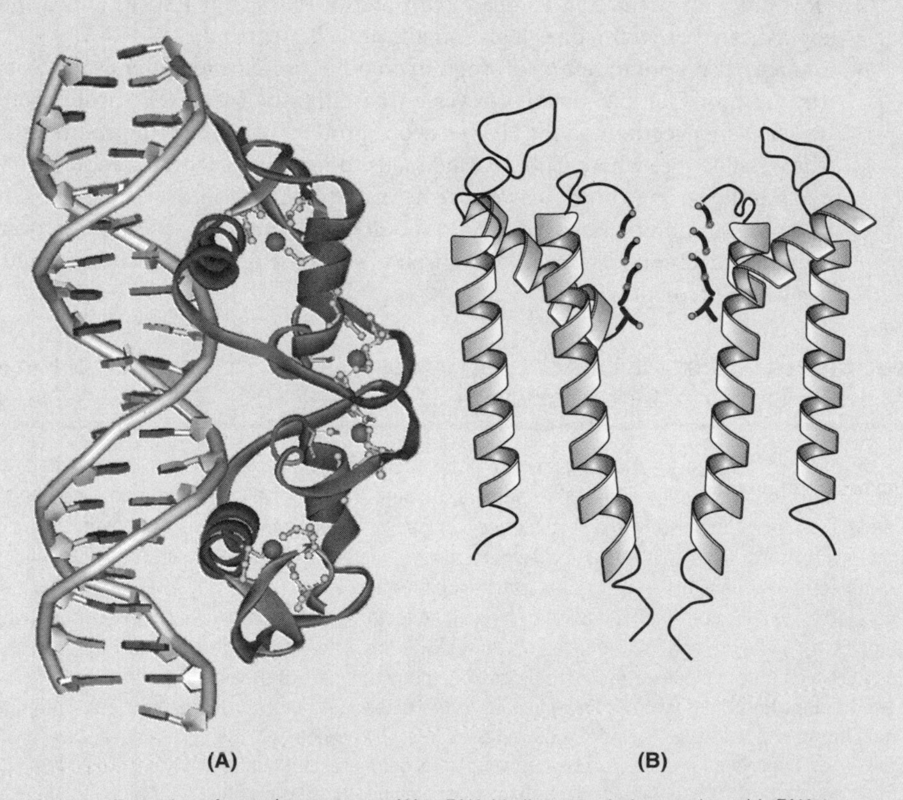

Figure 11.8 Examples of protein structure: (A) a DNA-binding protein interacting with DNA by means of two α-helices; (B) a channel protein that is anchored into a membrane by long helices creating a pore through which some transported molecule can pass. Part (A) structures from the RCSB PDB: PDB ID 1R4R (B. J. Luisi et al (1991) Crystallographic analysis of the interaction of the glucocorticoid receptor with DNA. Nature. 352:497-505).

think of protein conformation as being "encoded" in its gene in some sense, but folding is also influenced by the environment in which the protein folds (such as the cytoplasm or endoplasmic reticulum) and in some cases by interactions with other proteins.

Folding begins while the protein is still being synthesized, as soon as the amino-acid chain begins to emerge from the ribosome. Local interactions among amino acids, often driven by the instability of hydrophobic amino acids exposed to the surrounding watery environment, result in the formation of **secondary (2°) structures** (**Figure 11.9B**). The two most common forms of secondary structures are α-helices and β-sheets. In an α-helix, hydrogen bonds between amino

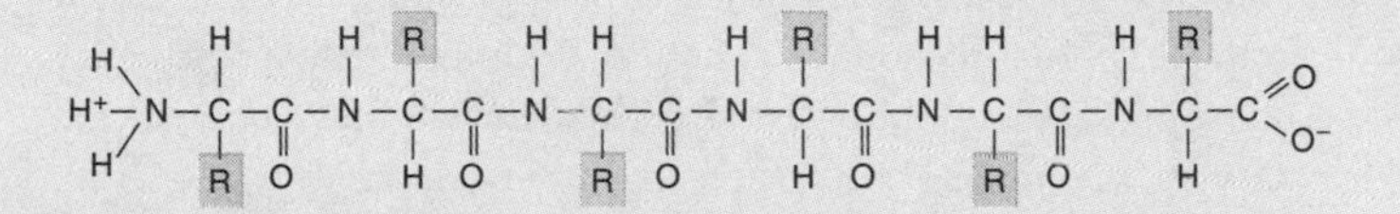

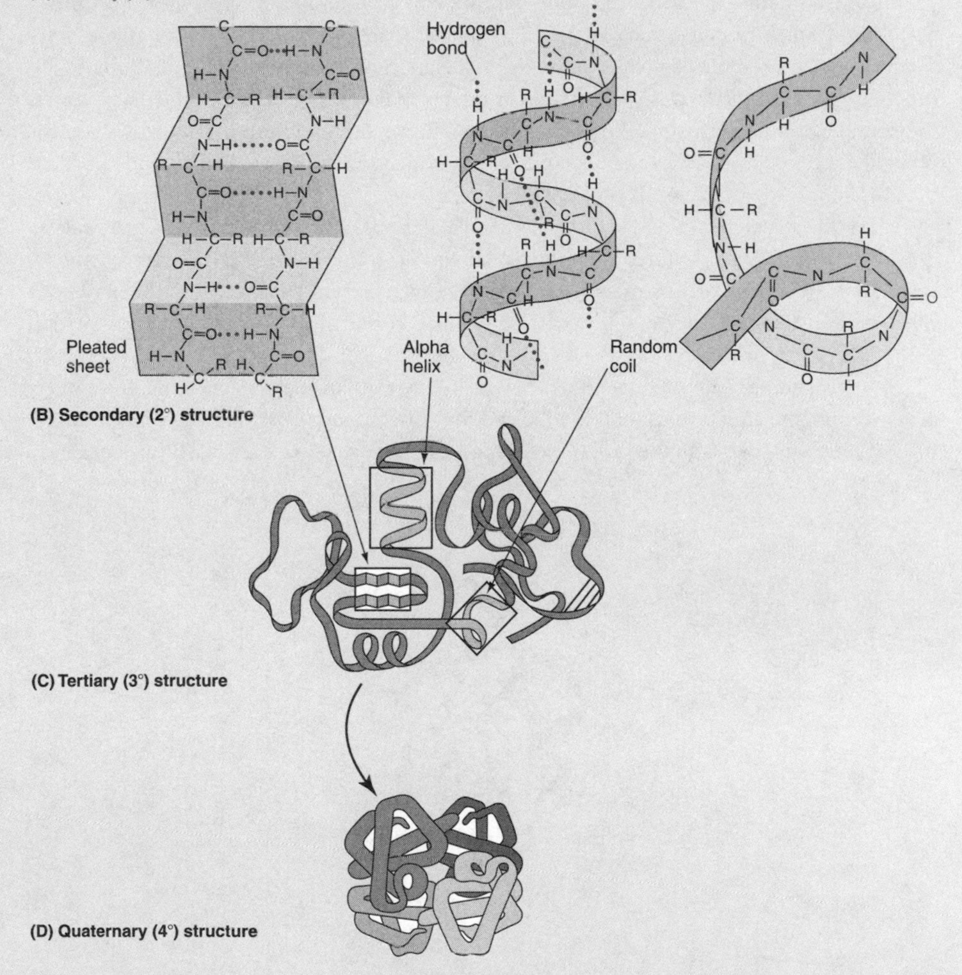

Figure 11.9 Folding of a protein: (A) primary structure, or the amino-acid sequence of the protein; (B) secondary structures formed by local interactions among amino acids: the β-sheet (or β-pleated sheet) and the α-helix; (C) tertiary structure, or the overall three-dimensional shape of the protein; (D) quaternary structure, or the association of two or more polypeptides to form a functional unit, necessary to the function of certain proteins.

acids spaced along a contiguous region form a regular, relatively rigid spiral-shaped structure. A β-sheet is formed by hydrogen bonds among extended, uncoiled stretches called β-strands; β-sheets create relatively flat surfaces in the folded protein. The β-sheet may form from β-strands that follow each other in the primary structure—if so, the strands are separated by hairpin **β-turns**—or may result from β-strands from different parts of the primary structure coming together. Stretches of amino acids with no particular secondary structure are referred to simply as **random coil** regions (Figure 11.9B).

As protein synthesis proceeds, secondary structures can interact with each other, folding the protein into an overall three-dimensional shape called its **tertiary (3°) structure (Figure 11.9C)**. Most proteins fold into a shape that is roughly spherical (globular), but some form long fibers or other configurations appropriate to their function. Within the tertiary structure of an enzyme, there is a binding pocket called the **active site** where the enzyme's substrate fits selectively, and there may also be binding pockets or clefts for other molecules that interact with the protein. Although any long amino-acid chain is commonly referred to as a protein, technically an amino-acid chain is a **polypeptide** and a protein as strictly defined is a *functional* unit. Some proteins, such as the CFTR protein, are composed of only a single polypeptide. However, some proteins require the association of multiple polypeptide subunits to function (**Figure 11.9D**); this is referred to as **quarternary (4°) structure**. The HIV protease, for example, is a **dimer**, composed of two identical polypeptide subunits. Hemoglobin, on the other hand, functions as a **tetramer** composed of two identical α-globin subunits and two identical β-globin subunits, four polypeptides in total.

When the structure of a folded protein is known, it can be represented in a variety of ways to quickly convey its major features to a viewer. A **ribbon** diagram (**Figure 11.10**) is a conventional way to represent the structure of a protein: Flat ribbons represent β-strands and coiled ribbons represent α-helices. Arrows point toward the protein's C-terminal end. A **cartoon** representation is very similar; here, the helices are shown as cylinders.

Proteins generally fold to reach their lowest energy state or most stable structure. Generally, hydrophobic amino acids fold into the interior of the protein, leaving hydrophilic ones on the outside to interact with the watery environment of the cytoplasm. Likewise, two negatively

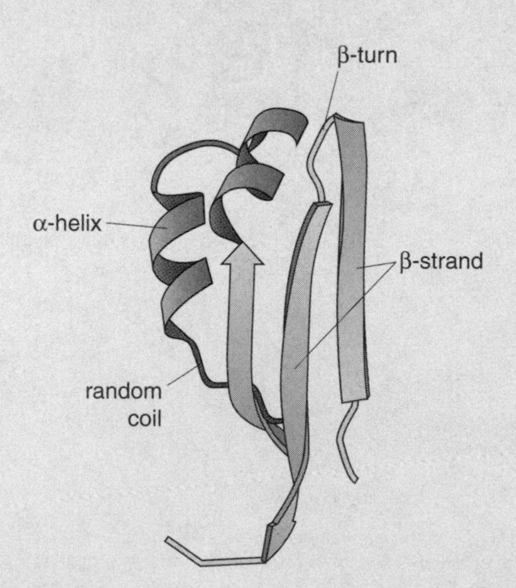

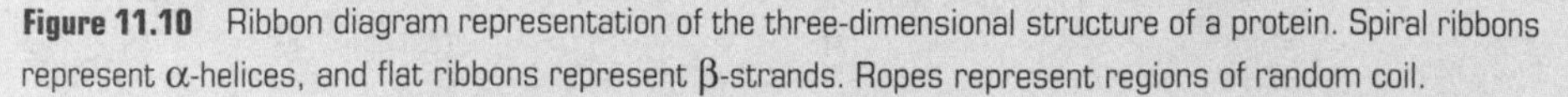

Figure 11.10 Ribbon diagram representation of the three-dimensional structure of a protein. Spiral ribbons represent α-helices, and flat ribbons represent β-strands. Ropes represent regions of random coil.

charged side chains fold to avoid each other and preferably interact with positively charged side chains. In practice, however, this process is constrained by factors such as the order of amino acids: If the first region of the protein folds as soon as it is synthesized to bring hydrophobic amino acids together, those amino acids are no longer available to interact with the next hydrophobic stretch. This reduces the number of possible folded structures for the real protein but tends to make computational prediction more difficult. Remember, too, that the interactions holding the folded structure together are generally weak and can be broken by increasing the temperature or changing the pH: We say this **denatures** the protein. We take advantage of this when we fry an egg, denaturing the watery, protein-rich goo into a more palatable form, or "perm" hair by chemically denaturing hair protein.

References and Supplemental Reading

Protein Folding, Misfolding, and Human Disease

Dobson, C. M. 2003. Protein folding and misfolding. *Nature* **426**:884–890.

Classic Papers on How Proteins Fold

Anfinsen, C. B. 1973. Principles that govern the folding of protein chains. *Science* **181**:223–230.

Pauling, L., and R. B. Corey. 1951. The polypeptide-chain configuration in hemoglobin and other globular proteins. *Proc. Natl. Acad. Sci. U.S.A.* **37**:282–285.

Chou-Fasman Algorithm

Chou, P. Y., and G. D. Fasman. 1974a. Conformational parameters for amino acids in helical, beta-sheet, and random coil regions calculated from proteins. *Biochem.* **13**:211–222.

Chou, P. Y., and G. D. Fasman. 1974b. Prediction of protein conformation. *Biochem.* **13**:222–245.

Chou P. Y., and G. D. Fasman. 1978. Prediction of the secondary structure of proteins from their amino acid sequence. *Adv. Enzymol. Relat. Areas Mol. Biol.* **47**:45–148.

HIV Protease and Protease Inhibitors

Louis, J. M., R. Ishima, D. A. Torchia, and I. T. Weber. 2007. HIV-1 protease: structure, dynamics, and inhibition. *Adv. Pharmacol.* **55**:261–298.

Wensing, A. M., N. M. van Maarseveen, and M. Nijhuis. 2010. Fifteen years of HIV protease inhibitors: raising the barrier to resistance. *Antiviral Res.* **85**:59–74.

Jmol

Hanson, R. M. 2010. *Jmol*—a paradigm shift in crystallographic visualization. *J. Appl. Crystallog.* **43**:1250–1260.

PSIPRED

McGuffin, L. J., K. Bryson, and D. T. Jones. 2000. The PSIPRED protein structure prediction server. *Bioinformatics* **16**:404–405.

Chapter 12

Nucleic Acid Structure Prediction:

Polymerase Chain Reaction and RNA Interference

Chapter Overview

Chapter 11 looked at computational methods to predict and analyze the structure of proteins; in this chapter, we'll look at the structure of nucleic acids. Far from being mere messengers in the information-transfer process, nucleic acids have attracted considerable attention in recent years due to new discoveries ranging from practical applications like PCR to such unexpected biological uses as ribozymes and RNA interference. In this chapter's Web Exploration, we see how nucleic acid structure prediction allows us to improve the design of PCR primers and consider the difficulty in predicting small RNA genes. In the Guided Programming Project, you will implement the Nussinov-Jacobson algorithm for nucleic acid secondary structure prediction, and in the On Your Own Project, you will apply your understanding of that algorithm to the problem of primer design.

Biological problems: PCR primer design, microRNA prediction

Bioinformatics skills: Nucleic acid secondary structure prediction (Nussinov-Jacobson algorithm), micro-RNA gene prediction

Bioinformatics software: Web Primer, Mfold, mirEval

Programming skills: Dynamic programming, recursion, windowing

Understanding the Problem:

Nucleic Acid Structure Prediction

*With the discovery of the structure of DNA and Francis Crick's subsequent declaration that DNA → mRNA → protein is the "Central Dogma" of molecular biology, genes became defined in both the popular and scientific consciousness as DNA segments that encode proteins, with messenger RNA (mRNA) in a supporting role as an intermediate template. Although it was known that ribosomes (***Figure 12.1***) had an RNA component (indeed, ribosomal RNA was the first ribosomal component to be purified), relatively little attention was paid to functional RNAs or the genes that encode them. With ensuing discoveries, however, functional RNAs have built momentum: in 1982, Thomas Czech discovered that some protist RNAs*

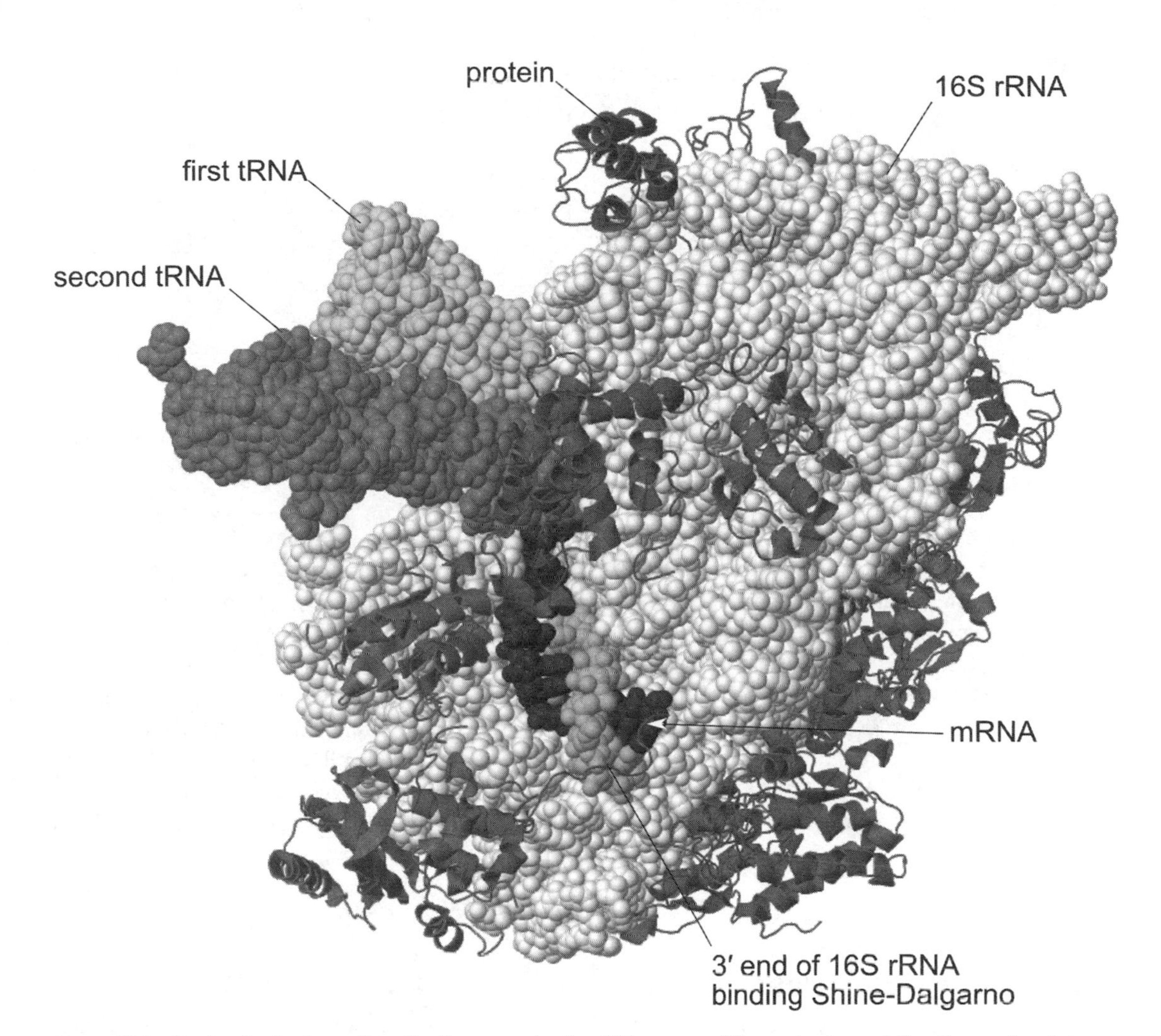

Figure 12.1 Small subunit of a prokaryotic ribosome, showing RNA as spacefilling molecules and the ribosomal proteins as cartoons. The 16S rRNA has a complex secondary structure and makes up most of the subunit. The first two tRNAs are shown bound to the mRNA, and the Shine-Dalgarno sequence of the mRNA can be seen interacting with the end of the 16S rRNA. Structures from the RCSB PDB (www.pdb.org): PDB ID 2QNH: A. Korostelev et al., Interactions and dynamics of the Shine Dalgarno helix in the 70S ribosome. *PNAS* 104:16840 (2007).

could catalyze their own splicing, and subsequently self-cleaving viral RNAs, RNAs used as catalytic subunits or templates in enzymes such as ribonuclease P and telomerase, and ***microRNAs (miRNAs****; see BioBackground) that influence gene expression have been identified. The function of each of these RNAs depends not merely on its sequence but on its unique three-dimensional structure. Nucleic acid structure has also become important as Kary Mullis' 1983 invention of the* ***polymerase chain reaction (PCR****; see BioBackground) revolutionized molecular biology with the ability to isolate any desired DNA molecule: Molecular biologists now routinely need small DNA primers they know will not form strong secondary structures.*

The essential three-dimensional structures of functional RNAs and the undesirable structures that may render PCR primers ineffective both result from base-pairing. By analogy to protein structure, we refer to base-pairing within a nucleic acid molecule (generally, a single-stranded nucleic acid molecule) as **secondary structure**. Such pairing results in loops and folds that can be important for function (for more detail, see BioBackground).

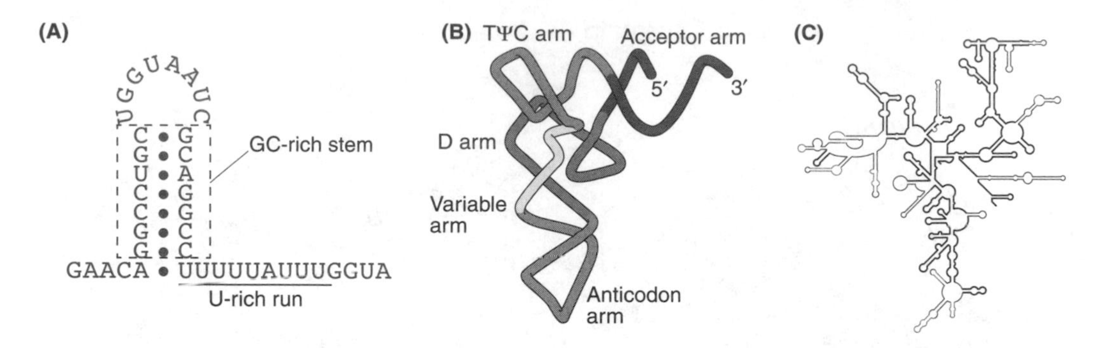

Figure 12.2 Examples of nucleic acids whose structure is critical to function. (A) Formation of a stem-and-loop structure with specific characteristics in mRNA leads to termination of transcription in prokaryotes. (B) Folded structure of a tRNA required for translation. (C) Secondary structure of 18S ribosomal RNA, the main RNA component of the small subunit of the eukaryotic ribosome.

Some examples include a "hairpin" loop in mRNA that leads to termination of transcription in prokaryotes; the compact, folded structure of transfer RNAs (tRNAs); and the complex folding of the ribosomal RNAs (**Figure 12.2**). Interactions among these folded structures then give an overall three-dimensional shape to the molecule. In addition, base-pairing between two nucleic acid molecules can have functional importance (e.g., tRNA binding mRNA or telomerase extending chromosomes) or create problems (two PCR primers binding each other). As with proteins, the primary structure (in this case, the nucleotide sequence) of the nucleic acid determines the possibilities for higher-order folding. This chapter explores bioinformatic methods to predict the secondary structure of nucleic acid molecules.

Bioinformatics Solutions:
Secondary Structure Prediction

The problem of nucleic acid folding is perhaps a little less daunting than that of protein folding, because there are only four nucleotides in an RNA or DNA molecule and the possible interactions are determined by the base-pairing rules: A pairs with T (or U in RNA) and G pairs with C. Nonetheless, this is a complex question because the number of possible pairings increases rapidly with sequence length: We need some way to decide which of the possible interactions is most likely. Furthermore, the strengths of the two pairs are not equal, and a structural prediction model needs to recognize that a G-C pair is stronger (and therefore more energetically favorable) than an A-T pair. Even more sophisticated prediction algorithms would account for the possibility of unusual base-pairing variations, such as Hoogsteen pairing, which can allow for the localized formation of a triple helix.

As with protein folding, we expect nucleic acids to fold in such a way as to maximize stability and minimize energy. Base-pairing stabilizes nucleic acid molecules—indeed, this is why a fully double-stranded DNA molecule is a highly stable structure—so conformations that maximize the internal base-pairing of a single-stranded RNA or DNA are those most likely to form, and long "stem" structures stabilize a folded nucleic acid more effectively than the same number of paired bases scattered throughout the molecule. Structure prediction algorithms thus attempt to maximize base-pairing and favor longer stretches of paired bases.

This chapter's projects use such structure prediction algorithms in several ways. For researchers studying functional RNAs, they can be used to predict secondary structure *ab initio* by maximizing base-pairing. However, secondary structure is not always desirable, as in the case of PCR primer design: The same basic structure prediction algorithm can be used in this situation to find primers that minimize pairing. Finally, we use structure prediction as one aspect of programs that extend gene prediction methods to genes encoding miRNAs.

BioConcept Questions

1. How are secondary and tertiary structures of nucleic acids similar to secondary and tertiary structures of proteins? How are they different?
2. What criteria might be used to decide whether one potential folded structure is better than another?
3. What are some kinds of structures that are commonly formed in a folded RNA?
4. If two PCR primers were able to base-pair partially with each other, what would you expect the result to be? How would the result be different if one of a pair of PCR primers were able to base pair internally?
5. Suppose you have a double-stranded DNA with one strand having the sequence 5′ `TATAGATCCGCG`. To amplify this sequence by PCR, you decide to use the primers 5′ `TATA` and 5′ `GCGC`. But your PCR reaction yields nothing. Why not?
6. In what two ways is the potential for miRNA to form secondary structure important to its function?

Understanding the Algorithm:
The Nussinov-Jacobson Algorithm

Secondary structures form when a single-stranded DNA or RNA folds over and base-pairs with itself, forming various types of loop structures (see BioBackground). The case of base-pairing between two separate nucleic acid molecules is conceptually similar. An algorithm published by Ruth Nussinov and Ann Jacobson in 1980 (see References and Supplemental Reading) predicts the locations of **hairpin** loops (**stem-loop** structures) within a nucleic acid strand. Stems form when base-pairing occurs between one or more pairs of RNA bases; unmatched bases form the loops. An *ad-hoc* approach to RNA folding might be to simply look for complementary bases where pairing could occur; however, in a long sequence, many such locations can be found and choices must be made among them. The Nussinov-Jacobson algorithm chooses the most stable (lowest energy) structure as the most likely one: This is the structure with the greatest number of paired bases. This algorithm is the basis of many nucleic acid folding algorithms in use today.

Figure 12.3 shows the possible base-pairing arrangements that would cause hairpin loops to form between two nucleotides, labeled i and j. Notice that a single loop may form between i and j when the nucleotide at location i pairs with a complementary nucleotide at location j (**Figure 12.3A**). Alternatively, a loop may form when the nucleotide at position $i + 1$ base-pairs with the nucleotide at position j (**Figure 12.3B**) or when the nucleotide at position i pairs with the nucleotide at position $j - 1$ (**Figure 12.3C**). The last example (**Figure 12.3D**) shows the possibility of *two* hairpin loops forming between positions i and j if a nucleotide k occurs between i and j such that k pairs with i and $k + 1$ base-pairs with j.

Determining all possible folds to choose the one with the greatest number of base pairs quickly becomes an intractable problem as the size of the sequence increases. This is the same problem we faced in aligning sequences (Chapter 3), so it may not surprise you that

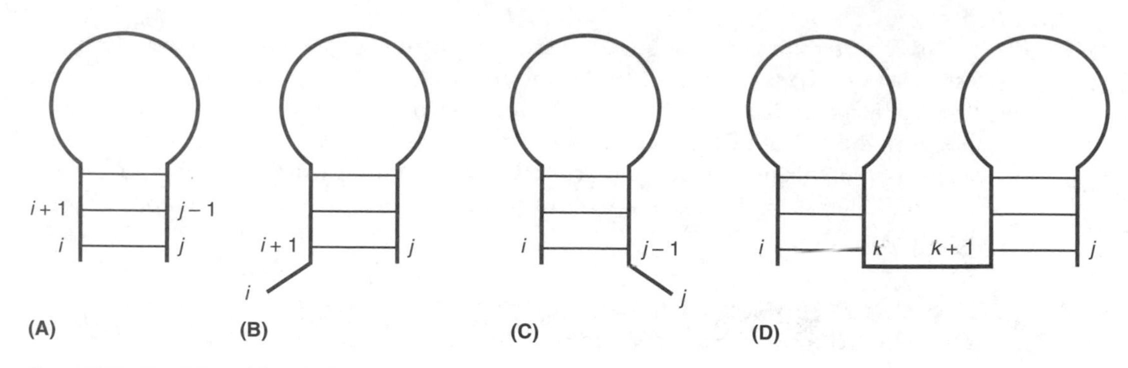

Figure 12.3 Possible positions for hairpin loops between nucleotide *i* and nucleotide *j* in an RNA molecule.

the Nussinov-Jacobson algorithm uses a dynamic programming technique similar to that used by Needleman and Wunsch in which a matrix represents the maximum number of base pairs that can occur between any two locations in the nucleic acid. It then uses a backtracking technique to determine which structures lead to an optimal solution.

Algorithm

Nussinov-Jacobson Algorithm

1. *Create an* n × n *matrix, where* n *is the length of the nucleic acid sequence.*
2. *Initialize the diagonal with zeros.*
3. *Fill in the matrix with the maximum number of base pairs (S) between each pair of nucleotides (i and j) by*

$$S(i,j) = \max \begin{Bmatrix} S(i+1, j-1) + w(i,j) \\ S(i+1, j) \\ S(i, j-1) \\ \max_{i<k<j}\left[S(i,k) + S(k+1, j)\right] \end{Bmatrix}$$

   where w(i, j) *is 1 if nucleotide* i *can base-pair with nucleotide* j *and 0 if it cannot and* k *is the location of a nucleotide between* i *and* j *as in Figure 12.3D.*

4. *Find the optimal (maximum) number of base pairs for the sequence as a whole in the upper right cell. Work back through the matrix to identify the base-paired structures that led to this number using the formulas in step 3. More detail on this backtracking step is given in the Guided Programming Project.*

Figure 12.4 shows the 13 × 13 matrix for the short sample sequence GGUUCCUUCCCAA. Looking at this sequence, notice it could base-pair in more than one way: For example, the two As could pair with either of the two UU sequences. You can see, however, that we get the most base pairs if two adjacent stem-loop structures form. To accomplish this algorithmically, for each nucleotide *i* in the sequence, we want to compute the maximum number of base pairs (*S*) that could form between that nucleotide and each other nucleotide, *j*. As shown in Figure 12.3, there are four possible scenarios, so *S*(*i*, *j*) is the greatest of four possible values calculated using the formulas above. However, notice that each cell's value

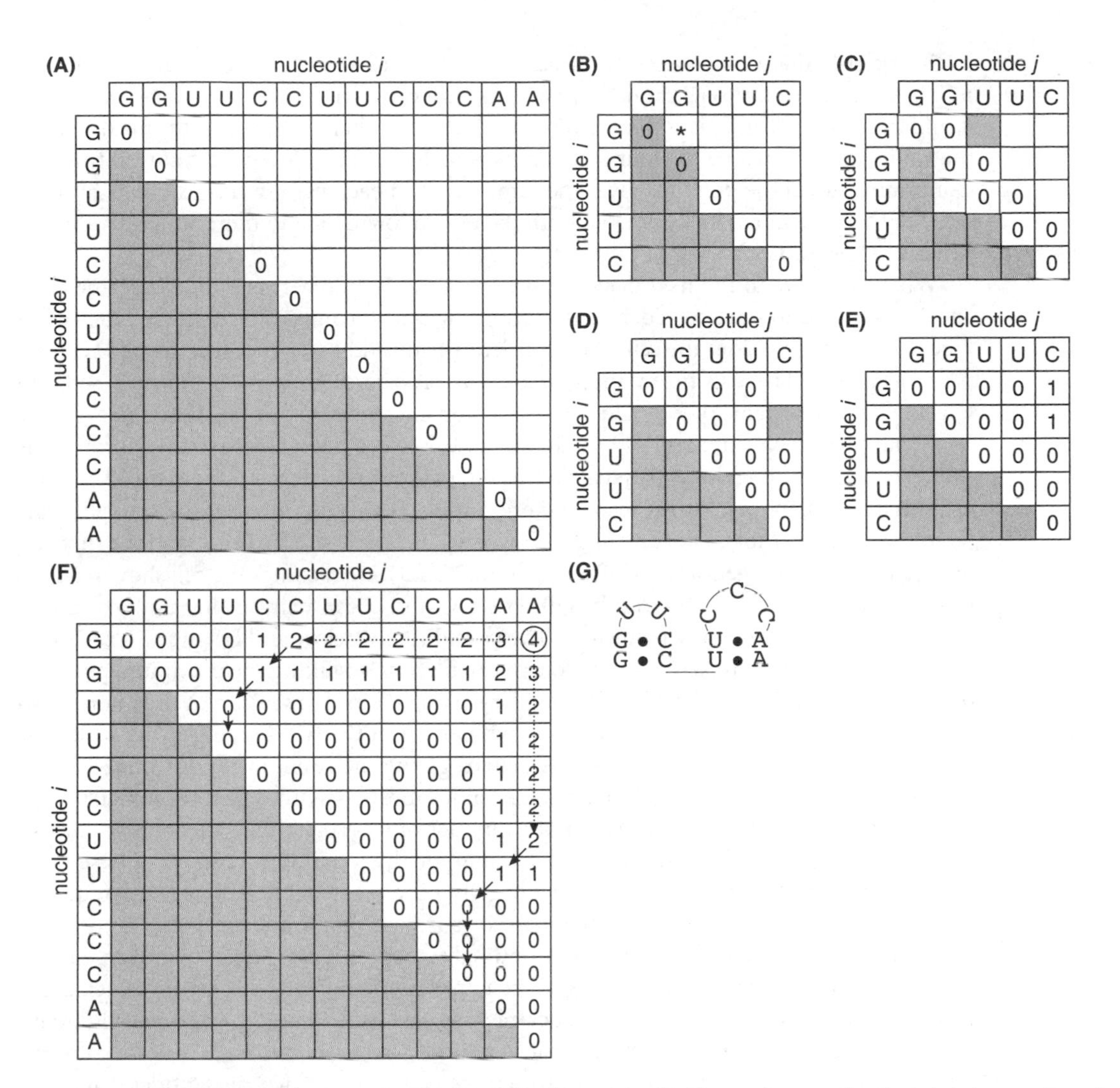

Figure 12.4 The Nussinov-Jacobson matrix: (A) after setting first values; (B–E) steps in computing the values for a portion of the matrix; (F) the backtrack path; (G) the predicted RNA structure.

is computed using the values of other cells in the matrix; we need to be sure those cells are filled in first. This means we need to initialize the matrix with some **first values** that we can calculate without depending on any other values. We know that a nucleotide cannot base-pair with itself, so we can initialize the diagonal to zeros (**Figure 12.4A**).

Now, we can compute the remaining values using the formulas just given: Each cell should contain the maximum of the four possible cases. Again, the value $w(i, j)$ is 1 if the bases at positions i and j can base-pair and 0 otherwise. Notice, however, that to find the values for the first three conditions, we need previously determined values for three other cells: $(i + 1, j - 1)$, $(i + 1, j)$, and $(i, j - 1)$. This is shown in **Figure 12.4B** (where for clarity only a small part of the matrix is shown), where the value of the cell at $i = 0, j = 1$ (marked by the asterisk; remember that computer scientists always count from zero) depends on the values of the three cells shown in blue. Therefore, we have to fill the cells in the proper order so these values are in place when needed: We need to fill the matrix diagonally. For the cell marked by the asterisk, there is no value for the cell at $(i + 1, j - 1)$, so the first formula becomes just $w(i, j)$, which is 0 because G cannot pair with G. The values of $(i + 1, j)$ and $(i, j - 1)$ are both 0, and because i is 0 and j is 1, there is no value k between them. Therefore, the maximum of the values determined by the four formulas is zero, and we can place that value in this cell. Continuing down the diagonal gives the same result in each case in this example, because none of the bases considered can pair with the one following it and there are no k values to consider (**Figure 12.4C**).

When we move to the next diagonal, there is a nucleotide between the i base and the j base, so we need to start considering the last of the four formulas. To fill in the blue cell in Figure 12.4C, we consider the cell diagonally left (0) and add $w(i, j)$; because G and U cannot pair, this is zero. The cells below and to the left are also zero. Then, $i = 0$ and $j = 2$, so there is a single possible value of k, 1, and we need to consider cells (i, k) or (0, 1) and $(k + 1, j)$ or (2, 2), outlined in blue in Figure 12.4C; in this case, both values are zero so the maximum is again zero. Continuing down this diagonal again gives zeros in every case (**Figure 12.4D**).

In the next diagonal, things become more complicated. There are now *two* possible values of k to consider for each i and j pair. However, the process is the same, and the first cell in the diagonal again gets a value of zero (Figure 12.4D). Next comes the cell shown in blue in Figure 12.4D, and here the result is different. G and C can pair, so the result of adding $w(i, j)$ to the cell diagonally below is 1. The cells directly below and left are still zero, and both values of k result in zeros for the fourth calculation, so taking the maximum puts 1 in this cell, representing a possible base pair. The final cell in this section of the matrix also gets a value of 1, because this C could pair with either of the two Gs (**Figure 12.4E**). This process continues through the rest of the matrix, giving the result in **Figure 12.4F**.

Like the Needleman-Wunsch matrix, the values in the Nussinov-Jacobson matrix represent partial solutions to the problem: the number of base pairs that can form as various segments of the RNA fold. The optimal solution, or the maximum number of base pairs that can form in this sequence (presumably giving the most stable structure), is found in the last column of the first row of the matrix: four base pairs. The final step is to determine *where* these base pairs occur in our optimal structure. This is done by traversing our base-pairing matrix backward, starting at the upper right and determining how each score was obtained. This is analogous to the backtracking step used to find the optimal alignment from the Needleman-Wunsch matrix but is a little more complex because of the fourth formula, which allows for base-paired structures *between* any two nucleotides i and j in the matrix. Computationally, this requires a recursive solution, which is discussed in the Guided Programming Project.

Figure 12.4F shows the results of the backtracking process for our sample sequence, and **Figure 12.4G** shows the optimal base-paired structure. Starting in the upper-right cell, we backtrack to the diagonal of the matrix by considering the path(s) that could have produced the score in that cell using the four formulas given. At each point we can make four possible moves: a diagonal move, a horizontal move, a vertical move, or a jump that causes a split in our path. A diagonal move (down and left) from a cell at position (i, j) indicates a base pair between the nucleotides in row i and column j (Figure 12.3A). A horizontal move (left) indicates the nucleotide in column j does not base-pair (Figure 12.3C). A vertical move (down) indicates the nucleotide in row i does not base-pair (Figure 12.3B). A jump indicates a split in the path, creating two new cells to consider: (i, k) and $(k + 1, j)$. Notice that for our sample sequence, a jump occurs immediately (Figure 12.4F). From each of the two new cells we move diagonally, indicating a base pair. Continuing to backtrack along these paths results in the two hairpin loops shown in Figure 12.4G. As with the Needleman-Wunsch algorithm, Nussinov-Jacobson can potentially identify more than one solution with the same optimal number of base pairs.

Test Your Understanding

1. Why does the Nussinov-Jacobson algorithm assume that the best structure is the one with the most base pairs?
2. Draw out the Nussinov-Jacobson matrix for the short sequence CCCUUGG. How many base pairs does the algorithm predict for the optimal secondary structure? Draw the predicted optimal structure.
3. The Nussinov-Jacobson algorithm cannot predict regions of the nucleic acid that form pseudoknots. Can you explain why these structures cannot be identified by this method?
4. In applying nucleic acid folding prediction to PCR primer design, the Nussinov-Jacobson algorithm can be used just as described here to test each primer for the possible formation of stable hairpins. However, it is also necessary to know whether the two primers can base-pair stably with each other. In what ways would this problem be different from the prediction of secondary structure within a molecule?

CHAPTER PROJECT:

Nucleic Acid Structure Prediction

The projects in this chapter apply the principles of *ab initio* nucleic acid structure prediction by the Nussinov-Jacobson algorithm and its derivatives to two different practical situations: the design of primers for PCR and the prediction of miRNA genes.

Learning Objectives

- Understand the importance of nucleic acid structure both in understanding biological function and in practical applications
- Use Web-based tools to predict nucleic acid secondary structure
- See how nucleic acid structure prediction is important to the successful design of PCR primers
- Understand how the Nussinov-Jacobson algorithm predicts secondary structure *ab initio*
- Appreciate the difficulty of identifying genes encoding functional RNAs and especially small RNAs

Suggestions for Using this Project

In the Web Exploration, students examine the secondary structure of a viral RNA, using a Web-based program that incorporates algorithms similar to the Nussinov-Jacobson algorithm. Related algorithms are used to find PCR primers that lack secondary structure and regions of complementarity. Students will also see how structure prediction can be one element of a program to predict miRNA genes within genomes. In the Guided Programming Project, students implement the Nussinov-Jacobsen algorithm for structure prediction. The On Your Own Project then examines how this algorithm can be modified for primer prediction.

Programming courses:

- Web Exploration: Use a Web-based structure prediction tool to fold a viral RNA and a primer-design tool to look for PCR primers suitable for amplification of its gene. Then use a small-RNA prediction program to look for potential miRNAs that might regulate herpes virus latency. Part III could be omitted to allow more focus on programming and structure prediction.
- Guided Programming Project: Implement the Nussinov-Jacobson algorithm to predict secondary structures in RNAs.
- On Your Own Project: Modify the Nussinov-Jacobson algorithm to develop a PCR primer-design program.

Non-programming courses:

- Web Exploration: Use a Web-based structure prediction tool to fold a viral RNA and a primer-design tool to look for PCR primers suitable for amplification of its gene. Then use a small-RNA prediction program to look for potential miRNAs that might regulate herpes virus latency.
- On Your Own Project: Consider modifications to the Nussinov-Jacobson algorithm to produce an algorithm suitable for PCR primer design.

■ **Web Exploration**: Applications of Nucleic Acid Structure Prediction

Oral and genital herpes result from infection with human herpesviruses 1 and 2 (HHV-1 and HHV-2, formerly known as HSV-1 and HSV-2, respectively). Herpes is a lifelong disease, because in addition to active replication in epithelial cells (producing the characteristic herpes lesions), these viruses can enter neurons where they remain **latent** (dormant) for long periods. The mechanisms involved in establishing and maintaining latency remain poorly understood, but an untranslated RNA called LAT (latency-associated transcript) has been implicated in down-regulating the other viral genes during latency. The RNA that is actually detected in infected neurons is in fact a stable intron spliced out of the original transcript of the LAT region; here, we refer to the stable intron as the LAT for convenience.

Part I: Secondary Structure in the LAT

Does the LAT have significant secondary structure and could that structure be important to its function? The LAT is believed to control the expression of genes that lead to active production of HHV-1 (keeping them turned off in neurons where the virus is latent) and possibly to block apoptosis (programmed cell death) of infected cells. The mechanisms by which it acts are largely unclear, but because it is a stable RNA maintained in the host cell, one might hypothesize that it has specific secondary structure that is of functional importance. We investigate its secondary structure using a Web-based implementation of mfold, a well-known secondary structure prediction program. This program uses an algorithm much like the Nussinov-Jacobson algorithm and adds a minimal energy calculation to further refine the predicted structures.

Start by finding the complete genome of HHV-1 in the NCBI **Nucleotide database** or in the **RefSeq database**. Search the text of the GenBank record for LAT (match the case as well as the text to make this easier) to identify a feature labeled as the stable LAT intron. Copy the nucleotide sequence for the intron. This is most easily done by using the numbered sequence at the bottom of the GenBank entry and then using the **Sequence Manipulation Suite** to clean up the copied DNA. Save your sequence to a text file in FASTA format. Now, because we want to look at the folding of the RNA transcribed from this region (and U-A base pairs are different in stability from T-A base pairs), replace all the Ts in the sequence with Us.

Now, find the RNA folding form at the **mfold server site**. Input your sequence. Notice many parameters can be changed; most can be left at their defaults for basic structure prediction. Be sure the RNA sequence is set to `linear`, and because of the length of this sequence, you will need to submit it as a batch job rather than an immediate job. Submit the job; when it is complete, you should see a link to a results page (which will also be emailed to you). The results page offers some options for downloading the folding data in various formats and lists all the possible structures found by mfold (up to whatever limit was set in the parameters). View a few structures by clicking the `pdf` link below each structure.

Web Exploration Questions

1. Does it appear that the LAT has significant secondary structure? Describe generally what the structures predicted by mfold look like and how extensive the base-pairing is.
2. How would you explain the fact that mfold found many possible structures, not just one optimal structure?
3. Are the different structures predicted by mfold generally similar or quite different?
4. Most RNAs are not very stable in the cytoplasm of cells: They are quickly broken down by RNAses that attack single-stranded RNA (relatively rapid mRNA turnover is necessary for genes to be turned off effectively). The LAT, however, is a highly stable RNA. Does its structure provide any clues to the cause of this unusual stability?

Part II: Designing PCR Primers to Amplify a Viral Gene

PCR (see BioBackground) is a sensitive way to detect a viral infection (the most sensitive HIV tests are PCR-based, for example) because of its ability to amplify as little as a single molecule of DNA to a level that can be detected by gel electrophoresis. Because HHV-1 is a virus with a DNA genome, you could use PCR to amplify an HHV-1 gene as a means of detecting infected cells. For this exercise, imagine you want to PCR-amplify the stable LAT intron to diagnose infection. You would first need to order primers, typically 20 nucleotides long, matching the two ends of the region you want to amplify. It is easy to choose 20 nucleotides from the right part of the HHV-1 genome, but not all potential primers work equally well: Self-complementarity within a primer could lead to generation of a stable hairpin structure that would impede binding to the HHV-1 DNA (**Figure 12.5A, B**) or the two primers could have a region of complementarity leading to the amplification of "primer dimers" in competition with amplification of the target sequence (**Figure 12.5C**). Thus, before ordering primers, molecular biologists use bioinformatic software to check for these undesirable characteristics; these programs have evolved to the point that now they commonly select appropriate primers given a desired sequence to be amplified.

Web Primer is a Web-based primer design program that gives good results and is very easy to use, requiring only an input sequence and selection of some basic parameters. Return to the complete genome of HHV-1 and again find the stable LAT intron. This time, copy the nucleotide sequence for the intron *plus* about 100 extra nucleotides at each end. Because it

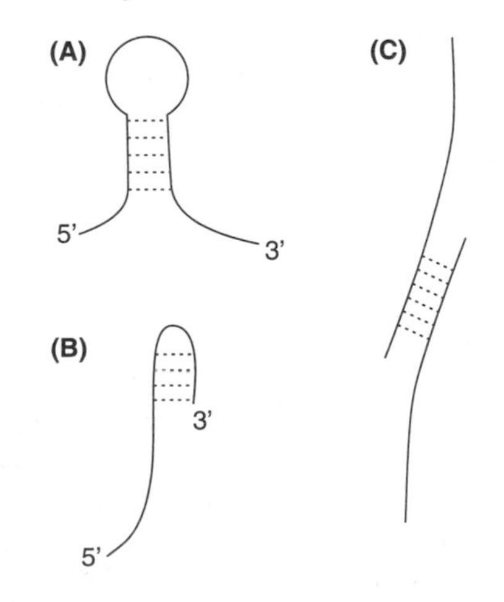

Figure 12.5 Base-paired structures formed by PCR primers that can interfere with annealing or DNA synthesis during PCR. (A) self-annealing, (B) 3′ self-annealing, (C) pair-annealing, or "primer dimer." Dotted lines represent base-pairing.

is unlikely that the ideal PCR primer will bind exactly to the ends of the sequence you want to amplify, it is helpful to have a larger region in which to find potential primers. Save this sequence to a text file.

Now, find **Web Primer** at the ***Saccharomyces* Genome Database** site. Paste your sequence into the input box (note that FASTA comments are not permitted here). On the next page you can set a number of parameters. You may choose to pick primers that exactly amplify the entire input sequence, but as noted this does not allow the program much latitude to pick the best primers. So, choose instead to let the program search within a 100-nucleotide region on each end.

The **melting temperature (T_m)** is the temperature at which the primer can no longer base-pair stably with the template—to be specific, 50% of the primer molecules should be unpaired if the primer is exactly at its T_m. This is critical in the PCR annealing step (see BioBackground), because the temperature chosen here must be low enough to allow the primers to bind but high enough to destabilize nonspecific base-pairing interactions. We usually want the annealing temperature to be about 5° C below the lower primer T_m and yet be at least 45–50° C. To make this work well, we also want the two primers to have very similar T_m. The default values for primer length and T_m reflect these considerations and can be used for now. The GC content of the primer also affects the stability of base-pairing with the template; use the default here for now as well.

Finally, the primer annealing parameters deal directly with secondary structure and interprimer pairing. The algorithm used by Web Primer (see References and Supplemental Reading) gives a simple score to reflect the likelihood of base-pairing within or between primers: 2 for an A-T pair and 4 for a G-C pair. **Self-annealing** refers to secondary structure within the primer (Figure 12.5A), with a separate category for secondary structure that includes the 3′ end (Figure 12.5B), because pairing here could prevent the 3′ from being used as the start point for DNA synthesis. **Pair-annealing** works the same way but refers to pairing between the two primers (Figure 12.5C). You can leave these values at their defaults for now; raising them allows for more base-pairing in your primers, whereas lowering them more stringently screens out pairing. When you run the analysis, Web Primer first searches for possible primers that meet the length, T_m, and GC criteria and then examines their potential for secondary structure and base-pairing interactions.

Web Exploration Questions

5. What is the initial result of the Web Primer analysis? Looking at your LAT intron sequence, can you see why this result was obtained?
6. Use the information from question #5 to decide how to adjust parameters appropriately to get a better result. What parameters did you have to change, and what values did you set?
7. What are the sequences (write them both from 5′ → 3′) of the best pair of primers found by Web Primer? How long a DNA fragment would they amplify?
8. What are the melting temperatures of the two primers in the best primer pair? What would be a good annealing temperature for your PCR reaction?
9. How much potential is there for secondary structure in the best primers? How much potential for interprimer interaction? Do these numbers change significantly as you go down the list of less optimal primer pairs?
10. If you were actually going to do this PCR reaction, what concerns might you have about the primers chosen by Web Primer? If you look at the list of all primer pairs that meet your criteria, do you see any others that might help with these concerns?

Part III: Predicting Genes Encoding Functional Small RNAs

The discovery of miRNAs (see BioBackground) touched off a great deal of interest in identifying small RNAs that function in gene regulation as well as the practical use of small RNAs to selectively decrease gene expression (RNA interference). Many such RNAs have now been found (see References and Supplemental Reading), including miRNAs encoded by viruses such as Epstein-Barr virus, HIV, and hepatitis C virus that are used to regulate viral or host functions. Databases such as **miRBase** catalog miRNAs and their functions. Because the LAT RNA appears to be important in HHV latency and because other known viruses use miRNAs for regulation, it is reasonable to investigate whether the LAT might include miRNA sequences that could be responsible for its function.

Given a DNA sequence, however, prediction of genes encoding functional RNAs is challenging: Although there may be identifiable promoter elements, the lack of a protein-coding sequence and the potential presence of introns make it difficult to predict the exact extent and sequence of a gene for a noncoding RNA. Secondary structure is one important component of predicting small RNA genes: Both miRNAs and small interfering RNAs, although different in origin and function, form hairpin structures recognized by the Dicer enzyme (see BioBackground). Secondary structure prediction is one element of the analysis done by **mirEval**, a Web-based miRNA prediction program. In addition to structure prediction, mirEval compares potential miRNAs with known miRNAs in miRBase, constructs alignments with other genome sequences (conserved regions are more likely to be functional), and examines flanking sequence for additional miRNAs (known miRNA genes often occur in clusters).

At the mirEval site, paste your sequence into the appropriate field. Select `Other species` as the organism. Because we are looking for miRNAs in HHV-1, not in the human genome, we can skip the conservation and cluster analysis: under `Select Analysis type`, check only `Structure` and `miRBase`. You can choose whether or not to run the `Advanced Structure` analysis (or try it both ways); this slows down the analysis, but on the short DNA sequence we are using, the time difference is not large. On the results page, green bars represent predicted miRNA precursor structures. The actual RNAs used in the RISC complex are cut from these precursor structures by Dicer and are 20–25 nt long.

Web Exploration Questions

11. How many potential miRNA precursors were identified?

12. Were there any similarities to known miRNAs that might strengthen the likelihood that these are genuine miRNAs?

13. HHV-1 and HHV-2 are closely related and behave similarly with regard to infection and latency. We might expect that any genuine miRNAs in HHV-1's LAT would also be found in the LAT of HHV-2. Use mirEval to identify potential miRNA precursors in the HHV-2 LAT; are there similarities that strengthen the case for authentic miRNAs?

More to Explore

Many primer-design programs are available. You may be interested in investigating this area of bioinformatics further by looking at **Primer 3**, which has the same goals as Web Primer but offers far more opportunity for the user to choose specific parameters. You might also want to look further into possible HHV-1 miRNAs by using mirEval to look at the entire genome instead of just the LAT.

Guided Programming Project: Structure Prediction with the Nussinov-Jacobson Algorithm

The Understanding the Algorithm section described an implementation of the Nussinov-Jacobson algorithm that can be used to find an optimal base-paired structure for a DNA or RNA sequence. You are already very familiar with the construction of matrices as a way to examine sequence data, and given your experience with the Needleman-Wunsch algorithm (Chapter 3), you should have little trouble implementing a program that accepts a single-stranded DNA or RNA sequence as input and constructs the Nussinov-Jacobson matrix (Figure 12.4) to determine an optimal number of base pairs in the folded structure. Remember to calculate first values before proceeding to evaluate the four formulae representing the four possible conditions existing between two nucleotides. In the Understanding the Algorithm section, we initialized the diagonal with zeros; however, in your code it will be easier to initialize the entire matrix to zeros so you do not have to worry about an empty cell value (as in Figure 12.4B) when applying the formulae.

The backtracking portion of the algorithm to identify the actual paired bases requires some additional discussion. Traversing the matrix in reverse is best implemented as a recursive algorithm. **Recursion** in a computer program occurs when a step in a function calls that same function. Recursion is a divide-and-conquer technique used to solve large problems by solving smaller and smaller subproblems, just the opposite of the dynamic programming technique used to build the matrix. The repetitive process of a recursive function is similar to a loop executing repeatedly until the loop's test condition fails; the steps of a recursive function are repeated until a **base case** is met that ends the recursive calls. At this point the current instance of the function is completed and execution returns to the previous call. Eventually, execution returns to the initial calling routine. The following steps should help you see how recursion can be used to traverse the Nussinov-Jacobson matrix.

Algorithm

Backtracking Algorithm for the Nussinov-Jacobson Matrix

1. *The optimum number of base pairs is found in the first row (row 0), last column (column* n-1, *for an* n × n *matrix). In the example in Figure 12.4F, this is row 0, column 12.*
2. *Call a function to look for a base pair, passing in these initial values for arguments* row *and* col.
3. *In the function:*

   If row >= col, *end the process. This is the base case and ends the recursive calls.*

   Else, if S(row, col) = S(row + 1, col), *recursively call the same function with* (row + 1, col) *as arguments.*

   Else, if S(row, col) = S(row, col–1), *recursively call the same subroutine with* (row, col–1) *as arguments.*

   Else, if S(row, col) = S(row + 1, col–1) + w(i, j), *recursively call the same subroutine with* (row + 1, col–1) *as arguments. If* w(i, j) = 1, *a base pair has been found: print values for* row *and* col.

   Else, *if none of the above cases is true, then find all values* k, *such that* S(row, col) = S(row, k) + S(k + 1, col). *For each value found, recursively call the same subroutine with* S(row, k) and S(k + 1, col). *We are now looking for base-paired structures that involve bases between* i *and* j *(Figure 12.3D).*

You saw that mfold incorporates drawing algorithms to produce a graphical representation of the folded nucleic acid. For our purposes, however, a text list of base-paired nucleotides is sufficient. For example, the matrix shown in Figure 12.4 would give the following output (using zero-based numbering for the nucleotides):

```
G(0) base pairs with C(5)
G(1) base pairs with C(4)
U(6) base pairs with A(12)
U(7) base pairs with A(11)
```

Additionally, we saw before that the Nussinov-Jacobson matrix can identify multiple solutions that give the same optimal number of base pairs. Your program need only find one solution for each sequence. The pseudocode that follows should allow you to implement this algorithm as described.

Algorithm

Nussinov-Jacobson Algorithm

Goal: To find the optimal number of base pairs given an RNA sequence

Input: An RNA sequence

Output: The optimal number of base pair locations

```
// Step 1: Initialization and Read in Sequence

open input file: infile
seq = ""
read and ignore first line of data in infile1
for each line of data in infile
    concatenate line of data to seq

// initialize matrix with zeros
N = length of seq
matrix = array of size [N,N]
for each i from 0 to N
    for each j from 0 to N
        matrix[i, j] = 0

for each diagonal from 1 to N
    for each row from 0 to (N-diagonal)
        col = row + diagonal
        w = isBasePair(seq[row], seq[col])
        v1 = matrix[row+1, col-1] + w
        v2 = matrix[row+1, col]
        v3 = matrix[row, col-1]
        v4 = getMaxK(matrix, row, col)
        matrix[row, col] = max(v1, v2, v3, v4)
findBasePairs(matrix, seq, 0, N-1)

// function to determine if parameters are base pairs
function isBasePair(a, b)
    if a == 'A' and b == 'U'
      or a == 'U' and b == 'A'
```

```
        or a == 'G' and b == 'C'
        or a == 'C' and b == 'G'
            return 1
    else
            return 0

// function to determine the maximum k value
function getMaxK(matrix, i, j)
    curmax = -1
    for each k from i+1 to j-1
        t = matrix[i,k] + matrix[k+1, j]
        if t > curmax
            curmax = t
    return curmax

// recursive function to find location of optimal base pairs
function findBasePairs(matrix, seq, r, c)
    if r >= c // ends base case
        return
    else if matrix[r, c] == matrix[r+1,c]
    findBasePairs(matrix, seq, r+1, c)
    else if matrix[r, c] == matrix[r, c-1]
    findBasePairs[matrix, seq, r, c-1)
    else if matrix[r,c] == matrix[r+1,c-1] + isBasePair(seq[r],seq[c])
        if isBasePair(seq[r],seq[c]) == 1
            print seq[r] + '(' + r + ') base pairs with' + seq[c] +
            '(' + c + ')'
        findBasePairs(matrix, seq, r+1, c-1)
    else
        for each k from r+1 to c-1:
            t = matrix[r, k] + matrix[k+1, c]
            if t == matrix[r, c]
                findBasePairs(matrix, seq, r, k)
                findBasePairs(matrix, seq, k+1, c)
```

Putting Your Skills Into Practice

1. Implement the Nussinov-Jacobson algorithm as described earlier and in Understanding the Algorithm. Your program should output a matrix to determine the maximum number of base pairs for any sequence and then list the nucleotides that base-pair along with their locations, as shown previously. Test your program with the sample sequence in Figure 12.4. Additional test data are provided on the *Exploring Bioinformatics* website.
2. Change the order of the conditional statements in the recursive backtracking routine. Does this change the structure your program predicts? Why? If your program found all possible solutions, would the order of statements matter?
3. Test your program with the HHV-1 LAT RNA sequence you downloaded. Does it produce a structure somewhat similar to that predicted by mfold? Discuss similarities and any significant differences.

4. The algorithm as described allows base pairs to occur between two nucleotides immediately adjacent to each other. This is actually not very realistic: Two bases that close together will be prevented from pairing by the geometry of the DNA or RNA molecule. It would be better for our algorithm to require at least one nucleotide between any two base-paired nucleotides. Modify your solution to incorporate this improvement.

On Your Own Project: PCR Primer Analysis with Nussinov-Jacobson

Understanding the Problem

As we saw in the Web Exploration, molecular biologists use bioinformatics software to help choose appropriate PCR primers, because not all potential primers work equally well to amplify regions of DNA. How could the analysis of secondary structure be applied to choosing PCR primers? For this project you are asked to consider how the Nussinov-Jacobson algorithm could be applied to a potential primer to look for the ability to form a hairpin.

As you saw for Web Primer, base-pairing, self-annealing, and 3′ self-annealing all affect a primer's ability to amplify a region of DNA. When choosing a primer, scores for these properties must be considered. Base pairs can be scored according to a simple system (e.g., Web Primer scores 2 for a weaker A-T pair and 4 for a stronger G-C pair). The user can decide how high a score to consider acceptable. (A more sophisticated algorithm would incorporate energy calculations to assess the stability of a secondary structure more precisely.) The special case of 3′ self-annealing can be addressed by using the same data and separately scoring base pairs starting at the 3′ end. This leaves two issues: the question of pair annealing, or pairing between the two PCR primers, and how (given an input DNA sequence to be amplified) to choose possible primer pairs to test. In this On Your Own Project, you will implement a program that determines potential primers for a given DNA sequence.

Solving the Problem

Start by first considering how to choose potential primer pairs given a DNA sequence and the location in the sequence to amplify. We assume that 20-nucleotide primers are desired (although a program could obtain a desired value from the user). Do not forget that the two PCR primers must bind to opposite strands of the DNA, with their 3′ ends facing the sequence to be amplified.

You want to eliminate any primer with a high self-annealing or 3′ self-annealing score; you can easily modify the Nussinov-Jacobson algorithm to assign a score to the base-paired structures it finds and separately score base-paired structures involving the 3′ end. You also need to set a threshold for how high a score is allowable; the default parameters you used with Web Primer would be a good starting point. Review Part II of the Web Exploration project for help in understanding these ideas.

Once you have your potential primers, consider how you could use the Nussinov-Jacobson algorithm to look for pairing *between* two potential primers. Although the algorithm works on a single sequence, consider how your two primers can be combined so that no change to the input requirements would be needed.

If you are in a nonprogramming class, list the steps of an algorithm that will (i) choose primers from a sequence; (ii) test for self-, 3′- and pair annealing; and (iii) choose the best pair. If you are in a programming class, proceed to implement the algorithm as described next.

Programming the Solution

Write a program that determines potential primers given an input DNA sequence and the starting/ending location of the area to amplify. Create a short test sequence that includes enough nucleotides around the amplification area to allow for at least several different possible primers. Use a windowing technique (see Chapter 10) to identify potential primers. Your results should list potential primer pairs (only those that met the criteria) and each primer's starting and ending location in the original sequence. The primer pairs with the best combined scores for self-annealing, 3′ self-annealing, and pair annealing should be listed first.

More to Explore

In addition to testing primers for undesirable base-pairing, Web Primer chose primers that matched T_m and GC content criteria. Melting temperature is particularly important: For PCR to work, it must be possible to choose an annealing temperature appropriate to both primers and not so low as to reduce the specificity of annealing. Calculating melting temperature can get quite complex, with the most accurate formulas accounting for the salt concentration of the solution and for interactions between adjacent nucleotide pairs as well as the base pairs themselves. However, a reasonable approximation of T_m can be calculated simply by adding 2° C for each A-T pair and 4° C for each G-C pair. Using this simple rule, modify your program so it calculates the T_m for each primer in a potential pair and discards the pair if (1) the two T_m values are more than 5° C apart or (2) the lower T_m value is less than 50° C (so the annealing temperature drops below 45° C). Then test only the remaining primer pairs for acceptable secondary structure.

BioBackground: Nucleic Acid Structure and Applications

Nucleic Acid Secondary Structure

Formation of secondary structure in a nucleic acid results from base-pairing within the nucleotide chain. Because this cannot occur in double-stranded regions, we usually think of secondary structure as a feature of single-stranded nucleic acids, notably RNA or the single-stranded DNA genomes of some viruses. However, even double-stranded DNA can have single-stranded regions in which secondary structure occurs.

Regions of paired bases in a single-stranded nucleic acid molecule are called **stems**; unpaired regions are **loops** (**Figure 12.6A**). Loops can occur at the ends of stems, forming **hairpins** (also called **stem-and-loop** structures), or within them, forming **internal loops** (**Figure 12.6B**); imperfect loops are known as **bulges**. In addition, interactions can occur between different loops in a molecule; these **pseudoknots** (**Figure 12.6C**) produce tertiary structure or an overall folded three-dimensional shape. As with proteins, RNA molecules fold to minimize the energy of the folded structure. Base-pairing protects "sticky" bases in a single-stranded molecule, so generally the more bases can be successfully paired, the more favorable the structure is. Structure prediction programs therefore generally seek to maximize base-pairing.

Secondary structure is critical to the function of many nucleic acids, notably in the increasing number of functional noncoding RNA molecules that have been shown to participate in various cellular processes. The importance of tRNA (Figure 12.2B) and rRNA (Figure 12.1 and Figure 12.2C) in translation is well known; however, more recent research suggests that a specific, highly structured segment of the RNA component of the ribosome, not one of the proteins, carries out the transfer of the growing peptide chain from the tRNA in the P site to the amino acid in the A site (peptidyltransferase activity). The rRNA thus acts as a **ribozyme**, or an RNA enzyme, and this is one piece of evidence in support of the hypothesis that RNA was both the original genetic material and the original catalyst (the "RNA world" hypothesis).

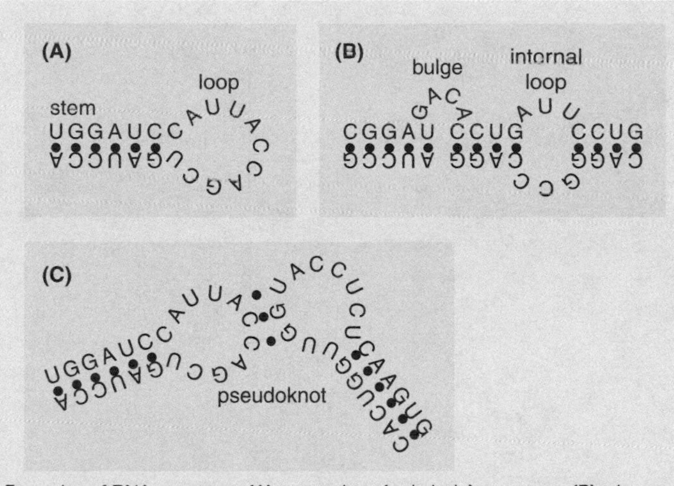

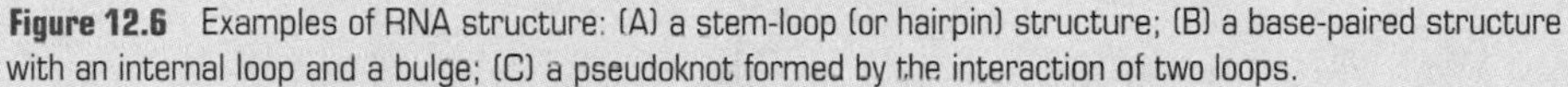

Figure 12.6 Examples of RNA structure: (A) a stem-loop (or hairpin) structure; (B) a base-paired structure with an internal loop and a bulge; (C) a pseudoknot formed by the interaction of two loops.

Additional ribozymes include RNAse P (involved in tRNA processing), self-splicing introns in some species, and a number of viral ribozymes. In all cases, secondary and tertiary structure are essential to ribozyme function. Folded RNAs are also important in noncatalytic roles. For example, in telomerase, the enzyme that adds repeated sequences to the ends of linear chromosomes so that incomplete replication of the ends does not damage genes, an RNA serves as the template for synthesis of the repeat DNA.

RNA Interference and microRNAs

In 2006 Andrew Fire and Craig Mello received the Nobel Prize for their discovery of a phenomenon called **RNA interference** (**RNAi**) in which small noncoding RNAs inhibit gene expression. The number of known natural functions of RNAi is rapidly increasing; these include antiviral responses and regulation of developmental genes. Additionally, RNAi has quickly become a valuable technique used to "knock down"—greatly reduce expression of—any desired gene. Eventually, we may see RNAi used therapeutically. Small RNAs with functional roles in bacteria have also been identified.

Although the terminology used for these functional RNAs is somewhat confusing, there are two major types of RNA important in RNAi: **miRNAs** and **small interfering RNAs**. The two have different origins and functions but share a common pathway. Double-stranded RNA or long base-paired RNA hairpins that result from viral infection (or in at least some cases from natural cellular pathways) are cleaved by an enzyme called Dicer into fragments 20–25 bp long (**Figure 12.7A**). One strand from these fragments then binds to a protein complex called RISC and then to mRNAs, resulting in "silencing" of translation through either cleavage or blockage of the mRNA. miRNAs are also cleaved by Dicer and bind to RISC to silence genes, but these are encoded in the genome and transcribed into folded RNAs with long base-paired regions (**Figure 12.7B**).

Polymerase Chain Reaction

PCR is a technique that revolutionized molecular biology by providing a means of isolating and amplifying any desired piece of DNA whose sequence is known. PCR amplification of DNA is useful for cloning genes, sequencing specific regions, measuring expression, assaying protein binding, and many other uses. The PCR technique harnesses DNA replication, using a DNA

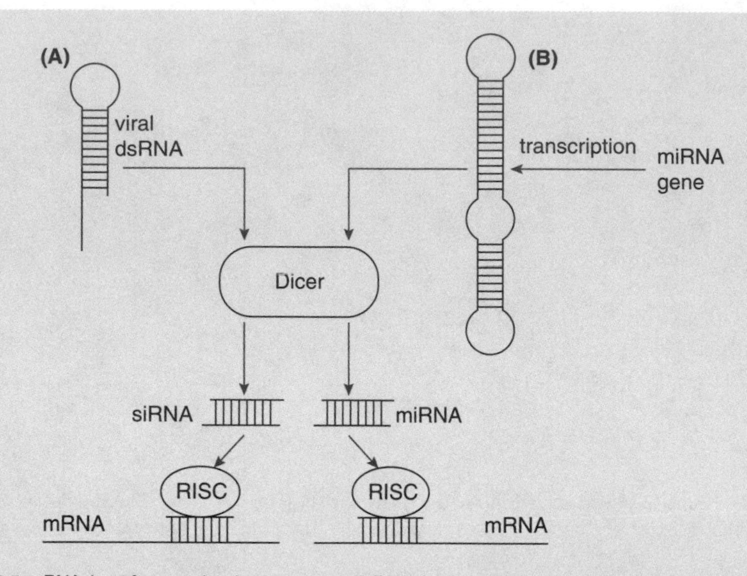

Figure 12.7 RNA interference involves structured RNA molecules. (A) siRNA is generated from viral or cellular dsRNA or hairpin RNA structures; (B) miRNA is generated from folded RNAs transcribed from the genome. In either case, Dicer generates short double-stranded fragments and RISC binds a single strand and facilitates binding to mRNA and inhibition of translation.

polymerase (*Taq* polymerase or a similar enzyme) isolated from bacteria that live in hot springs and can thus withstand high temperatures.

The location at which the DNA copying occurs is controlled by **primers**, single-stranded nucleic acids roughly 20 nucleotides long (a short nucleic acid is also called an **oligonucleotide**, or **oligo** for short). A primer of any desired sequence can be purchased inexpensively from a commercial supplier. The primers are chosen so that one base-pairs with the DNA sequence at one end of the region to be copied and the other with the sequence at the other end (**Figure 12.8**). Specifically, one must bind to each strand of the DNA, and both must bind so that their 3′ ends (the end from which DNA polymerase begins copying) "face" the sequence to be copied. The primer that binds the "left" end of the sequence as the researcher looks at it is referred to as the **forward primer** and the other is the **reverse primer**.

The DNA to be copied (the template) is mixed with the two primers, the DNA polymerase, and a supply of all four DNA nucleotides (referred to as dNTPs). The mixture is heated in a

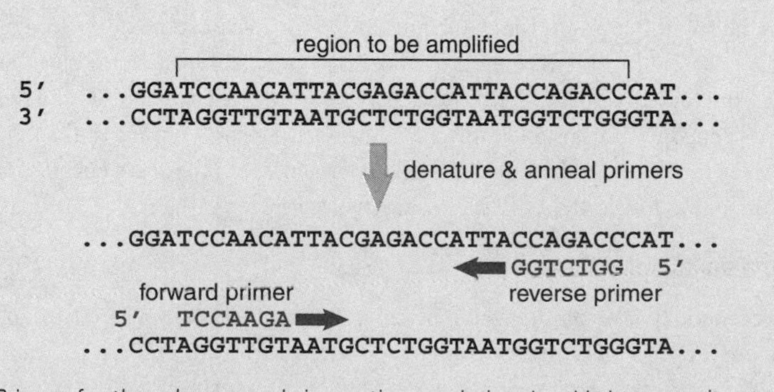

Figure 12.8 Primers for the polymerase chain reaction are designed to bind to opposite strands of the DNA at opposite ends of the region to be amplified. Arrows show direction of extension by DNA polymerase.

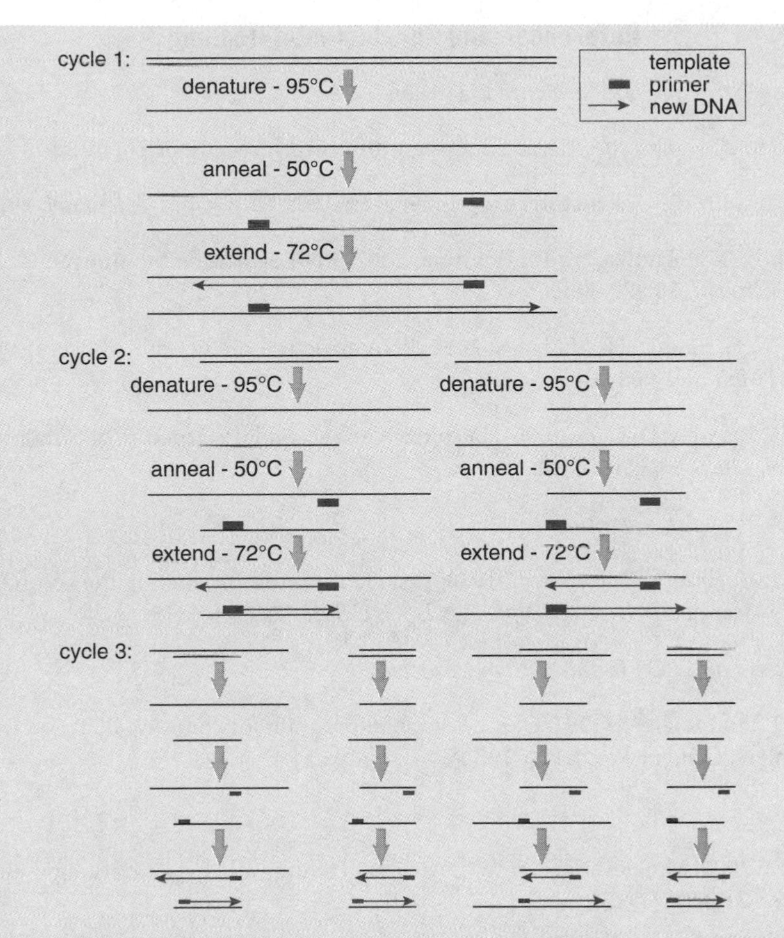

Figure 12.9 The polymerase chain reaction. In each cycle (three are shown here), DNA is denatured, primers are annealed, and DNA polymerase extends from the primers. With each cycle, more of the correct-sized product is synthesized. The process is repeated for 30 cycles to produce a large amount of specific amplified product.

thermal cycler (an instrument that can precisely control and rapidly change temperature) to 95° C, at which point the hydrogen bonds holding the two DNA strands together are broken and the DNA denatures or "unzips" (**Figure 12.9**). Now the temperature is reduced to the **annealing temperature,** a point just cool enough for the primers to base-pair with their targets (usually 45–60° C). The DNA polymerase starts at the 3′ ends of the primers and begins copying the DNA, copying efficiently as the temperature is raised to an extension temperature of about 72° C. As there is nothing to stop it other than the biochemical limits of its efficiency, the polymerase will copy beyond the desired region so that only one end of each new strand is in the correct place.

Now, the temperature is raised to the denaturing temperature again and a new cycle begins. After annealing and extension, some of the new strands are now the correct length (cycle 2 in Figure 12.9). A third cycle then begins, and now some double-stranded correct-length DNA is seen (cycle 3 in Figure 12.9). The process of denaturing, annealing, and extending is repeated over and over, and by the end of 30 total cycles (assuming complete efficiency), more than a billion copies of only the desired sequence are produced from a single template molecule.

References and Supplemental Reading

Secondary Structure and Structure Prediction

Eddy, S. R. 2004. How do RNA folding algorithms work? *Nat. Biotechnol.* **22**:1457–1458.

Felden, B. 2007. RNA structure: experimental analysis. *Curr. Opin. Microbiol.* **10**:286–291.

Jossinet, F., T. E. Ludwig, and E. Westhof. 2007. RNA structure: bioinformatic analysis. *Curr. Opin. Microbiol.* **10**:279–285.

Seetin, M. G., and D. H. Mathews. 2012. RNA structure prediction: an overview of methods. *Methods Mol. Biol.* **905**:99–122.

Svoboda, P, and A. Di Cara. 2006. Hairpin RNA: a secondary structure of primary importance. *Cell. Mol. Life Sci.* **63**:901–908.

Nussinov-Jacobson Algorithm

Nussinov, R., and A. B. Jacobson. 1980. Fast algorithm for predicting the secondary structure of single-stranded RNA. *Proc. Natl. Acad. Sci. U.S.A.* **77**:6309–6313.

Web Primer and PCR Primer Design Principles

Hillier, L., and P. Green. 1991. OSP: a computer program for choosing PCR and DNA sequencing primers. *Genome Res.* **1**:124–128.

mfold

Zuker, M. 2003. Mfold web server for nucleic acid folding and hybridization prediction. *Nucleic Acids Res.* **31**:3406–3415.

Small RNAs and miRNA Prediction

Griffiths-Jones, S., R. J. Grocock, S. van Dongen, A. Bateman, and A. J. Enright. 2006. miRBase: microRNA sequences, targets and gene nomenclature. *Nucleic Acids Res.* **34**:D140–D144.

He, L., and G. J. Hannon. 2004. MicroRNAs: small RNAs with a big role in gene regulation. *Nat. Rev. Genet.* **5**:522–531.

Ritchie, W., F. X. Théodule, and D. Gautheret. 2008. Mireval: a web tool for simple microRNA prediction in genome sequences. *Bioinformatics* **24**:1394–1396.

Appendix

Introduction to Programming

Programming Basics

Programming consists of writing a **program**—a set of statements that instructs a computer to perform actions to solve a particular problem. This includes developing a set of logical steps that can solve the problem (the **algorithm**) and then "translating" these steps into statements in a particular programming language (such as Java, C++, or—most commonly in bioinformatics applications—PERL or Python). Each language has a specific **syntax**, and just as the rules of English grammar must be followed to create a coherent sentence, a computer can correctly interpret a programming statement only if the syntactical rules of the programming language are followed. These statements, the source code, are eventually translated into the computer's internal language (which is independent of the programming language), allowing the computer to execute (or run) the program.

Learning to program includes learning the syntax of one or more languages so program statements can actually be written, but, even more importantly, it means learning how to problem-solve. Solving the problem is what programming is all about, and no amount of syntactical knowledge will save you if you cannot come up with an algorithm that actually solves the problem. The goal of this text is to help you develop the skills to solve problems in bioinformatics by working on real-world problems through a series of hands-on projects. Although the ability to program is not necessary to learn about bioinformatic algorithms and complete the Web Exploration and On Your Own projects, learning to program can (1) enhance your learning, because implementing the algorithm in code is the real test of how well you understand it; (2) provide a basis for making better choices of programs and parameters when faced with a bioinformatic problem; (3) give you the power to write programs to deal with repetitive or tedious tasks you would otherwise have to do by hand; and (4) give you a marketable skill that can open doors later on.

This appendix is not intended to substitute for an introductory programming course. However, if you have limited programming background and want to try (or teach) some of the programming exercises in the text, it can help you with fundamental concepts of programming and the basic constructs used in the text's pseudocode examples.

Five Main Programming Constructs

Programming languages allow you to write five main types of statements, often called the five constructs of programming languages: (1) **assignment** statements, (2) **input** statements, (3) **output** statements, (4) **decisions** (or conditionals), and (5) **loops** (or iterations). Each construct tells the computer to perform a specific kind of action, as illustrated in **Figure A.1**.

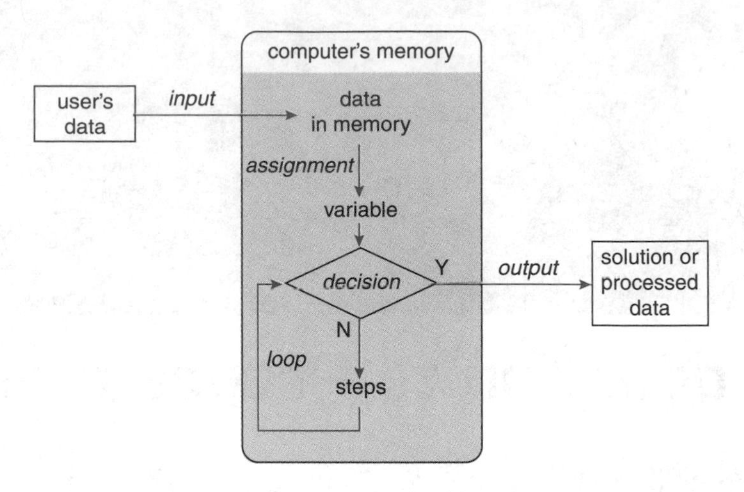

Figure A.1 How the five main programming constructs are used to build a program: Input statements bring data into memory where it can be acted upon by means of assignment statements, decisions, and loops. Results of processing are returned to the user via output statements.

The first three constructs control the movement of data. Data (from the keyboard or a file) can be moved into the main **memory** of a computer using input statements, out of the computer's main memory (to a printer, monitor, or file) using output statements, and manipulated in memory using assignment statements (which identify a piece of data with some variable name). The remaining two constructs control the flow of execution of a program's statements. A program is executed in a linear, sequential order unless instructed otherwise. Decisions and loops are statements used to alter this sequential flow.

These five constructs are the basis of most programming languages. To write a useful program, however, we also need things like **operators** (allowing us to manipulate variables in assignment statements and compare them in decision statements), **functions** (steps to be executed repeatedly), and **data structures** (convenient types of variables to hold different kinds of data, such as arrays that can hold matrices). Next, we examine these five constructs more closely, with examples in **pseudocode**. Pseudocode resembles the syntax of a program statement but is not specific to any particular language, showing how an algorithm is implemented in programming statements without requiring you to understand any particular syntax. Because pseudocode closely resembles programming syntax, you should be able to see how to easily transfer any pseudocode statement to the language you are using. This is often how a programmer starts writing the algorithm to solve a complex problem. The pseudocode examples that follow are based on solutions in the Guided Programming Projects in this text.

Assignment Statements

Assignment statements are responsible for moving data from one location in main memory to another. We can better understand the need for these statements if we understand how data are stored in a computer's memory. A piece of data, such as a number or a **string** (a set of alphanumeric characters), must be stored at some specific location (or "address") within memory; to access it, the placement location must be identified in some way. A **variable** is just a named storage location. Some programming languages require you to specify the kind of data (or **data type**) associated with each variable (e.g., string, or text, integer, decimal number, etc.), whereas others are more flexible.

The general syntax for an assignment statement is `variable = expression`. The expression may consist of another variable, a constant, an arithmetic expression, or a function call. The expression is evaluated, and its result is stored at the storage location referenced by the variable name. Any data already stored with that variable name are overwritten. Although the right-hand side of an assignment statement (part to the right of the = symbol) can be quite complex, the left-hand side (part to the left of the = symbol) is always a variable. This is because the purpose of an assignment statement is to alter the contents of a storage location.

Throughout the text you will find many examples of pseudocode representing assignment statements. Some examples are as follows:

```
currentrow = n − 1
add 'H' to dstring
subtract 1 from currentcol
add element m to key
iSmallest = i's key
```

Each element indicates a change to some storage location. When converting the pseudocode to actual programming syntax, each statement needs to be converted to an assignment statement in the syntax of the chosen programming language. For example, the first statement might simply be written `currentrow = n − 1`, but the syntax of some languages requires `currentrow=n−1;` (with the semicolon representing the end of the statement). Some identify the variable name with a special character (`$currentrow = n − 1`), and some allow shortcuts like `currentrow−−`. The second example given might be translated to something like `dstring = dstring + 'H';`, using an appropriate operator to add an H onto the end of an existing text string named `dstring`.

Input Statements

Input statements are responsible for moving data into memory from the keyboard or from external storage devices (files stored on a hard drive or data being collected by some instrument, for example). This process is also called reading data. When data are read in, the variable that holds the data must be specified. The syntax for reading in data from a file is similar to the assignment statement syntax, except the input file (or external device) must be indicated in some way. Here are some examples of pseudocode representing input statements used in the text:

```
Input the number of clusters: numClusters
Input the two sequences: s1 and s2
```

Output Statements

Output statements are responsible for moving data from main memory to the screen, to a file, or to a storage device. This process is also called writing or printing data (even if it is not actually printed on paper). In this text you will find pseudocode statements like these that represent output statements:

```
Output "no mismatches found – strings are identical"
Write seq1[i] to outfile
```

The term output here indicates an output statement directed to the console (screen), whereas write is used to indicate output directed to a file.

Decision Statements

The statements of a program execute in sequential order. That is, each statement executes in its entirety before moving on to the next statement. However, there are times when the flow of execution may need to change—for example, when you want your program to choose one path if one event occurs and a different path in response to a different event. Decision statements allow for this choice.

A decision consists of a conditional test, statements that execute if the conditional test is true, and potentially alternative statements that execute if the conditional test is false. Almost all languages use an `if` statement for decision making; we have therefore indicated decision statements with `if` in the pseudocode examples. The basic structure of an `if` statement is as follows:

```
if conditional expression
    statement or block
else
    statement or block
```

The conditional expression is the test to be evaluated, such as `if x > 2`. If the condition is true (if x is actually greater than 2), then the statement or block of statements that follows the `if` is executed. The statements associated with the `if` statement are always indented to show their dependency on the conditional expression. The `else` block is optional and is used if there are alternative statements that should be executed if the condition is *not* true. Many languages also allow for one or more `else if` conditions: for example, if x is not greater than 3, we could use `else if` to test whether x is greater than 2 and execute some statements if so as well as an `else` marking statements to execute if x is neither greater than 3 nor greater than 2. It is also possible to nest `if` statements: to have one `if` statement within another `if` statement. The following pseudocode examples show two independent `if` statements:

```
if (s1[i] not equal s2[i])
    add 1 to difCtr
if (dstring[dirpos] equals "D")
    align s1[seq1pos] and s2[seq2pos]
    subtract 1 from seq1pos and seq2pos
else if (dstring[dirpos] equals "V")
    align s1[seq1pos] and a gap
    subtract 1 from seq1pos
else
    align s2[seq2pos] and a gap
    subtract 1 from seq2pos
```

Loop Statements

Loops are used when a section of code needs to be repeated, either for a specified number of times or until some specified condition is met (tested with a decision statement). There are three requirements when using loops: (1) there must be a test condition that will eventually fail (ending the loop), (2) the test condition must be initialized, and (3) the test condition must change. These three requirements occur at different locations based on the type of loop used. Programming languages support three main loop types: for, while, and do-while.

Loop statements consist of a conditional test and a loop body. The conditional test is evaluated, and, if true, the statements in the loop body execute. This process repeats until the conditional test fails. The `for` and `while` loops perform the conditional test first, whereas the `do-while` loop performs the conditional test for the first time after the loop body executes once. Thus, for the `for` and `while` loops, the statements in the loop body may never execute, but the `do-while` loop always executes its statements at least once. The `for` loop statement generally includes a variable that is incremented (or decremented) automatically with each trip through the loop, but the other two must change their variables explicitly in the loop body. The following pseudocode examples represent the three kinds of loops. Indentation is used to indicate the statements in the loop body.

The `for` loop:

```
for each i from 0 to length of s1 – 1
    coverage[i] = 0

for each line of data in infile
    clusterNames[i] = first value in line
    distances[i] = remainder split using space as delimiter
```

The `while` loop:

```
while (i < coverage length and met is true)
    if coverage[i] < fold
        met = false
    add 1 to i
```

The `do-while` loop:

```
do
   randLen = random number between fMin and fMax, inclusive
   randSt = random number between 0 and length of s1
   frags[numFrags] = s1.substring(randSt, randSt+randLen)
   add 1 to numFrags
   coverage = true
   for each i from randSt to (randSt + randLen – 1)
      add 1 to coverage[i]
      if coverage[i] less than coverageMin
        coverage = false
while (coverage is false)
```

Data Structures

As mentioned, variables hold data values in memory. **Scalar** variables hold a single piece of information: name, phone number, DNA sequence, and so on. Sometimes, however, we want a variable to allow us to work on a set of related data. Consider, for example, how you would store a matrix of values. You could create a scalar variable for each element of the matrix, but it would be much easier to have a single variable represent the entire matrix. Data structures are storage structures that can hold data more complex than simple scalar variables.

The following sections discuss the two data structures used extensively throughout this text: arrays and hash tables. Note that these structures can have different names in different programming languages: Arrays are called lists in some languages, and hash tables may be called maps or dictionaries.

Arrays

Arrays are used to hold multiple pieces of related information. Arrays can be single dimensional (having a single row) or multidimensional (having rows and columns). Therefore, you might have an array named `firstnames` that holds the elements `"Jim"`, `"Sue"`, `"Bob"`, and `"Alice"`. Subscript notation is used to indicate a specific element in an array; it is important to remember that the first element is always numbered 0, not 1: `matrix[0]` represents the first element in an array named matrix, whereas `matrix[2]` represents the third element in the same array. If the array is a two-dimensional array, the row and column must be indicated using two subscript values, so `matrix[2,4]` represents the element in the third row and fifth column (again, subscripts start at 0).

Hash Tables

Like arrays, hash tables allow multiple values to be stored using a single variable name. In a hash table, however, a value can be accessed by its name instead of by its position in the array. This is very useful when instead of a matrix of data you want to look up items from a list. For example, in this text, hash tables are used to store the amino acids encoded by each of the 64 possible mRNA codons so we can "look up" a codon and get its corresponding amino acid.

Entries in a hash table consist of two components: a **key** and a data **value**. The key is the name used to access (or look up) the data value, so it only makes sense to use this structure when there is a clear relationship between two sets of variables. For the amino-acid hash table, the key `"ATG"` would be associated with the value `"Met"` (or `"M"` if the table used the single-letter code). Hash tables are used throughout this text to improve the efficiency of the algorithmic solution. Pseudocode for storing the value `"Trp"` in a hash table named `aminoAcid` under the key `"UGG"` would look like this:

```
aminoAcid['UGG'] = 'Trp'
```

Another example of using a hash table is in the substitution matrix for a protein alignment. Here, we need to look up two keys, because we need a score for when, for example, Ala is substituted by Gly that is different from the score when Ala is substituted by Phe or Trp. To accomplish this, we can nest one hash table within another so the value for the key `"Ala"` is actually a whole hash table of values for various amino acids that might substitute for Ala. This allows us to quickly access a substitution value based on a pair of amino acids. Pseudocode to store a value in a nested hash table would look like this:

```
a1 = 1st amino acid
a2 = 2nd amino acid
hashTable[a1, a2] = score
```

Functions

When confronted with a large problem, it is helpful to break the problem down into manageable pieces that can be solved more easily than the original larger problem. In a program, we may therefore have some code that needs to be executed repeatedly, once for each piece of a bigger problem. Or, we may just have some steps that need to be done at several points in the program or even in several different programs. For example, many programs discussed in this text require conversion of a DNA sequence to a complementary (other strand) sequence. In all these cases, a **function** (called a subroutine in some languages) can be used to advantage. A function is a set of statements outside the main flow of program execution. It can be **called** from a statement within the main program, its statements execute, and then execution goes back to the main program (often with some changed variables).

Thus, if you have written a function to find the complement of a DNA strand, you could write the code once and call it multiple times from within the program, passing it a different nucleotide sequence each time. Or, you could copy that piece of code into another program and, because it is a function, it would not require changes in the main program. Using functions thus allows us to modularize code. In fact, you can think of a program as a set of modules: a required main set of statements and optional functions. When a program executes, the main block of code executes and calls any functions that are needed at various points.

A function consists of a header and a block of code designated as the function body. The header consists of the name of the function and any **parameters** (information passed into the function by the calling routine). In some programming languages it is necessary to specify the data type for the data to be returned. A function has its own memory, so parameters and return values are used to share information between a function and its calling routine. Function pseudocode in this text is formatted as follows, with indentation used to indicate the statements in the body of the function:

```
function functionname (parameter 1, parameter 2, . . .)
    body of function
```

The following example shows a function that determines whether a minimum coverage value was met, which would need to be done repeatedly in a sequence assembly program. The function returns a value that is either true or false to the routine that called the function.

```
function coverageMet(coverage, fold)
    i = 0
    met = true // assume coverage met
    while (i < coverage length and met is true)
        if coverage[i] < fold
            met = false
        add 1 to i
    return met
```

Using this function would simplify the do-while example shown previously, because now this section of main program code merely needs to call the function (notice that the call is in the `while` statement), adding readability as well as reducing the code required:

```
do
    randLen = random number between fMin and fMax, inclusive
    randSt = random number between 0 and length of s1
    frags[numFrags] = s1.substring(randSt, randSt+randLen)
    add 1 to numFrags
    for each i from randSt to (randSt + randLen – 1)
        add 1 to coverage[i]
while coverageMet(coverage, fold) is false
```

A Sample Program

As an example of how these ideas are put together, let's construct a simple program that finds the complementary strand for any input DNA strand. First, consider what the steps would be in an algorithm to solve this problem.

Algorithm

Complementary Strand Algorithm

1. *Accept an input string from the user that represents the nucleotides in one DNA strand.*
2. *Read one character from the string.*
3. *Output the correct character for the complementary strand using the following rules:*
   - *If the character is A, output T.*
   - *If the character is C, output G.*
   - *If the character is G, output C.*
   - *If the character is T, output A.*
4. *If there are no more characters, stop; otherwise, go back to step 2.*

Now, that algorithm needs to be converted into a series of programming steps, using the five programming constructs described. Here, the program is presented in pseudocode, but of course in reality it would have to be implemented in a programming language:

```
prompt user to input a nucleotide sequence
input the DNA sequence: dnaSeq
output "Complementary strand:"
for each i from 0 to length of dnaSeq
    dnaBase = character from dnaSeq at position i
    compNuc = getNuc(dnaBase)
    output compNuc

function getNuc(dnaBase)
    if dnaBase = "A"
        comp = "T"
    else if dnaBase = "C"
        comp = "G"
    else if dnaBase = "G"
        comp = "C"
    else
        comp = "A"
    return comp
```

Glossary

–10 sequence A DNA sequence similar to 5'TATAAT that along with the nearby –35 sequence makes up the bacterial promoter sequence recognized by RNA polymerase.

3′ One end of a DNA strand. Synthesis of nucleic acids is from 5′→3′, so this is the last end synthesized. At this end, there is a hydroxyl (OH) group attached to the 3′ carbon of the last nucleotide.

–35 sequence A DNA sequence similar to 5'TTGACA that along with the nearby –10 sequence makes up the bacterial promoter sequence recognized by RNA polymerase.

454 sequencing See *pyrosequencing*.

5′ One end of a DNA strand. Synthesis of nucleic acids is from 5′→3′, so this is the first end synthesized. At this end, there is a phosphate group attached to the 5′ carbon of the first nucleotide.

5′ cap A structure (actually a G nucleotide added by a "backward" 5′-5′ linkage) added to the 5′ end of a messenger RNA after transcription and before translation. The cap is used by the ribosome to identify the 5′ end of the mature mRNA.

α-helix A secondary structure commonly found in proteins in which a region of amino acids is arranged in a spiral held together by hydrogen bonds.

β-sheet A common secondary structure in protein in which short stretches of linear chains of amino acids (β-sheets) lie parallel to each other, also called a β-pleated sheet.

β-strand A stretch of amino acids within a protein in a linear, extended conformation that makes up part of a β-sheet.

β-turn A bend between two consecutive β-strands in the secondary structure of a protein.

***ab initio* structure prediction** Prediction of the folded structure of a protein (or nucleic acid) given only its amino-acid (or nucleotide) sequence. Also called *de novo* structure prediction.

accession number A unique identifier assigned to a DNA or protein sequence in a database such as GenBank.

active site The region of an enzyme that binds one or more molecules in such a way that a chemical reaction occurs between them.

affine gap penalty Used to score gaps in an alignment. An affine gap penalty (see also linear gap penalty) assigns a higher penalty (lower score) for creating a new gap than for extending an existing gap. This is intended to account for the fact that a single indel event might span multiple bases.

agglomerative clustering A hierarchical clustering method for grouping related items from a data set by means of a repetitive process in which at each iteration the most closely related items are merged into a group.

algorithm A set of specific, well-defined steps that describes how a problem can be solved.

alignment Comparison of two nucleotide (DNA, RNA) or amino-acid (protein) sequences so that individual residues from one sequence are matched with identical or similar residues in the other.

alignment-based gene prediction Identification of genes within a longer DNA sequence by looking for regions which are conserved over evolutionary time.

allele A specific form of a gene, arising when a mutation produces a distinct DNA sequence. Generally, an allele is referred to as resulting in a particular characteristic, such as an allele that produces a genetic disease.

alpha helix See α-*helix.*

alphabet In a hidden Markov model, symbols representing possible states.

ambiguous nucleotide A position in a nucleotide sequence where any one of two or more nucleotides can be a possible match. For example, the TATA box sequence within the eukaryotic promoter can have either T or A as its last nucleotide.

amino acid The chemical subunit of a protein. Hundreds or thousands of amino acid subunits, with various chemically distinct side-chains, are joined together to make a protein.

amino terminus The end of a protein that is synthesized first; when the protein is complete, the amino group of the first amino acid in the chain will be at the amino terminus of the protein.

ancestral state Sequence of a gene or protein from an evolutionary ancestor of the gene or protein under consideration.

annealing temperature In a polymerase chain reaction, temperature at which primers are allowed to base-pair with the template.

annotated database Repository of genomic or sequence information which includes details about a gene or protein sequence as well as the sequence itself.

annotation See *genome annotation.*

antiparallel Arrangement of two base-paired nucleic acid strands in which the two strands run in opposite directions, so that the 5′ end of one strand is opposite the 3′ end of the other.

assembler A computer program that carries out assembly of DNA sequences.

assembly See *sequence assembly.*

AUG See *start codon.*

base Informal term for a nucleotide.

base-calling In DNA sequencing, the process of examining (manually or by computer) an electropherogram and deciding which nucleotide is represented at each point in the sequence.

base condition A conditional test used in a recursive function that ends the function.

base case The condition tested in order to determine when a recursive process ends.

Bayesian statistics A field of statistics where degrees of belief are used to describe probabilities of events occurring based on prior knowledge and updated as new data is obtained.

beta sheet See β-*sheet.*

beta strand See β-*strand.*

beta turn See β-*turn.*

bioinformatics A discipline at the interface of the biological and computer sciences in which computerized algorithms are used to solve biological problems, especially in evolutionary biology, molecular genetics, and genomics.

BLAST Basic Local Alignment Search Tool. A very popular algorithm used to find sequences in a database (such as GenBank) that are similar to an input DNA or protein sequence and generate alignments between the database sequences and the input sequence. As the name suggests, this algorithm makes local alignments.

BLOSUM BLOcks SUbstitution Matrix. A substitution matrix commonly used to determine similarities among amino acids in making alignments of amino-acid sequences.

branch site A sequence within an intron where one end of the intron is chemically joined after the intron is removed from the mRNA, during splicing, preventing inadvertent translation of the spliced-out intron.

bulge In an RNA secondary structure, a point within a base-paired stem where one or more bases are not base-paired with bases on the opposite side of the stem.

call The point in a computer program at which program execution is transferred to a subroutine or function.

C-terminus See *carboxyl terminus.*

carboxyl terminus The end of a protein that is synthesized last; when the protein is complete, the carboxyl group of the last amino acid in the chain will be at the carboxyl terminus of the protein.

carrier An individual who harbors a recessive allele that could potentially produce a genetic disease or other genetic characteristic but who also has a dominant allele preventing that allele from showing its effect.

cartoon Representation of a protein structure using ribbons with directional arrows to represent secondary structures.

casein The major protein in milk.

CBI See *codon bias index.*

cDNA A DNA molecule synthesized by reverse transcriptase so that it has a nucleotide sequence complementary to that of a particular RNA molecule.

centroid linkage In agglomerative clustering, a method of determining the distance between one cluster and a second cluster (or data point) in which the average distance among all items within the clusters is taken as the distance between the clusters.

character-based method An algorithm for constructing a phylogenetic tree in which indicators of genuine biological similarity are valued over simpler quantitative measures of distance between two species.

chromosome A distinct segment of an organism's genome, composed of a single, long DNA molecule (including many genes) and associated proteins.

clade In a phylogenetic tree, an ancestral species or group and the species or groups that descend from it.

cladist An evolutionary biologist who favors character-based methods and places greatest importance upon biological similarity in phylogenetic studies.

cladogram A representation of a phylogenetic tree showing the relationships among species or groups but in which the branches are not proportional to distance.

cluster A group of related items in a data set.

clustering algorithm A method for grouping related items from a data set.

coding sequence The portion of a gene specifying the amino acids to be assembled into a protein.

codon A group of three nucleotides within an mRNA molecule that together specify one amino acid that should be added to a growing protein chain during translation.

codon bias index A measure of how much the codon usage in a particular sequence of interest varies from the average codon usage for the organism in question.

common ancestor A species from which two or more related modern species are descended by evolution.

complement Given a DNA strand written from 5′ to 3′, its complement is the strand whose nucleotides could base-pair with the original strand, written from 3′ to 5′ (so, the complement of 5′-`GAT` is 3′-`CTA`).

complementary Describes the two strands of a DNA molecule that are not identical but are related by the base-pairing rules: e.g., where one strand has `A`, and the complementary strand has `T`.

complete linkage In agglomerative clustering, a method of determining the distance between one cluster and a second cluster (or data point) in which the distance between the two most distant items is used as the distance between the clusters.

complex See *multifactorial.*

computational biology The application of computer-science methods to the solution of biological problems.

conformation The three-dimensional structure of a functional protein.

consensus sequence A functional set of nucleotides that can be identified by their sequence; the sequence is determined by looking for commonalities among several known sequences with a particular function.

conservation The tendency of a DNA or protein sequence to remain unchanged over evolutionary time; a conserved residue is one that has not changed between two species being compared.

conservative substitutions A mutation resulting in replacement of one amino acid in a protein with another amino acid of similar size, shape, hydrophobicity, and so on, so that the function of the protein is unlikely to be affected.

constraint A restrictive condition or limit placed upon a computer program, often an assumption made in order to simplify a complex problem.

content-based gene prediction The identification of genes within a DNA sequence by evaluating (often statistically) the properties of the sequence, such as codon usage or high frequencies of CpG pairs.

contig A DNA sequence made by assembling multiple short, overlapping sequences.

core promoter The part of the promoter region of a eukaryotic gene absolutely required for any transcription to take place, usually including a TATA box sequence and an Inr sequence.

coverage In a DNA sequencing project, the number of times a particular region of a DNA molecule has been sequenced.

CpG A C (cytosine) nucleotide followed by a G (guanine) nucleotide; the "p" refers to the phosphate molecule between them.

CpG island A region within an organism's genome that has an unusually high concentration of CpG nucleotide pairs.

C-terminus See *carboxyl terminus.*

data mining A field of study in computer science concerned with finding patterns in large data sets.

ddNTP See *dideoxy nucleotide.*

***de novo* structure prediction** See *ab initio structure prediction.*

deep sequencing Use of next-generation sequencing techniques to generate DNA sequence with a high level of coverage.

deletion A mutation resulting in loss of one or more nucleotides from a DNA sequence.

denaturation Unfolding of a protein so that its three-dimensional conformation is lost and it is merely a chain of amino acids; denatured proteins are usually unable to function.

dideoxynucleotide A nucleotide that lacks a hydroxyl (-OH) group in the position required for joining another nucleotide onto the chain; dideoxynucleotides are used in DNA sequencing to halt synthesis of new DNA, creating fragments from which the sequence can be read.

dideoxy sequencing Determination of DNA sequence using dideoxy nucleotides to terminate DNA synthesis, producing a set of fragments of all possible lengths with a known or measurable nucleotide at the 3′ end; also called Sanger sequencing.

dimer A functional protein made up of two polypeptide subunits.

directed sequencing A strategy for sequencing of a large DNA molecule (such as a chromosome or an entire genome) in which segments of the molecule are sequenced in order, minimizing the problem of assembly.

distance metric A means of quantitating the number of differences between two DNA or protein sequences, usually used in constructing phylogenetic trees.

distance-based method An algorithm for constructing a phylogenetic tree that is based on a relatively simple quantitative measure of distance between DNA or protein sequences (distance metric).

distributed computing A technique for solving very large problems by breaking them up into small units that can be distributed to a large network of ordinary computers rather than requiring a supercomputer.

DNA Deoxyribonucleic acid, the genetic material of all living organisms, consisting of a very long, double-stranded chain of nucleotides that includes sequence encoding instructions for synthesizing proteins and functional RNAs.

DNA polymerase An enzyme that synthesizes DNA; in order for a cell to divide, DNA polymerase must make a faithful copy of all of its DNA.

DNA sequencing Determination of the sequence of nucleotides making up a particular DNA molecule.

DNA trace An electropherogram.

dNTPs Abbreviation referring to all four possible DNA nucleotides; for example, dNTPs would be added to a sequencing or PCR reaction for a DNA polymerase to use in copying DNA.

domain A functional region of a protein, such as a region that binds DNA or inserts into a membrane.

dominant allele An allele whose effect is detectable even if the individual's genome contains only a single copy of that particular allele (see also *recessive allele*).

dynamic programming A programming technique in which a problem that would otherwise be computationally expensive can be solved by breaking it into subproblems that are much easier to address.

e-value A statistical measure used by BLAST to evaluate the probability that a match between a query sequence and a database sequence is due to chance. Lower e-values represent better matches.

electropherogram The output of an automated DNA sequencing instrument, showing the intensity of fluorescence of each fragment that passes through a reader, used to reconstruct the nucleotide sequence of the original DNA; also called a DNA trace.

emission probability In a hidden Markov model, the probability that a particular result (in bioinformatics, often a nucleotide) is produced for a particular (hidden) state.

enhancer A region of DNA near the promoter of a eukaryotic gene to which activator proteins or other transcription factors bind, increasing (generally) the efficiency of transcription from that promoter.

enzyme A protein that catalyzes (increases the rate of) a particular desirable chemical reaction taking place in a cell.

eukaryote An organism whose cells contain nuclei (specialized structures enclosing the DNA) and other complex functional structures; animals, plants, fungi, and protists are all eukaryotes.

evolution Descent of one species from an ancestral species with modification (mutation) over time.

exon A segment of a gene containing part of the sequence that encodes a protein; in eukaryotes, segments of the coding region are typically separated by introns that are transcribed but must be spliced out of the mRNA in order for it to be translated.

expression See *gene expression*.

first values An initial set of data that can be readily determined and which a computational algorithm can then use in solving a complex problem.

forward primer In the polymerase chain reaction, the short nucleic acid used to define one end (the "left" end as the researcher looks at the sequence) of a template sequence to be amplified.

frameshift A mutation in which one or more nucleotides are inserted or deleted within a protein-coding sequence, causing the codons to be read incorrectly.

frequency matching The process of determining the number of occurrences of a particular pattern in a set of data.

frequency value In the Chou-Fasman algorithm for predicting protein secondary structure, the likelihood that a particular amino acid will be found in a β-turn.

function An independent segment of a computer program that is used for a specific task and can be called repeatedly whenever it is necessary to perform that task; referred to in some languages as a subroutine.

functional constraint Limitation on the likelihood that a mutation occurring in a particular region of a DNA molecule will become fixed; DNA segments that encode parts of a protein crucial to its function are much less likely to show fixed mutations than other regions of the DNA molecule.

gap An insertion or deletion introduced by the alignment algorithm into one of two aligned genes or proteins relative to the other in order for adjacent residues to align optimally.

gap penalty In alignments of two genes or proteins, the score given in places where it is necessary to introduce a gap.

gene An inheritable unit, comprising a specific region of a DNA molecule and generally thought of as encoding either a protein or a functional RNA.

gene discovery The identification of DNA segments within a genome that are transcribed and encode proteins (or functional RNAs).

gene expression The process of transcribing and translating a gene so that its protein product is made in a cell.

gene prediction See *gene discovery.*

gene therapy Treatment of genetic disease by adding functional genes or removing defective ones.

genetic code The set of rules showing the correspondence between a nucleotide (DNA or RNA) sequence encoding a protein and the amino-acid sequence of the protein itself; the genetic code tells which of the 20 amino acids is represented by each of the 64 possible codons in DNA and RNA.

genome The complete set of genetic material of an organism; all of the organism's standard DNA sequence, including both genes and regions that appear to have no functional significance.

genome annotation The process of describing the genes (their functions, when and where they are expressed, etc.) in an organism's genome once their sequences are known.

genome-wide association study A large-scale experiment attempting to identify known genetic variations with a particular genetic disease or trait, often a complex trait.

genotype An organism's genetic makeup with regard to a particular gene or trait; the combination of alleles a particular individual has for a particular gene.

germ-line gene therapy Treatment of a genetic disease or condition by changing the genetic makeup of an individual's germ-line cells (sperm- or egg-producing cells in the testes or ovaries) so that the individual can no longer pass on an undesirable trait or will pass on a desirable trait.

global alignment A technique for alignment of two genes or proteins that attempts to match every residue in one sequence with a corresponding residue in the other, or with a gap.

graph A data structure used to show relationships among elements.

greedy algorithm An algorithm that attempts to solve a complex or computationally expensive problem by making a locally optimal choice from among a limited set of alternatives at each stage, thus reducing the number of calculations needed to determine an overall optimal solution. This type of algorithm does not guarantee an optimal solution.

guide tree A phylogenetic tree, usually developed by a simple distance measure, used to determine the order with which sequences should be introduced into a multiple sequence alignment.

GWAS See *genome-wide association study.*

HA See *hemagglutinin.*

hairpin Region of nucleic acid secondary structure in which two adjacent stretches of nucleotides base-pair; also called a stem-and-loop structure.

hemagglutinin Protein on the surface of an influenza virus that interacts with a cellular receptor (sialic acid) to allow entry of the virus into a human cell.

heuristic A "shortcut" method that simplifies a complex computational problem by making assumptions; it typically finds a good solution rather than taking the time and resources to find the best solution.

HGT See *horizontal gene transfer.*

hidden Markov model A means of predicting the probability with which an unobservable state of interest occurs given a pattern of observations (used in this text as a means of gene prediction).

hierarchical clustering A method of grouping related data items in which each new group (or cluster) is based on the clusters generated earlier.

high-scoring pair In a BLAST search, a match between a query sequence and a database sequence of sufficient quality as to exceed some threshold.

HMM See *hidden Markov model.*

homodimer A functional protein made up of two identical polypeptide subunits.

homology modeling Determination of a protein structure by comparison of its amino-acid sequence to that of a similar protein whose structure is already known.

horizontal gene transfer (HGT) Movement of genetic information from one organism to another, rather than by inheritance to its offspring.

HSP See *high-scoring pair.*

hydrophilic The ability to dissolve in or interact freely with water.

hydrophobic Lacking the ability to dissolve in or interact freely with water; oily or greasy.

ideogram A schematic representation of a chromosome showing its arms, centromere, and pattern of bands and often the location of genes relative to those bands.

indel An insertion or deletion mutation.

initiator sequence (Inr) A DNA sequence similar to `5'YYCARR` found within the promoter region of a eukaryotic gene and usually containing the start site of transcription.

inprocessing A means of data manipulation in computer programming in which the algorithm implemented by the program examines and modifies the data as needed by the algorithm. This process is often used to detect possible errors in the data at run time.

Inr sequence See *initiator sequence.*

insertion mutation A mutation in which one or more nucleotides are inserted within a DNA sequence.

internal loop In nucleic-acid secondary structure, a region of unpaired bases on each strand occurring between two base-paired stems.

internal node The point within a phylogenetic tree where two branches join, usually representing the common ancestor of the species which branch from that point.

intron A noncoding DNA segment inserted between portions of a coding sequence and removed from the mRNA by splicing in order for the mRNA to be translated correctly.

inverse Given a DNA strand written from 5′ to 3′, its inverse is the same sequence written from 3′ to 5′ (so, the inverse of 5′-`GAT` is 3′-`TAG`).

inverse complement Given a DNA strand written from 5′ to 3′, its inverse complement is the strand whose nucleotides could base-pair with the original strand, written from 5′ to 3′ (so, the complement of 5′-`GAT` is 5′-`ATC`).

iteration One repetition of a repetitive process within a computer program.

latent In a viral infection, the state in which a virus can infect human cells but remain dormant, not producing new virus particles.

linear gap penalty System of scoring gaps in an alignment in which each individual gap receives the same score, regardless of whether there are multiple individual gaps or a group of multiple consecutive gaps.

linkage method A method of determining the distance between clusters in an agglomerative clustering algorithm.

local alignment An alignment between two DNA or protein sequences that seeks to find conserved regions of the sequences even when the degree of similarity of the two sequences as a whole may be low.

log-odds ratio A statistical measure of probability of a particular event, generally the log of the likelihood of that event.

loop (1) A segment of a computer program that is executed repeatedly until some condition is met, or (2) A region of unpaired bases at the end of a base-paired stem in nucleic-acid secondary structure.

match bonus See *match score.*

match score In scoring of a DNA or protein alignment, the score given when two aligned nucleotides or amino acids are identical (or similar, in the case of a protein alignment).

mate-pair read DNA sequences read from two ends of the same fragment, used to check the accuracy of assembly.

matrix A set of data arranged in a rectangular table of rows and columns; matrices can be added or multiplied using mathematical rules.

mature mRNA An mRNA molecule that has been processed by addition of a 5′ cap and 3′ poly(A) tail and by splicing and is ready for translation.

maximum likelihood One of several commonly used character-based method for constructing a phylogenetic tree.

melting temperature The temperature at which 50% of single-stranded nucleic acid molecules would be able to base-pair with complementary or partially complementary strands.

merge step In an agglomerative clustering algorithm, grouping of closely related data elements or clusters into a single cluster.

messenger RNA (mRNA) An RNA molecule produced by transcription whose nucleotide sequence is complementary to the DNA sequence of a single gene (sometimes several sequential genes, in prokaryotes) and which is used in synthesis of a protein (translation).

metadatabase A database which brings together information from multiple primary databases related to each of its entries.

metagenome DNA sequence information obtained from a clinical or environmental sample, including genes or genomes from all the organisms present in the sample.

metagenomics The obtaining of metagenome sequence information and its analysis.

microbiome The community of microorganisms living in a particular environment, especially a particular environment within the human body or within an animal or plant.

microRNA A small, non-coding RNA molecule that functions in gene regulation.

miRNA See *microRNA*.

mismatch penalty See *mismatch score*.

mismatch score In scoring of a DNA or protein alignment, the score given when two aligned nucleotides or amino acids are not identical (or similar, in the case of a protein alignment).

missense mutation A mutation with the protein-coding region of a gene which causes one amino acid in the protein to be substituted by a different amino acid.

molecular clock A DNA or protein sequence that is used in measuring evolutionary distance; the more changes that have occurred in the sequence between two species, the longer the time since they diverged from a common ancestor.

molecular evolution The use of DNA and protein sequences and other molecular genetic tools to study evolutionary relatedness.

monophyletic group In phylogenetics and evolutionary biology, a group whose members are all genuinely descended from a common ancestor.

motif A short, functional region within a DNA or protein sequence, usually recognizable by a sequence or structure pattern.

mRNA See *messenger RNA*.

multifactorial In genetics, a genetic disorder which is inherited or has a genetic component but which results from the effects of multiple genes rather than a single one; also called a complex disorder.

multiple sequence alignment An alignment made to compare more than two nucleic-acid or protein sequences (see also pairwise alignment).

mutant An organism or cell resulting from a mutation.

mutation A change in a DNA sequence, such as the replacement of one nucleotide by another, or the insertion or deletion of one or more nucleotides.

N-terminus See *amino terminus*.

NA See *neuraminidase.*

natural selection The tendency for individuals in a population who are genetically better adapted to some environmental condition (climate, predation, nutrient availability, mate selection, etc.) to have a greater chance of reproducing, thus increasing the representation of their alleles in future generations.

Needleman-Wunsch algorithm An algorithm utilizing dynamic programming to obtain the optimal global alignment between two DNA or protein sequences.

neighbor-joining An algorithm for constructing unrooted phylogenetic trees without having to assume that the evolutionary distance from each of two species to their common ancestor is equal.

neural network An algorithm based on the functioning of networks of nerve cells and designed to predict an outcome by integrating the states and relative importance of various inputs.

neuraminidase A protein on the surface of an influenza virus which assists the progeny viruses in freeing themselves from the host cell.

Newick format A shorthand form of representing a phylogenetic tree in which nested parentheses are used to represent groups of related organisms; for example, ((A, B), C), (D, E).

NJ See *neighbor-joining.*

NN See *neural network.*

nonsense mutation A mutation occurring within a protein-coding sequence that changes a codon encoding an amino acid to a stop codon, resulting in premature termination of the protein.

non-template strand The DNA strand that is not read by the RNA polymerase during transcription; the sequence of the non-template strand is the same as the sequence of the mRNA, except that it has `T` rather than `U` nucleotides.

nosocomial An infection acquired in a hospital environment.

N-terminus See *amino terminus.*

nucleotides The chemical subunits that make up DNA and RNA molecules; DNA is composed of two complementary nucleotide chains, using nucleotides adenine (`A`), guanine (`G`), cytosine (`C`), and thymine (`T`), whereas RNA is usually single stranded and is composed of nucleotides A, `C`, `G`, and `U` (uracil).

nucleus The membrane-bounded region of a eukaryotic cell containing the genetic information (DNA) and where transcription takes place.

odds ratio The ratio of two probabilities, often used as a measure of increase or decrease from an expected frequency.

oligo See *oligonucleotide.*

oligonucleotide A DNA molecule composed of only a few nucleotides.

open reading frame (ORF) A DNA sequence containing a start codon, a number of codons encoding amino acids and a stop codon. An ORF can potentially encode a protein and thus is a candidate to be a gene.

operon A group of two or more genes (usually with related functions) located close together on the chromosome of a bacterial cell (operons are found only in prokaryotes) and transcribed on one piece of mRNA from a single promoter.

optimal Best, such as the optimal (best possible) alignment of two DNA sequences.

ORF See *open reading frame.*

ortholog A gene that is similar in DNA sequence to a gene in another species. Orthologs result from mutation of a single gene over time in two different species that have diverged from a common ancestor.

outgroup In construction of a phylogenetic tree, a species or group accepted as clearly unrelated to the species or groups under consideration which can be used to find the root of the tree relative to the other species.

pair-annealing Base-pairing occurring between two PCR primers.

pairwise alignment Alignment of two nucleic-acid or protein sequences.

PAM Point Accepted Mutation matrix. A substitution matrix commonly used to determine similarities among amino acids in making alignments of amino-acid sequences.

pandemic A disease outbreak which reaches multiple continents.

paralog A gene that is similar in DNA sequence to another gene in the same species. Paralogs result from mutation over time of two genes derived from a single original gene that became duplicated.

parameter A value that helps define the scope of an algorithm.

pathogen A disease-causing organism, usually a bacterium or virus.

pattern-matching algorithm An algorithm that searches for a particular pattern in a set of data.

PCR See *Polymerase chain reaction.*

pheneticist An evolutionary biologist who favors distance-based methods and places greatest importance upon quantitative measures of evolutionary distance in phylogenetic studies.

phenotype The visible or measurable result of an organism's genetic makeup (see genotype).

phylogenetic tree A representation of the evolutionary relationships among organisms, illustrating the descent of modern species from common ancestors.

phylogenetics The study of evolutionary relationships among organisms.

phylogram A phylogenetic tree drawn so as to show not only the relationships among species or groups but also the distance or evolutionary time that separates them from their common ancestors.

poly(A) tail A large number of adenine (A) nucleotides added to the end of a messenger RNA to increase the time before enzymes attacking its 3′ end can affect its coding sequence.

polyadenylation site The site at which an mRNA is cleaved and its poly(A) tail added after transcription.

polymerase chain reaction (PCR) A laboratory technique for making multiple copies of (amplifying) a desired region within a complex DNA molecule.

polymorphism Variation at a particular genetic site.

polypeptide A chain of amino acids encoded by a particular gene.

polyphyletic In evolutionary biology and phylogenetics, a group that includes species not all truly descended from a common ancestor.

polyprotein A polypeptide translated as a single amino-acid chain and then cleaved into individual functional proteins (used for example in synthesis of HIV proteins).

postprocessing A means of data manipulation in computer programming in which the data are examined and modified after execution of the program. This process is often used to format data for specific user needs or to fix errors that may still result after execution of the program.

pre-mRNA A messenger RNA as it is translated from the DNA, before splicing, capping, and the addition of the poly(A) tail. See also mature mRNA.

preprocessing A means of data manipulation in computer programming in which the data are examined and modified prior to sending them to the program. This process is often used to detect and fix, if possible, noise or noisy data.

primary (1°) structure The amino-acid sequence of a protein, its most basic structure before any folding occurs.

primary database A database of genetic or genomic information in which data obtained directly by a researcher (DNA sequence, amino-acid sequence, protein structure, etc.) is deposited.

primer A short piece of DNA (or, in natural DNA replication, RNA) used as a starting point for DNA polymerase to copy a longer DNA template, such as in DNA sequencing or PCR.

program A sequence of instructions executed by a computer to perform a specific task (implement an algorithm).

progressive alignment A multiple sequence alignment algorithm which builds the alignment by adding individual sequences in the order specified by a guide tree.

prokaryote An organism whose cells lack nuclei and other complex functional structures; bacteria are prokaryotes.

promoter The region of a DNA molecule in which RNA polymerase recognizes an appropriate site for transcription to begin.

propensity value In the Chou-Fasman algorithm for predicting protein secondary structure, the likelihood that a particular amino acid will be found in an α-helix or β-strand structure.

protease An enzyme which cleaves or breaks down proteins.

pseudocount A default starting value used to initialize a variable.

pseudoknot A structure found in folded RNA molecules in which secondary structures from different parts of the molecule come together.

pyrosequencing A form of next-generation sequencing in which the addition of nucleotides to a growing chain is detected by the emission of light when reagents react with the pyrophosphate released in the polymerase reaction. Also called 454 sequencing.

quarternary (4°) structure A level of structure found in some proteins in which two or more folded protein subunits come together to make up the final, functional unit. Hemoglobin, for example, is made up of four distinct protein subunits, two each of the products of two different genes, which associated to form the functional hemoglobin molecule.

random coil A region of a protein with no specific secondary structure.

rational drug design The process of developing molecules for therapeutic use based on identifying a specific target and then developing a pharmaceutically useful molecule to bind that target.

RBS See *ribosome binding site.*

read length The number of nucleotides that can be determined by one set of DNA sequencing reactions.

reading frame A set of nonoverlapping codons as they would be read in translation. A reading frame is established by a start codon, and subsequent nucleotides are read in groups of three.

recessive allele An allele whose effect is not detectable unless the individual's genome contains two copies of that particular allele (see also *dominant allele*).

recursion A computer programming technique in which a subroutine calls itself repeatedly until some base-case condition is met.

resistance The ability of bacteria to avoid being killed by an antibiotic.

reverse primer In the polymerase chain reaction, the short nucleic acid used to define one end (the "right" end as the researcher looks at the sequence) of a template sequence to be amplified.

ribbon In a representation of a protein structure, a thickened line used to represent an α-helix or β-strand.

ribonucleic acid A long, usually single-stranded chain of nucleotides which functions in gene expression (mRNA), translation (rRNA, tRNA), gene regulation (miRNA), or other cellular processes.

ribosomal RNA Non-coding RNA molecule that contributes to the structure and enzymatic activity of the ribosome.

ribosome A large complex of protein and RNA where translation of an mRNA to produce a protein is carried out.

ribosome binding site (RBS) See *Shine-Dalgarno sequence.*

ribozyme An enzyme whose active component is RNA, not protein.

RNA See *ribonucleic acid.*

RNA interference Negative regulation (whether natural or artificial) of a gene by a small, non-coding RNA (miRNA or siRNA).

RNA polymerase An enzyme that carries out transcription, reading one strand of a DNA molecule and synthesizing a complementary strand of RNA.

RNA splicing The process of removing introns from an mRNA and rejoining the exons to form an unbroken coding sequence; occurs in the nucleus after transcription and before the mRNA can be translated.

RNAi See *RNA interference.*

RNA-seq Determination of which genes in a cell are being actively transcribed by making cDNA copies of all the cellular mRNA and then sequencing it using a next-generation sequencing method.

rooted A phylogenetic tree in which the location of the common ancestor of all the species or groups in the tree can be determined.

rRNA See *ribosomal RNA*.

Sanger sequencing See *dideoxy sequencing*.

scoring matrix In gene or protein alignment algorithms, a matrix representing various ways in which two sequences might be aligned and computing the scores for those possible alignments.

scoring metric A system for scoring matches, mismatches, and gaps in an alignment.

secondary (2°) structure For proteins, folding of local regions of a protein into three-dimensional shapes, notably α-helices and β-sheets. For DNA or RNA, folding of a single-stranded molecule or region of a molecule by formation of base pairs between different regions of the same strand.

secondary database A database of genetic or genomic information in which data from a primary database is combined with other relevant information.

selectively toxic Ability of an antimicrobial drug (e.g., antibiotic) to kill or inhibit a microbe while leaving the human host unharmed.

self-annealing Base-pairing (secondary structure) within a PCR primer (or, more generally, within any single nucleic-acid molecule).

semiglobal alignment An alignment between two nucleotide or amino-acid sequences, particularly when one is much shorter than the other, in which gaps at the ends of the sequences are not penalized—for example, when searching for a single gene within an entire genome.

sequence assembly In DNA sequencing, the process of joining individual short DNA sequences into the complete nucleotide sequence of a gene, plasmid, chromosome, or genome.

sequence logo A visual means of representing the frequency of nucleotides found at each position of a consensus sequence.

sequence read In DNA sequencing, an individual stretch of nucleotide sequence read by means of a single sequencing reaction.

sequence-based gene prediction The identification of genes within a DNA segment based on sequence patterns such as start and stop codons, consensus sequences for promoters, etc.

Shine-Dalgarno sequence A sequence similar to `5'AGGAGG` that is found in mRNA molecules in prokaryotes and to which the ribosome binds in order to identify the correct start codon. Also called the ribosome binding site.

shotgun sequencing A DNA sequencing technique in which a large DNA molecule (such as an entire chromosome or genome) is broken up into random small fragments whose sequences are determined and then assembled using computer software.

side chain The portion of an amino acid that produces the distinctive chemical properties of that type of amino acid.

silent mutation A mutation which changes a nucleotide within the coding sequence of a gene but does not change the amino acid encoded.

similarity The relatedness of two nucleotide or amino-acid sequences.

simple-nucleotide polymorphism Variation at a particular genetic site in which different alleles differ only by a single-nucleotide substitution or a very small insertion or deletion.

single linkage In agglomerative clustering, a method of determining the distance between one cluster and a second cluster (or data point) in which the distance between the two closest items is used as the distance between the clusters.

siRNA See *small interfering RNA*.

sliding window A computer programming technique in which one segment ("window") of data is examined, then the focus shifts to an adjacent, overlapping segment.

small interfering RNA A double-stranded RNA or an RNA with secondary structure resulting from a viral infection and leading to negative regulation of viral gene expression (a form of antiviral defense).

small nuclear ribonucleoprotein A complex of proteins and small non-coding RNAs functioning in the nucleus in the RNA splicing mechanism.

Smith-Waterman algorithm An algorithm used to find an optimal local alignment between two DNA or protein sequences.

SNP See *simple-nucleotide polymorphism.*

somatic-cell gene therapy Treatment of a genetic disease or disorder by changing the genetic makeup of the affected cells but not of the germ-line cells so that offspring are unaffected.

space-filling A representation of a protein structure in which atoms are shown as spheres to indicate the space they occupy.

spliceosome A large complex of protein and RNA in the nucleus of eukaryotic cells that carries out splicing of mRNA.

splicing See *RNA splicing.*

start codon The codon signaling the beginning of a protein coding sequence in mRNA, almost always an AVG (methimine) codon occurring in a context recognized by the ribosome.

stem Within DNA or RNA secondary structure, a base-paired region.

stem-and-loop Within DNA or RNA secondary structure, a base-paired region with an unpaired "loop" region at its end. (Also known as a hairpin structure.)

stop codon A codon (UAG, UGA, or UAA in mRNA) that represents no amino acid and that signals the end point of protein synthesis.

strain A variant population within a species that has distinct genetic characteristics.

strand One of two long chains of nucleotides that makes up a complete DNA molecule.

subroutine See *function.*

substitution Informally, any mutation that results in the replacement of one nucleotide by a different nucleotide. More specifically, in molecular evolution, a fixed mutation that creates a difference in the genome sequence of one species relative to another.

substitution matrix A set of values indicating scores that should be applied based on degrees of similarity when various pairs of amino acids are aligned in making a protein alignment.

substitution rate A measure of the frequency with which fixed mutations have occurred over evolutionary time.

superstring A string consisting of a single set of common characters constructed from a set of overlapping substrings; in DNA sequencing, a superstring resulting from sequencing of multiple small fragments is called a contig.

TATA box A DNA sequence similar to `5'TATA(A/T)A(A/T)` found in the promoter region of most eukaryotic genes and required for transcription to occur.

template A strand of DNA from which a complementary strand is to be synthesized ("copied") by DNA polymerase.

template strand A DNA strand read by DNA polymerase during replication or RNA polymerase during transcription; the polymerase synthesizes a strand complementary to the template strand.

terminal gap A gap introduced at one end of a DNA or protein sequence in the process of aligning it with another, often longer, sequence.

terminal node The end of a branch on a phylogenetic tree, often an extant (modern) species.

terminator A DNA sequence signaling the end point for transcription of a gene.

tertiary (3°) structure Folding of a protein into an overall three-dimensional conformation.

tetramer A functional protein composed of four polypeptide subunits.

TF See *transcription factor.*

thermal cycler An instrument that can heat and cool solutions in small tubes quickly and precisely, used for PCR.

threading Protein structure prediction by comparing a segment of amino-acid sequence to known common structural units.

Tm See *melting temperature.*

track In a genome browser, the display of a specific category of information about a DNA sequence: expression, similarity, locations of SNPs, etc.

training set A set of data used to fit a model (train an algorithm).

transcript RNA molecule produced by transcription of a DNA sequence.

transcription Synthesis of an RNA molecule, especially an mRNA, from a DNA template.

transcription factor (TF) A protein that binds to a DNA sequence in the promoter region of a gene (especially in eukaryotes) that affects the frequency or efficiency of transcription.

transcription unit A transcribed segment of a DNA molecule, consisting of a promoter, a protein coding sequence or functional RNA sequence, and a terminator sequence, resulting in transcription of a single mRNA or other RNA molecule.

transfer RNA Non-coding RNA used in translation to carry amino acids to the ribosome.

transformed distance In construction of phylogenetic trees, a method which corrects for unequal rates of evolution along two branches of the tree.

transition A mutation converting one purine to the other purine (A to G or G to A) or one pyrimidine to the other (C to T or T to C).

transition probability In a hidden Markov model, the probability of moving from one state to the next.

translation The process of synthesizing a protein based on an mRNA template.

transversion A mutation converting a purine to a pyrimidine (A to C, A to T, G to C, or G to T) or a pyrimidine to a purine (C to A, C to G, T to A, or T to G).

traveling salesperson problem (TSP) A famous mathematical problem seeking to determine the shortest path a salesperson can take in order to visit all the cities on his or her route; solutions to this problem have many applications in finding optimal solutions to programming problems.

tRNA See *transfer RNA*.

ultrametric A method for calculating distances in a phylogenetic tree that assumes a constant rate of evolution for all species in the tree.

UPGMA Unweighted Pair-Group Method with Arithmetic mean. A linkage method used in constructing phylogenetic trees in which distances between groups of related organisms are constructed based on the average distance between organisms in the group.

vertical gene transfer Inheritance of a gene from parents by offspring. Contrast with *horizontal gene transfer*.

virome The community of viruses living in a particular environment within the human body or within an animal or plant.

virulence factor A structure, protein, toxin, or other characteristic produced by a microbial pathogen that contributes to its ability to cause disease (such as by adherence, breaking down of tissues, uptake of nutrients, etc.).

wild-type The allele of a particular gene that is the most prevalent in the population, or an individual exhibiting the characteristics associated with the wild-type allele.

x-ray crystallography A technique used to precisely determine the three-dimensional structure of a protein by examining the patterns resulting from scattering of x-rays by a crystal of the protein.

Index

Note: Page numbers followed by "*b*", "*f*", or "*t*" indicate material in boxes, figures, or tables, respectively.

H

I

J

K

L

M

Q

R

S

U

V

W

X